An Introduction to Modeling of Transport Processes Applications to Biomedical Systems

Organized around problem solving, this book gently introduces the reader to computational simulation of biomedical transport processes, bridging fundamental theory with real-world applications. Using this book the reader will gain a complete foundation to the subject, starting with problem simplification, implementation in software, through to interpretation of results, validation, and optimization. Ten case studies focusing on emerging areas such as thermal therapy and drug delivery, with easy-to-follow step-by-step instructions, provide ready-to-use templates for further applications. Solution process using the commonly used tool COMSOL Multiphysics is described in detail, useful biomedical property data and correlations are included, and background theory information is given at the end of the book for easy reference. A mixture of short and extended exercises make this book a complete course package for undergraduate and beginning graduate students in biomedical engineering, biological engineering, and other engineering curricula, as well as an invaluable self-study guide.

Ashim Datta is a Professor in the Department of Biological and Environmental Engineering at Cornell University, where he has developed and taught modeling of biomedical transport processes as a course since 1996. He is recipient of the Michael Tien '72 Excellence in Teaching award from the College of Engineering, and he has authored a textbook, authored and co-authored over 85 technical papers and book chapters, and also co-edited three books on biological heat and mass transfer.

Vineet Rakesh is a Research Scientist in the Computational Medicine and Biology Division of a biomedical research company. He received his Ph.D. in Biological Engineering from Cornell University. He has also worked as a teaching assistant for the biomedical process modeling course at Cornell for three years and has been presented with the Outstanding Teaching Assistant award. His research has included modeling of airflow in the upper airway and drug transport in cancer therapy.

Cambridge Texts in Biomedical Engineering

Series Editors
W. Mark Saltzman, *Yale University*
Shu Chien, *University of California, San Diego*

Series Advisors
William Hendee, *Medical College of Wisconsin*
Roger Kamm, *Massachusetts Institute of Technology*
Robert Malkin, *Duke University*
Alison Noble, *Oxford University*
Bernhard Palsson, *University of California, San Diego*
Nicholas Peppas, *University of Texas at Austin*
Michael Sefton, *University of Toronto*
George Truskey, *Duke University*
Cheng Zhu, *Georgia Institute of Technology*

Cambridge Texts in Biomedical Engineering provides a forum for high-quality accessible textbooks targeted at undergraduate and graduate courses in biomedical engineering. It covers a broad range of biomedical engineering topics from introductory texts to advanced topics including, but not limited to, biomechanics, physiology, biomedical instrumentation, imaging, signals and systems, cell engineering, and bioinformatics. The series blends theory and practice, aimed primarily at biomedical engineering students, it also suits broader courses in engineering, the life sciences, and medicine.

An Introduction to Modeling of Transport Processes

Applications to Biomedical Systems

Ashim Datta and **Vineet Rakesh**
Cornell University

CAMBRIDGE
UNIVERSITY PRESS

University Printing House, Cambridge CB2 8BS, United Kingdom

Cambridge University Press is part of the University of Cambridge.

It furthers the University's mission by disseminating knowledge in the pursuit of education, learning and research at the highest international levels of excellence.

www.cambridge.org
Information on this title: www.cambridge.org/9780521119245

© A. Datta and V. Rakesh 2010

This publication is in copyright. Subject to statutory exception and to the provisions of relevant collective licensing agreements, no reproduction of any part may take place without the written permission of Cambridge University Press.

First published 2010

A catalogue record for this publication is available from the British Library

Library of Congress Cataloguing in Publication data

Datta, Ashim K.
    An introduction to modeling of transport processes: applications to biomedical systems / Ashim Datta and Vineet Rakesh.
    p. ; cm. – (Cambridge texts in biomedical engineering)
    Includes bibliographical references and index.
    ISBN 978-0-521-11924-5 (Hardback)
    1. Biological transport–Computer simulation. I. Rakesh, Vineet. II. Title. III. Series: Cambridge texts in biomedical engineering.
[DNLM: 1. Biological Transport. 2. Computer Simulation. 3. Models, Biological. 4. Software. QU 120 D234i 2010]
  QH509.D38 2010
  571.6´40113–dc22

                                                                               2009035316

ISBN 978-0-521-11924-5 Hardback

Additional resources for this publication at www.cambridge.org/9780521119245

Cambridge University Press has no responsibility for the persistence or accuracy of URLs for external or third-party internet websites referred to in this publication, and does not guarantee that any content on such websites is, or will remain, accurate or appropriate.

To my sister, Namita, and my brother, Ashis
                                    Ashim

To my sisters, Smita, and Namrata
                            Vineet

Contents

Preface   xiii
Acknowledgments   xvii
Introduction and overview   xix
List of symbols   xxv

I Essential steps   1

1 Problem formulation   3
1.1   Context: biomedical transport processes   3
1.2   What is problem formulation?   5
1.3   Steps in problem formulation   7
1.4   Defining goals for problem formulation   9
1.5   Simplify, simplify, simplify   11
1.6   Geometry: setting the computational domain   11
1.7   Governing equations   21
1.8   Boundary and initial conditions   27
1.9   Material properties   32
1.10   Other input parameters   35
1.11   Summary   35
1.12   Problems   37

2 Software implementation 1   50
2.1   Choosing a software   50
2.2   Software is not to be used as a blackbox   51
2.3   Organization of a typical CAE software: preprocessing, processing and postprocessing   51
2.4   Some general guidelines to preprocessing   54
2.5   Introduction to preprocessing in a computational software (COMSOL)   55
2.6   Geometry and analysis type   56
2.7   Geometry creation   60

2.7	Geometry creation	60
2.8	Governing equations	65
2.9	Boundary conditions	72
2.10	Initial conditions	77
2.11	Material properties	77
2.12	Miscellaneous implementation aspects	82
2.13	Problems	85

3 Software implementation 2 — 87

3.1	Which numerical method to use	87
3.2	Items needed in specifying the solution methodology	88
3.3	How to discretize the domain: mesh	88
3.4	How to choose a time step	98
3.5	How to choose a solver to solve the system of linear equations	103
3.6	Problems	106

4 Software implementation 3 — 108

4.1	Useful information in a biomedical context	109
4.2	Obtaining data at a particular location	111
4.3	Plotting transient data at one or more points, line or surface as a function of time	113
4.4	Obtaining surface/contour plots (in 2D problems) for observing variation within a region	117
4.5	Obtaining a surface plot in a 3D problem	121
4.6	Obtaining average values at a particular time or as a function of time	125
4.7	Obtaining arbitrary functions of computed variables	129
4.8	Creating animations	129
4.9	Dedicated plotting and postprocessing software	131
4.10	Relating to the goals of the simulation: guidelines for postprocessing	131
4.11	Analysis of data obtained from postprocessing	133
4.12	Presenting the simulation results to others	133
4.13	Problems	134
4.14	Appendix	136

5 Validation, sensitivity analysis, optimization and debugging — 139

5.1	Types of errors and error reduction	140
5.2	Estimating error: validation of the model	144
5.3	Reducing discretization error: mesh convergence	146

5.4	Estimating uncertainty and relating to design: performing sensitivity analysis	151
5.5	Objective functions: simple optimization	155
5.6	When things don't work: debugging	157
5.7	Problems	161

II Case studies   177

6 Case studies   179

6.1	Introduction	179
6.2	How to use the case studies	182
6.3	Additional case studies from the work of students at Cornell University	183
	Case Study I: Thermal ablation of hepatic tumors	186
	Case Study II: Cryosurgery of a wart	201
	Case Study III: Drug delivery from a patch	221
	Case Study IV: Drug delivery in therapeutic contact lenses	243
	Case Study V: Elimination of nitrogen from the blood stream during deep sea diving	251
	Case Study VI: Flow in human carotid artery bifurcation	258
	Case Study VII: Radioimmunotherapy of metastatic melanoma	270
	Case Study VIII: Burn injury in blood-perfused skin	280
	Case Study IX: Radiofrequency cardiac ablation	288
	Case Study X: Laser irradiation of human breast tumor	303
6.4	Problems	321

III Background theory   333

7 Governing equations and boundary conditions   335

7.1	Conservation of mass: the continuity equation	336
7.2	Conservation of momentum: governing equation for fluid flow	338
7.3	Conservation of thermal energy: governing equation for heat transfer	342
7.4	Governing equation for heat conduction with change of phase	346
7.5	The bioheat transfer equation for mammalian tissue	349
7.6	Conservation of a mass species: governing equation for mass transfer	353
7.7	Non-dimensionalization of the governing equations	357
7.8	Coupling of governing equations	360
7.9	Summary: governing equations	362

7.10	Boundary conditions: general comments	362
7.11	Boundary conditions: fluid mechanics	362
7.12	Boundary conditions for heat transfer	364
7.13	Boundary conditions for mass transfer	367
7.14	Governing equations in various coordinate systems	373
7.15	Problems	378

8 Source terms — 379

8.1	Heat source terms due to metabolism and blood flow	380
8.2	A generic form for the heat source term	380
8.3	Heat source term for electromagnetic heating	382
8.4	Microwave heating and its heat source term	384
8.5	Radiofrequency heating and its heat source term	389
8.6	Ferromagnetic heating and its heat source term	392
8.7	Infrared heating and its heat source term	393
8.8	Laser heating and its heat source term	395
8.9	Ultrasonic heating	401
8.10	Mass source terms	405
8.11	Summary	408
8.12	Problems	411

9 Material properties and other input parameters — 414

9.1	What material property data and input parameters do we need?	414
9.2	Where do we get data?	415
9.3	How accurate should the data be?	416
9.4	What to do when accurate data is not available	416
9.5	Anatomical and physiological parameters	417
9.6	Rheological properties	424
9.7	Thermal conductivity	427
9.8	Specific heat	429
9.9	Density	433
9.10	Thermal diffusivity	433
9.11	Thermal properties of related materials	434
9.12	Latent heat of fusion and evaporation	434
9.13	Radiative properties	434
9.14	Equilibrium vapor pressure	435
9.15	Properties of an air–water vapor mixture	436
9.16	Mass diffusivity	436
9.17	Partition coefficient	439
9.18	Diffusive permeability and transmissibility	439
9.19	Reaction rate constants	443

9.20	Other parameters	444
9.21	Summary	450
9.22	Problems (short questions)	453

10 Solving the equations: numerical methods — 455

10.1	Flexibility of numerical methods	456
10.2	Finite difference method (FDM)	458
10.3	FDM: converting the 1D heat equation to algebraic equations	460
10.4	FDM: stability (limitations in choosing step sizes)	464
10.5	FDM: summary	464
10.6	Finite element method (FEM)	465
10.7	FEM: converting the 1D heat equation to algebraic equations	465
10.8	FEM: solving the linear system of algebraic equations	478
10.9	FEM: choice between linear solvers	483
10.10	FEM: linearization of non-linear equations	483
10.11	FEM: error in the finite element method and its reduction	485
10.12	FEM: convergence of the numerical solution as the mesh is refined	487
10.13	FEM: stability of the numerical solution	488
10.14	FEM: generalization of methodology to more complex situations	488
10.15	FEM: summary	490
10.16	Problems	491

Index — 497

Preface

Simulation is an important component of the engineering design process in many sectors. The integration of simulation into undergraduate engineering education in an appropriate manner, so that it enhances the fundamentals, and also provides students with a cutting-edge tool, has been in the forefront of education thinking, as evidenced by the interest in a recent workshop at Cornell University (ISTEC, 2008), and in a report by the National Science Foundation (NSF, 2006). The tremendous growth in biomedical engineering over the last 10–15 years has encouraged increased quantitative treatment of biomedical product, process and equipment design, and design of treatment procedures. Such quantitative treatment has made simulation into a useful tool in biomedical applications as well. The synergy between increased use of simulation and the availability of improved interfaces has brought down the barriers to the use of simulation, from only specialized modelers, to just about anyone who has the necessary prerequisite of the physical process (engineering science content such as heat transfer or mass transfer). The increased need in industry and research, and the lower barrier to modeling can be integrated further by having all the essential information under one umbrella – which is the goal for this book. This introductory book walks a person without any prior knowledge in modeling through all of the necessary steps thus helping them to join the modeling community, and thereby enabling a productivity tool for design and research.

Although more widespread use of modeling in practical terms means that a typical modeler will now have a less formal background training than in the past, training material that introduces modeling to the less well prepared is elusive. We have made our primary goal in this book to provide a smooth, easy-to-follow introduction which takes the newcomer through the entire model development, formulation from physical to mathematical, required biomedical data on all parameters, complete implementation in software, validation of the solution, making sense of the data and, finally, sensitivity analysis and optimization. It is designed for someone with a transport-phenomena background as a self-study tool, in order that they can walk through the entire process without any external help; thus making it suitable in both industry and introductory classroom situations. In academia, although the material was used as a follow-up course to heat transfer or heat and mass transfer, the gentle approach of the book will allow, without additional instruction, students to perform small projects and thus enhance an introductory transport process course.

The needs of the introductory modeler, without an elaborate background in modeling, were far from obvious when this course was started in 1996 (Datta, 2009). It is through observing the students (3rd and 4th year engineering) and the questions they asked, that we learned where they needed the most help. For example, problem formulation from a physical process to a mathematical one, was clearly somewhere they needed significant help – this led to the development of a complete chapter (Chapter 1). The instructors learned from the process just as much. For example, problem formulation is not an explicit topic in any typical curriculum, however, it made us think more about this topic, which also happens to be at the center of the engineer's ability to solve problems. Likewise, much of the book is centered around the precise information that the students needed to complete their modeling projects. Since its inception, the students on this course have completed close to 100 projects, most of which have been learning experiences for the instructors as well. This knowledge has been distilled and incorporated into the book.

Obviously, the book intends to develop a modeling background which is not meant to replace more specialized training, typically graduate coursework in computational fluid flow, heat transfer and so on (computational fluid dynamics or CFD). Rather, the intent here is that the user of the book should be able to perform less-complex simulations and, for more complex simulations, be a more effective team member. Also, the person will be sufficiently familiar with simulation methodology to be competent to take part in discussions involving simulation.

To summarize, we feel the novelties of this book are three-fold: (1) introducing modeling to an audience with less formal and specialized training in modeling, and to undergraduates; (2) bringing the context of emerging biomedical processes; and (3) to achieve this without making the process a complete blackbox. Our hope is that as modeling in biomedical processes and modeling in general become more ubiquitous, this book will serve the need for gentle introduction for the beginner and a resource that can be put to use right away in the classroom situation, academic research or industrial design and research.

References

Datta, A. K. Computer-Aided Engineering: Application to Biomedical Processes. Course home page. Dept. of Biological and Environmental Engineering, Cornell University. 1 Jan. 2009 <http://courses.cit.cornell.edu/bee4530/>

ISTEC. 2008. Integration of simulation technology into the engineering curriculum: A University – Industry Workshop, July 25–26, 2008, at Cornell University, Ithaca, New York. On the web at http://www.mae.cornell.edu/swanson/workshop2008/.

NSF. 2006. A Report of the National Science Foundation Blue Ribbon Panel on Simulation-Based Engineering Science. On the Web at http://www.nsf.gov/pubs/reports/sbes_final_report.pdf

Acknowledgments

We are indebted to the students of BEE 453 from 1996 onward at Cornell University, working with whom has led to the development of this book. The course could not be taught (and therefore the book would not exist) without the great teaching assistants in this course who have helped us to think about how to introduce simulation to a young audience and, most importantly, how to get their many projects to complete when we ran into problems, as we often did. The graduate teaching assistants include Haitao Ni, Hua Zhang, Jifeng Zhang, Shrikant Geedipalli, Amit Halder, Ashish Dhall, JMR Apollo Arquiza and Vineet Rakesh himself. We are also indebted to Sarah Snider, Siddharth Khasnavis, Gwen Owens, Frank Kung, Jackie Arenz, Amit Halder, Ashish Dhall, JMR Apollo Arquiza, and Kunal Mitra for various comments on portions of the manuscript and/or other input. The departmental and university support that made the teaching of this course possible through supporting the hardware, software and teaching assistants is also gratefully acknowledged. In particular, support from the Innovation in Teaching Program at Cornell University helped brainstorm improvements in presenting the material to a young audience. The help of Ms. Valorie Adams in obtaining the many copyright permissions needed for this book is much appreciated. Finally, Ashim Datta would like to express his gratitude to his wife Anasua, and daughters Ankurita and Amita, for sacrificing their quality time during the preparation of this book.

Introduction and overview

What does modeling involve? Why should we care about modeling? Do I have to be a math whiz to succeed? How can this book help me to get started in modeling (in case I do care)? These are some of the questions we try to answer in this introductory chapter. This chapter is critical and should not be skipped.

What is modeling?

By modeling, we mean developing a replica on a computer of a physical process that interests us so that we can manipulate the process on the computer. In contrast with a computer-aided design (CAD) model which deals mostly with geometric or solid modeling, shading, etc., we must include the detailed physics of the system in order to evaluate its performance. In short, such a model involves simplifying the geometry and physics of a real situation and solving the simplified equations that describe the physics, using a software that is primarily an equation solver.

You have modeled before. As a child, you learned that the area of a trapezoid is the height multiplied by the average of the two bases. If we program this on a calculator, so that we only have to input the height and the two bases and the calculator spits out the area, we have a model for area calculation. Most real areas will not be exact trapezoids, but you will have to use your intuition in simplifying the geometry to that of a trapezoid so that you can use the program you wrote for the calculator (see O'Connor and Spotila, 1992). In this book we develop models that involve more complex calculations. We have to use our understanding of the physics to simplify the geometry and the process and we tell the computer to solve for this situation. This procedure is worthwhile if the computer can quickly solve a problem that is close to reality. The ongoing revolution in computer processing speed makes it possible now to do just that.

Why model?

You don't have to model. But let's look at two scenarios – you are trying to design either medical equipment or a medical procedure. In designing medical equipment, your choice is to build a prototype, test it, and evaluate its shortcomings to guide you to an improved prototype. Building and testing physical prototypes is expensive and *time consuming*. While this procedure cannot be totally eliminated, it sure would be nice to reduce the time and money involved in prototype building. In today's industrial setting, saving time (also called development cycle time) alone can justify alternatives. With a realistic computer model, we can check out the expected performance of the equipment for different scenarios on the computer before we build the first prototype. This is often faster than building and testing physical prototypes.

Another scenario to consider would be the development of a new medical procedure. The comments made above about time and resources do apply to this case but, perhaps more importantly, the safety of the new procedure becomes more critical. Again, a realistic computer model can allow us to reduce the amount of experimentation that is typically needed to optimize a new procedure. Using a computer model, we can improve the performance and safety of the procedure before it is actually carried out on a patient.

Thus, models primarily provide vastly *improved understanding* of a process that is often not possible to achieve in other ways. The other major advantage stems from the fact that, once a model is built, it is much easier to modify. By repeated use of the model, we can check "what if" scenarios, making it much easier to *optimize* a process. Models can often (but not always) provide improved understanding and optimization in less time and with fewer resources. It is also inherently safer.

Many other engineering applications have embraced modeling as part of the design process for industries such as automotive, aeronautical, manufacturing, etc. A major push has occurred in recent years in the quantitative understanding of biomedical systems; modeling is therefore becoming a standard biomedical engineering tool.

What preparation do I need?

To develop the modeling skills presented here, one needs to have a relevant background in the fundamental physics of the processes under consideration. For the types of processes considered in this book, this means having a background in fluid flow, heat transfer, and mass transfer. In the past, one also had to learn about numerical methods and computer programming (not to mention all the time that would be involved in actually writing the programs needed). Now the time and experience needed for this has been reduced dramatically as computers continue to run faster and become more user friendly. Insofar as it is possible to model problems today that involve less complex physics, we can now almost avoid these procedures altogether. Thus, much of the mathematics is hidden and it is no longer necessary to bring knowledge of numerical methods and computer programming to the task of modeling.

So what level of preparation do I need in fluid flow, heat transfer and mass transfer? The answer to this, like that for any other subject, depends on the complexity of the physics that you want to study. For the types of problems mentioned in this book, introductory engineering or science courses in these topics are considered sufficient. The goal of this book is to help you solve problems yourself that are not overly complex. For more complex problems, this book will prepare you to work effectively in a group having members with more specialized or advanced background in those topics.

With the userfriendliness of software, one may wonder if training in the fundamental physics can be skipped since "the computer is taking care of the solution." As an example, this is akin to skipping the understanding of the sine curve since a calculator is available that can easily compute a sine curve with the touch of a button. The understanding of what a sine curve is must precede the rapid calculation of the same. Otherwise, it is hard to make sense of what is computed and one would never know if different types of error had crept in (for example, you punched a wrong button!). This analogy very much applies to modeling, as it is discussed here. Depending on the physics that your problem involves, having a background in the subject matter, such as fluid flow, cannot be avoided. For, if we do, we would not know the difference between a solution and garbage, the latter being common due to the very nature of computer (numerical) solutions.

How does modeling fit into design?

Typical steps in a biomedical design procedure (e.g., see King and Fries, 2009) are shown in Figure 0.1. As noted in this figure, several of the design steps can benefit from modeling through trying out various possibilities on the computer, or pre-optimizing.

Overview of the text

This text is intended to provide a quick start to modeling and Figure 0.2 shows how this is organized. Thus, the first five chapters walk the reader through the essential steps in modeling. *Chapter 1* deals with problem formulation, the all-important topic of how to take a real situation and simplify it in terms of equations that describe the process. *Chapters 2, 3 and 4* simply show how to tell a typical software

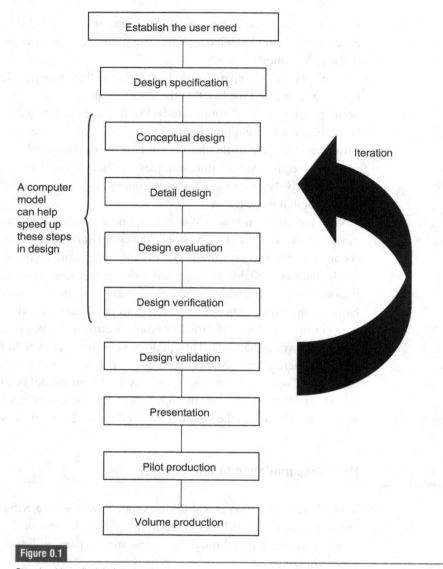

Figure 0.1

Steps in a biomedical design procedure.

application to solve the equations that we have selected (in problem formulation) and how to view the solution that we obtain. *Chapter 5* covers two important topics – validating a solution that we have just obtained, and extracting trends from the solution by changing parameters, also called sensitivity analysis, on our road to optimization of the process. *Chapter 6* provides case studies that show many examples of modeling, applying them to a number of biomedical situations. This completes the quick start to modeling. *Chapters 7–10* are provided as reference

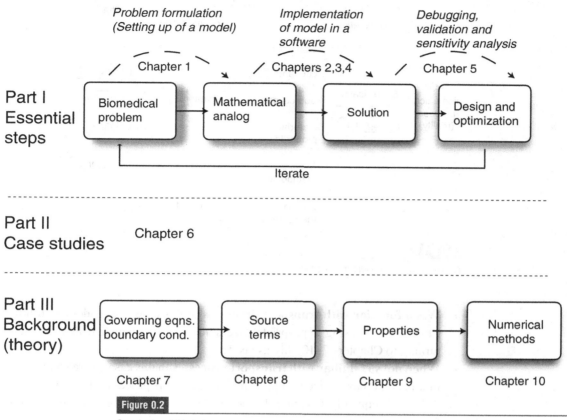

Figure 0.2

Organization of this text.

material that one can dig into, as needed. These chapters are also intended to provide, in a simple manner, the theory on which the computations are based – this should encourage the user to treat the software less as a black box. *Chapter 7* provides a quick and simple overview of how governing equations are derived. *Chapter 8* discusses how to include in the model the various modes of biomedical heating processes. *Chapter 9* is important as it discusses how one obtains the input parameters, particularly the properties, needed for the model. Finally, *Chapter 10* gently introduces the mathematical details behind numerical modeling.

Possible ways of using this text

The book is written to provide do-it-yourself training. Figure 0.3 shows another representation of the organization of this text. Depending on the user's familiarity with transport processes, two obvious approaches to using the book are:

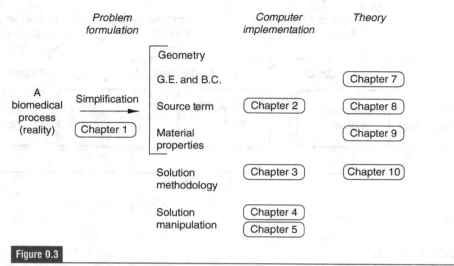

Figure 0.3

Another look at the organization of this text.

- **When familiar with transport processes** In this case, the reader can follow Chapters 1–5, using appropriate case studies from Chapter 6 as examples, turning to Chapters 7–10 only as needed.
- **When not so familiar with transport processes** In this case, the reader should first become familiarized with, or review, transport processes using Chapters 7–10. Subsequently, Chapters 1–5 are followed with appropriate case studies from Chapter 6 as examples.

Reference

King, P. H. and R. C. Fries. 2009. Design of Biomedical Devices and Systems. CRC Press, Boca Raton, Florida.

O'Connor, M. and J. R. Spotila. 1992. Consider a spherical lizard: Animals, models, and approximations. *American Zoologist*, **32**(2): 179–193.

List of symbols

a_w	water activity, dimensionless
A	area, m^2
Bi	$= hL/k$, Biot number, dimensionless
Bi_m	$= h_m K^* L/D_{AB}$, mass transfer Biot number, dimensionless
BMR	basal metabolic rate, kcal/kg/day
c	total concentration (sometimes abbreviated c_A), kg/m^3
c	speed of light in vacuum, m/s
c_*	speed of light in a medium, m/s
c_A	concentration of component A, kg of A/m^3
$c_{A,s}$	concentration of component A at a surface, kg of A/m^3
$c_{A,\infty}$	concentration of component A in the bulk fluid, kg of A/m^3
c_{av}	average concentration, kg/m^3
C_p	specific heat at constant pressure, kJ/kg·°C
C_{pa}	apparent specific heat, kJ/kg·°C
$CEM43$	Cumulative number of equivalent minutes at 43°C
D	diffusivity, as a general reference
D_{AB}	diffusivity of species A in species B, m^2/s
$\vec{E}$	electric field vector
E_a	activation energy, J/mole
f	force per unit volume, N/m^3
f	fraction of ice in an ice-water system
F	energy flux, W/m^2
F	force, N
Fo	$= \alpha t/L^2$, Fourier number, dimensionless
g	gravity, m/s^2
h	convective heat transfer coefficient, W/m^2·°C
h_r	radiative heat transfer coefficient, W/m^2·°C
h	Planck's constant, 6.625×10^{-34} J·s
h_m	convective mass transfer coefficient, m/s
H	enthalpy per unit mass, kJ/kg
H	Henry's law constant, atm/mole fraction

$\vec{H}$	magnetic field vector
I	intensity or energy flux, W/m²
I_0	incident energy flux, W/m²
k	thermal conductivity, W/m·°C
k_a	absorption coefficient
k''	reaction rate constant
K	partition coefficient, units vary
L	half-thickness of a slab or characteristic length, m
m	consistency coefficient; also mass fraction
m	mass, kg
M	molecular weight, kg
M	metabolic rate, W
$\dot{m}$	mass flow rate, kg/s
n	flow behavior index, dimensionless
n	order of reaction, dimensionless
$n_{A,z}$	mass flux of species A in the z direction, kg/m²·s
Nu	Nusselt number, dimensionless
p_A	partial pressure of component A, N/m² or Pa
P	total pressure, N/m² or Pa
P^*	non-dimensional pressure
P	power, W/m²
Pe	$= Re \cdot Pr$, Peclet number, dimensionless
Pr	$= \mu C_p/k$, Prandtl number, dimensionless
q_x	heat flow in the x-direction, W
q_x''	heat flux in the x-direction, W/m²
Q	volumetric heat generation, W/m³
Q_m	metabolic heat generation per unit volume of tissue, W/m³
Q^*	volumetric heat generation, non-dimensional
r	radial direction
r_A	rate of generation of A per unit volume, kg of A/m³·s
R	universal gas constant = 8.315 kJ/kmol·K
Ra	$= Gr \times Pr$, Rayleigh number, dimensionless
Ra_m	$= Gr_{AB} \times Sc$, mass transfer Rayleigh number, dimensionless
Re	$= \rho u_\infty L/\mu$, Reynolds number, dimensionless
Sc	$= \mu/\rho D_{AB}$, Schmidt number, dimensionless
Sh	$= h_m L/D_{AB}$, Sherwood number, dimensionless
t	time, s
T	temperature; stress
T_s	surface temperature
T_i	initial or inlet temperature

T_f	freezing point
T_R	reference temperature
T_∞	fluid or ambient temperature
u	velocity in the x-direction, m/s
u^*	dimensionless velocity in the x-direction, m/s
u_∞	free stream velocity in the x-direction, m/s
U	thermal energy per unit volume, J/m^3
U	overall heat transfer coefficient, W/m$^2 \cdot$°C
U_m	overall mass transfer coefficient, m/s
X	source term in momentum equation, m/s^2
v	velocity in y-direction, m/s
v_x, v_y, v_z	velocities in x, y and z-directions, respectively, m/s
$\dot{V}_b^v$	volumetric flow rate of blood per unit volume of tissue
V	volume, m^3
w	moisture content, kg of water/kg of dry solids
x	x-coordinate, m
x_A	mole fraction of species A in liquid phase
y	y-coordinate, m
z	z-coordinate, m

Greek Letters

α	thermal diffusivity, m^2/s
α	absorption coefficient (ultrasound)
β	coefficient of thermal expansion
β	non-linearity coefficient (ultrasound)
δ	penetration depth of microwaves
Δ	finite change
ϵ	emissivity, dimensionless
ϵ_0	permittivity of free space, $= 8.86 \times 10^{-12}$ F/m
ϵ'	relative dielectric constant, dimensionless
ϵ''_{eff}	relative dielectric loss, dimensionless
ϕ	electric potential, volts
κ	Boltzmann's constant, 1.380×10^{-23} J/K
λ	latent heat, kJ/kg
λ_f	latent heat of fusion, kJ/kg
λ_{vap}	latent heat of vaporization, kJ/kg
λ	wavelength, m
λ_0	wavelength in free space, m

μ	viscosity, kg/m·s
μ_a	absorption coefficient, m^{-1}
μ_s	scattering coefficient, m^{-1}
ν	kinematic viscosity, m^2/s
ω	$= 2\pi f$, angular frequency (f is frequency, in Hz)
Ω	extent of injury in a thermal burn
ρ	density or mass concentration, kg/m^3
σ	Stefan–Boltzmann constant, 5.676×10^{-8} W/m^2·K^4
σ	electrical conductivity, Siemens
σ_{AB}	collision diameter, Å
$\Omega_{D,AB}$	dimensionless function, defined in Chapter 9
τ	stress, Pa
τ	thermal relaxation time
θ	non-dimensional temperature

Essential steps

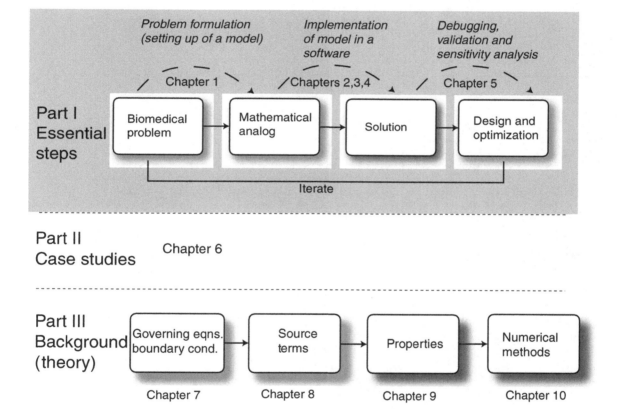

1 Problem formulation
From reality to realistic computer representation

In this chapter we will learn about problem formulation, which is the first step in developing a mathematical model of a physical (such as a biomedical) process, as illustrated in Figure 1.1. In problem formulation, we take reality, make assumptions thereby simplifying it, and apply universal physical laws to generate the equations (*the model*) which describe the real physical process. It is critical to see that everything we will learn from our model depends on how we have formulated the problem. This chapter provides the big picture of problem formulation, with additional details available in theory chapters (7–9). As shown in Figure 1.1, problem formulation in this chapter will be followed by subsequent chapters on the solution process.

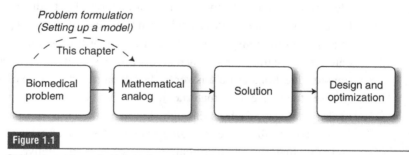

Figure 1.1
Problem formulation as the first step in modeling.

1.1 Context: biomedical transport processes

Transport processes, that is fluid flow, heat transfer and mass transfer, often underlie a biomedical process, perhaps the most common examples being drug delivery and thermal therapy. The relevance of heat and mass transfer to biomedical processes is now introduced.

1.1.1 Heat transfer and thermal therapy

Heat transfer refers to movement of thermal energy due to conduction, convection or radiation. Thermal therapy is any treatment or technique that elevates or decreases cell/tissue temperature for some length of time with an ultimate therapeutic goal. Thermal therapy can include hyperthermia, tissue coagulation and ablation as well as ultrasonic, laser, radiofrequency and microwave heating to destroy tissue, plus cryosurgery, burn therapy, bone growth stimulation, wound healing and thermally mediated gene therapy. Clinical applications include deep heating for various musculoskeletal diseases (rheumatoid arthritis, osteoarthritis, fibrositis and myositis), deep heating for many neuromuscular disorders (muscular dystrophy, progressive muscular atrophy), treatment of various eye disorders (iritis, postoperative uveitis), dental problems (swelling and trismus following extractions, toothache), elevating body temperature following hypothermia surgery and cancer therapy using hyperthermia (40–50 °C). Thus, modeling of thermal therapy would require modeling of heat transfer.

1.1.2 Mass transfer and drug delivery

Mass transfer refers to movement of a material due to diffusion and convection. Drug delivery can be described as the process of delivering a drug to the site of action. Drug transport, i.e., drug diffusion or flow, is intimately related to drug delivery. Even traditional drug delivery processes, such as oral intake, require mass transfer questions to be answered. For example, how do we design a tablet that releases the active drug material at a rate that is nearly constant? Newer methods of drug delivery, such as through skin as in the case of a patch placed over the skin, often require somewhat greater understanding of mass transfer through skin and other materials. In a critical process such as drug delivery in the brain, a mass transfer model that includes diffusion and elimination of drug can provide valuable insight into the complex process. Thus, modeling of drug delivery processes would require modeling of mass transfer.

1.1.3 Quantification of goals in a biomedical process

The goals of a thermal therapy or a drug delivery process need to be stated quantitatively. Examples of quantitative measures of several biomedical processes are shown in Table 1.1.

Table 1.1 Examples of quantitative measures of success in various biomedical processes.

Process	Design goal in quantitative terms
Radiofrequency heating of tumorous tissue	Reach temperatures in the tumorous tissue in the region of 43–45 °C.
Laser ablation	Reach temperatures in the region to be ablated above 300 °C.
Cancer therapy	Cumulative number of equivalent minutes of heating at 43 °C, defined by Eqn. 8.39, needs to reach a certain value.
Cryosurgery	Tissue to be destroyed needs to reach below −45 °C.
Drug delivery (design of a coated tablet, design of a patch)	Rate of drug release over time has to meet certain criteria such as a minimum dose; rate needs to stay near constant over time; release needs to be over a certain time period.
Drug delivery (placement in tissue, as in a brain tumor)	Penetration distance into the tissue (area of coverage) has to meet certain criteria. Additional criteria can be from those listed under tablet and patch.

1.2 What is problem formulation?

Problem formulation is creating an equivalent mathematical formulation of a physical problem, i.e., coming up with equations which describe the physical process or processes that constitute the problem (and therefore virtually replace such a process or processes). It is the first step in modeling. Consider, for example, the whitening strip in Figure 1.2 which is placed over teeth to remove unwanted stains. Hydrogen peroxide from the strip diffuses into the teeth and reacts with the stain (an organic material), thus removing the stain and whitening the teeth. We would like to know the rate of diffusion of the hydrogen peroxide into the teeth over time which will provide the time needed to whiten them. This is the problem in the physical world, as shown on the left side of the figure. In order to simulate this physical problem on the computer, we need to describe the physical process in terms of mathematical equations (i.e., develop a problem formulation).

The right side of Figure 1.2 shows the mathematical equivalent, or analog, that is achieved after many simplifications of the real process. The mathematical analog consists of a geometry (computational domain), governing equation, boundary

6 Problem formulation

A biomedical process →*Problem formulation*→ Its simplified mathematical equivalent

A plastic strip with a film containing hydrogen peroxide is placed on the teeth to whiten them (see Figure 1.22 for a section of a tooth).

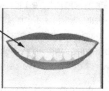

Goal
To find out how far into the teeth the whitening progresses and how that changes with time

Goal
To find the concentration of hydrogen peroxide in the teeth as it varies with position and time

Schematic

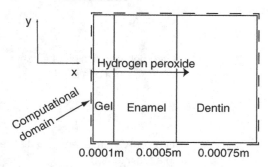

Governing equation

$$\frac{\partial c}{\partial t} = D\frac{\partial^2 c}{\partial x^2} + r_A$$

Boundary conditions

$$\left.\frac{\partial c}{\partial x}\right|_{x=0} = 0 \quad c(x=l) = 0 \quad \left.\frac{\partial c}{\partial y}\right|_{y=0} = 0 \quad \left.\frac{\partial c}{\partial x}\right|_{y=h} = 0$$

Initial condition

$$c = \begin{cases} 2.8 \times 10^{-5} \text{ g/liter in gel} \\ 0 \text{ elsewhere} \end{cases}$$

Properties and Parameters

$$\text{Diffusivity} = \begin{cases} 1.3 \times 10^{-9} \text{ m}^2\text{/s (gel)} \\ 7.8 \times 10^{-11} \text{ m}^2\text{/s (enamel)} \\ 7.8 \times 10^{-11} \text{ m}^2\text{/s (dentin)} \end{cases}$$

$r_A = 1.871 \times 10^3$ g/s

Figure 1.2

Illustration of problem formulation where a real process (whitening of teeth) is transformed into its computational model consisting of a goal, a computational domain, governing equation, boundary conditions, initial condition and material properties. Properties data taken from Bermudez *et al.* (2004).

conditions, material properties and other parameters that define the real process. The computational domain is the region over which computations will be performed. The governing equation represents the conservation of mass species, H_2O_2 in this case. Boundary conditions are the conditions imposed by the surroundings on the computational domain. In this figure, the first condition, $\partial c/\partial x = 0$ at $x = 0$, implies that H_2O_2 cannot escape into the air, while the second condition, $c(x = L) = 0$, means that H_2O_2 cannot penetrate very far into the teeth and therefore, at a far away place given by $x = l$, the concentration stays at $c = 0$. This mathematical analog can now be used in a computer to simulate the problem in the physical world. For the purpose of understanding and optimizing of the physical process, the mathematical analog can now be a substitute.

Where does this mathematical analog come from? A mathematical analog uses the fundamental laws of the physical world in a mathematical form. The fundamental physical laws that are used in the problems of interest to us are: (1) conservation of total mass (continuity equation); (2) conservation of a mass species (mass transfer equation); (3) conservation of momentum (fluid flow or Navier–Stokes equations); and (4) conservation of thermal energy (heat transfer equation). These are called the governing equations – they are presented in Section 1.7 and are derived in Chapter 7.

1.3 Steps in problem formulation

Problem formulation is perhaps the most critical activity in modeling a process. On the one hand, this step can be made overly complex by retaining a lot of unnecessary details in the model that will increase the computational complexities greatly. On the other hand, if simplification is not done carefully, the main physics of the process can be lost, leading to a worthless model for the purpose of simulating the physical process of interest. For these reasons, all of the steps in problem formulation require simplification, based on understanding of the process in terms of the fundamental physics, such as fluid flow, heat transfer or mass transfer. Problem formulation is also critical because the results we will obtain simply reflect how the model has been set up.

Depending on how we proceed in problem formulation, we can end up with different sets of governing equations and boundary conditions to solve. Analytical solutions to these equations, of the kind found in first courses in transport processes, are possible only for simpler forms of these equations which force a more drastic simplification of the physics. As numerical solutions, the kind that will be used in this book are considerably more flexible than analytical solutions, many more details of the physical process can be included, making the formulated

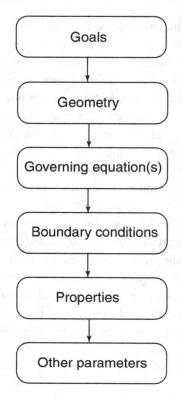

Figure 1.3

Steps in problem formulation. Each of the steps requires some simplification, as discussed throughout this chapter. For an example, see Figure 1.2.

mathematical problem considerably more realistic, i.e., closer to the actual physical situation. As we go through the various steps of problem formulation, we can always consider more realistic alternatives.

An example of problem formulation is shown in Figure 1.2. The process of arriving at the mathematical formulation can be divided into several steps, as shown in Figure 1.3. We first decide on the eventual goal of the simulation – this sets the stage for approaching the steps that follow. Next, we decide on the geometry or the region over which simulation will take place. This is followed by the choice of the governing equations and boundary conditions that will describe the process. To actually solve these equations for specific cases, we need the material properties and parameters corresponding to the situation.

Guidelines on performing the primary steps of problem formulation (geometry, governing equations, boundary conditions and properties) are provided in this chapter, Chapters 7–9 on theory and various software implementations in Chapters 2, 3 and 6. This interrelationship between chapters is illustrated in

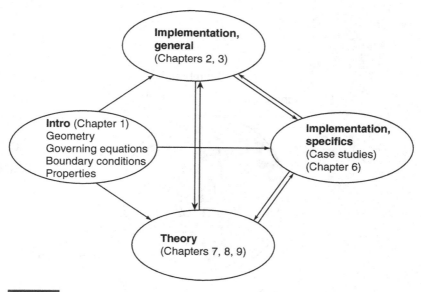

Figure 1.4

Relationship between the components of problem formulation and their software implementation, as presented in various chapters. Cross-references between chapters on a topic are highlighted throughout the chapters. For example, boundary conditions are introduced in Chapter 1, its general implementation in COMSOL is discussed in Chapter 2, more specific implementations discussed in Case studies in Chapter 6 and its theoretical description is provided in Chapter 7.

Figure 1.4 and is highlighted through cross-referencing in the individual chapters. Chapters 4 and 5 provide ways to visualize and further analyze the results.

1.4 Defining goals for problem formulation

Although the general goal of modeling is to improve understanding and facilitate optimization, it is critical to define the specific goals of a problem formulation clearly at the outset, as the formulated problem very much depends on what exactly we want to achieve from it. For example, Figure 1.5 shows two different formulations for the physical situation of drug delivery through skin, using a patch. In the first formulation the primary goal is to look at the effect of penetration enhancers on diffusion through skin. Details of the patch construction are probably unnecessary for this goal.

The second has the goal of finding the effect of using three different materials for the microporous membrane (with correspondingly varying diffusivities) on the rate of drug delivery. Obviously, the details of the membrane would have to

10 Problem formulation

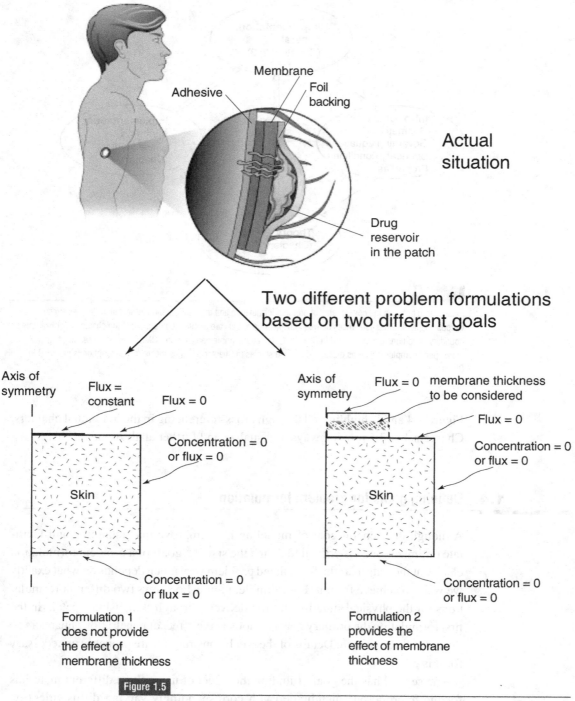

Figure 1.5

Two different goals lead to two different problem formulations for approximately the same physical problem. In Formulation 1, the goal is to study the drug transport primarily in the skin region. In Formulation 2, the goal is to study the drug transport inside the patch as well as in the skin.

be included in this problem formulation. A list of possible goals in biomedical problem formulation is shown in Table 1.1.

1.5 Simplify, simplify, simplify

We always simplify to achieve the least complex problem formulation possible, keeping in mind the important phrase "as simple as possible, but no simpler." Sometimes we simplify because the software cannot handle the required complexity. We simplify at other times to be within the limits of computing resources (memory and cpu speed) available. Even if we need to eventually solve the more complex problem, it is often instructive to solve a simpler problem first as simplification enhances our own understanding of the problem. By dissecting the problem into distinct processes, we can focus on a very specific goal within the more complex physics. Starting simple also makes it easier to debug. Some possible simplifications are: (1) starting with a 2D instead of a 3D geometry, or a 1D instead of a 2D geometry; (2) starting with no heat transfer (isothermal formulation) even though the physical situation is non-isothermal and therefore heat transfer will eventually need to be added; (3) starting with constant properties instead of properties varying with temperature or concentration, etc.

1.6 Geometry: setting the computational domain

The computational domain is the chosen region of the physical domain (actual geometry) where computations will be performed. Generally speaking, the larger the computational domain, the more computation is required. Thus, deciding on the computational domain is a very critical step in problem formulation. Although today's numerical methods can handle various shapes and sizes, and computers have significant speed and memory, prudent choices must be made in simplifying the actual geometry; otherwise meshing can be difficult or we can run into cpu speed and/or memory limitations.

In the example given in Figure 1.2, the geometry can have several possibilities, as shown in Figure 1.6. Depending on the geometry chosen, the amount of computation increases dramatically, but the question we have to answer is: are we learning anything new as we move from 1D to 3D? If we think of the 3D geometry in terms of r, θ and z directions, it seems reasonable to assume that most of the diffusion of H_2O_2 will be in the r direction (since the boundary concentration is uniform in the θ and z directions, respectively). Thus, a 1D computation, as shown in Figure 1.6(c), should suffice. Sometimes, however, the software does not allow a 1D geometry. In this case, the 2D geometry can be made to represent a 1D physics

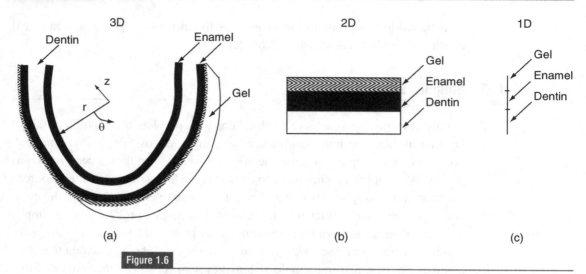

Figure 1.6

Various choices of geometry for the problem in Figure 1.2, showing 3, 2 or 1-dimensional approximations, respectively, of the same physical situation. Since computational complexity increases, often dramatically, as we go from 1D to 3D, we have to ask the question whether going to a higher dimension will truly provide additional information.

by using no flux boundary conditions at the two ends, as shown in Figure 1.6(b) (explained more fully with boundary conditions later).

Four of the important decisions that must be made in deciding on the computational domain are:

(1) What regions (entities or materials) need to be included?
(2) How much of a very large region should be included?
(3) How many dimensions are needed?
(4) How can symmetry be used to reduce the domain?

These decisions are inseparable from the physics of the problem, which needs to be considered in making the decisions.

1.6.1 What regions need to be included?

There are often two or more regions in a problem due, for example, to two different solid materials or a solid and a fluid (see Figure 1.7). Not all regions need to be included in the computational domain. This is illustrated in Figure 1.7 where we have two regions – fluid (blood) and solid (arterial tissue surrounding the blood). Depending on the physical process in which we are interested, it may be possible to stay completely in the fluid, with the solid as the boundary condition. Alternatively,

1.6 Geometry: setting the computational domain

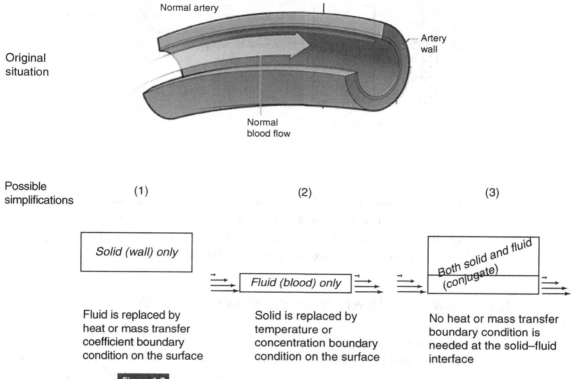

Figure 1.7

How a problem involving both solid and fluid may be simplified to consider only solid (1), or only fluid (2). An appropriate boundary condition at the solid–fluid interface is needed for such simplification. If such simplification is not possible, both regions are kept, as in (3). Artery figure reproduced from NHLBI (part of the National Institute of Health and the US Department of Health and Human Services) (http://www.nhlbi.nih.gov/).

it may be possible to stay completely in the solid with the fluid as a boundary condition (such as when using a heat or mass transfer coefficient). Of course, if neither is possible, we have to consider both solid and fluid simultaneously, which becomes more challenging. If we have two or more regions in the computational domain, the individual regions are also referred to as computational subdomains.

Also, depending on the situation, some equations might apply only to a certain region(s) and not others. For example, the flow equations will not be solved for a solid region, obviously. The exact implementation of governing equations and boundary conditions in the subdomains depends on the software.

Connectivity between the regions In drawing a geometry with multiple regions, such as for multiple materials as shown in Figure 1.8 for gel, enamel and dentin (extracted from Figure 1.2), one needs to make sure that the line drawn at an

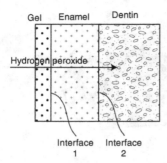

Figure 1.8

While drawing geometry with multiple regions, care needs to be taken so that the regions (i.e., gel and enamel, or enamel and dentin) stay connected in the software.

interface is shared by the two regions on either side of the interface. This means the line at the interface maintains that continuity between the two regions so that the fluxes of energy or mass on either side of the interface are the same. Often, this is automatic in the geometry program, but not always.

1.6.2 How much of a very large region should be included?

The larger the computational domain, the greater the requirement for computation memory and time. The computation time often increases disproportionately with the number of nodes (discrete locations in the computational domain where computations are performed), and thus one needs to be careful in deciding how much of a large domain needs to be included.

An infinite domain is an assumption made in developing analytical solutions (remember the error function solution of a heat equation in a semi-infinite region?). A numerical solution (and software) cannot simulate infinite regions literally (as it would need an infinite number of nodes to cover the region). The computational domain would simply need to be larger than the region over which the variables (such as temperature and concentration) undergo any appreciable changes. This may seem like an impossibility since the region over which variables will change is not known initially. In practice, one starts with an approximate size for the domain. For example, this could be based on a very approximate analytical solution, experimental data or, more commonly, intuition. Once the computations are made, the results can show whether the choice of domain size was appropriate.

An example of choosing a computational domain size for an effectively infinite domain is shown in Figure 1.9. The computational domain on the top of Figure 1.9(a) is much larger than necessary, since the variation in temperature for the three minutes of heating has stopped long before the far boundary, and therefore the effective infinite situation has been reached long before reaching the far

1.6 Geometry: setting the computational domain

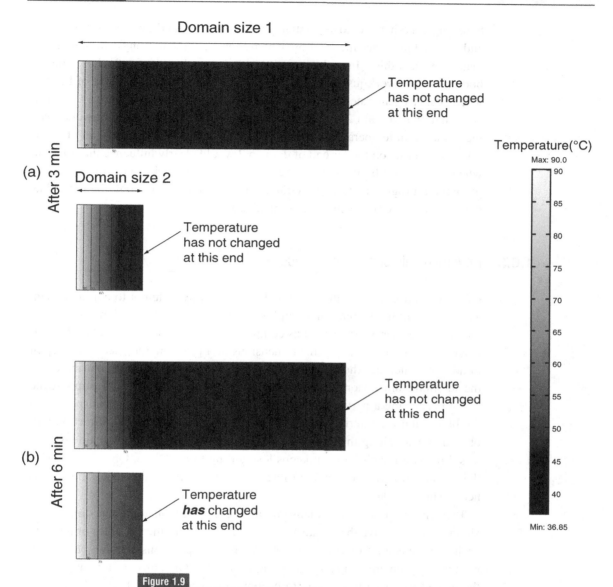

Figure 1.9

Two different computational domain sizes (larger on the top and smaller on the bottom) were chosen for the same material in a 1D transient heat transfer problem (Case study I in Chapter 6). Computations demonstrate that small domain size is indeed sufficient for a three-minute heating period (the cooler end stays at the initial temperature). However, the smaller domain is not sufficient for a six-minute heating period (the cooler end increases temperature over its initial value, as illustrated by the contours for domain size 2 in (b).

boundary. Another way to understand this is that the boundary condition at the far end does not influence the solution, which should be true in a physical sense if the material is very thick. In such a situation, the shorter domain size 2, chosen on the bottom, would work just fine. This situation changes for six minutes of heating when heat has penetrated further into the material, as shown in Figure 1.9(b). The computational domain 2 is now smaller than required and would not work since the variation in temperature has reached its far end. In other words, any boundary condition set on the far end of domain 2 would clearly influence the solution, which should not be the case if the material is very thick. Thus, in a transient problem, a progressively larger computational domain size would be needed as the temperature or concentration front advances.

1.6.3 How many dimensions are needed?

Of course, all problems are in 3D and we can always attempt to compute them as full 3D, but the increase in complexity due to the additional dimension often increases the user effort as well as computation time disproportionately. Thus, it makes sense to try to reduce the dimensions of a problem. One needs to keep in mind that sometimes the very physics of the process of interest is contributed by the additional dimensions. For example, with reference to Figure 1.7, suppose the blood is carrying a particular drug and we are interested in drug transport from the blood into the arterial tissue. The drug transport in the radial direction will obviously vary along the length of the artery, as the blood will contain less and less drug as it travels, since it keeps losing drug to the tissue. Obviously, due to the physics involved here and our interests, both the radial and axial dimensions need to be considered.

To decide on how many dimensions are needed, we look to find the dimensions along which we have the greatest number of changes in the variables of interest, such as temperature or concentration. As an example, consider simulation of a burn injury problem, as illustrated in Figure 1.10(a). Of course, heat transfer from the heated disc into the skin is truly three-dimensional, i.e., heat will move in all three directions into the skin. However, for the expected small duration, since the temperature change into the skin (in the negative z direction) will be over a small distance compared to the diameter of the disc, unless we are particularly interested in the region near the disc edge, temperature changes in the x and y directions are likely to be small. Thus, the temperature variation would be primarily in the z direction and therefore it may be possible to consider the process as one-dimensional in the z direction. If the disc is not very large, the edge effects can be significant and a 1D approximation may not be sufficient. As explained

1.6 Geometry: setting the computational domain

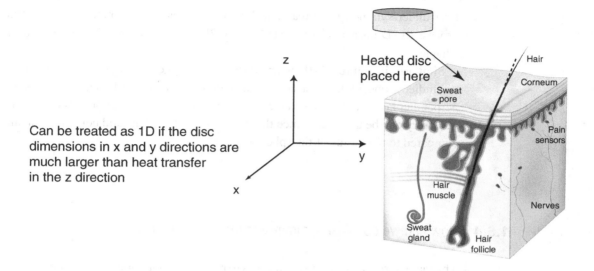

(a) Cross-section of a skin with a heated disc placed on it to simulate a burn injury

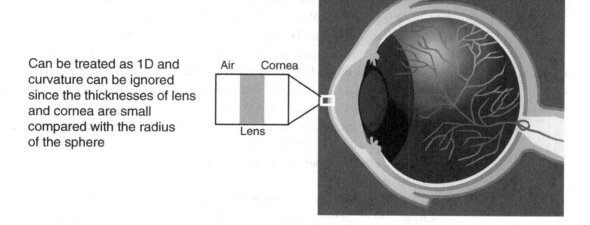

(b) Oxygen diffusion through a contact lens into the cornea

Figure 1.10

Examples of situations that can be simplified to 1D.

later under symmetry, we may still be able to consider symmetry, reducing the process to 2D instead of a full 3D (see the 2D formulation in Case study VIII in Chapter 6).

Similarly, in Figure 1.10(b) where diffusion of oxygen through a contact lens is to be studied, concentration variation can be considered primarily one-dimensional, perpendicular to the lens surface. Also, curvature can be ignored (i.e., planar geometry can be assumed) since the thicknesses of the lens and cornea are small compared to the radius of the sphere (eye).

1.6.4 How can we consider symmetry to reduce the domain?

The main purpose of using symmetry is to reduce computation by reducing the size of the computational domain. Figure 1.11 shows three examples where the computational domains can be made much smaller than the actual physical domains, by the use of symmetry. Note that *symmetry* about a line is different from *axisymmetry* which is more restrictive and means symmetry about any plane passing through the axis. In Figure 1.11(a), symmetry about the x, y, and z axes of the 3D rectangular block (the physical domain) is used to reduce the corresponding computational domain to a 2D square of only a quarter of the cross-section of the original physical domain. When a spherical domain is axisymmetric about one axis and symmetric about the other, as illustrated in Figure 1.11(b), only a quarter circle (2D geometry) needs to be used in a computation. When the same spherical domain is axisymmetric about any axis, only a line (1D geometry) needs to be used in a computation, as shown in Figure 1.11(c). Note that in all cases, symmetry refers to that in *physics, geometry, boundary and initial conditions*.

Conversely, when we draw a 2D geometry in the software as the computational domain, but treat it as either 2D or axisymmetry in the analysis (e.g., heat or mass transfer), we effectively simulate various 3D situations in the physical domain, as illustrated in Figure 1.12.

For a problem to be treated axisymmetrically, the physical domain, the boundary conditions, the properties and the initial condition all need to be axisymmetric. The same requirements apply to symmetry about a line.

Depending on our region of interest, these requirements can be relaxed. For example, Figure 1.13 shows a situation where axisymmetry can be assumed even when the original geometry is not axisymmetric. This is when the patch is large enough and we are interested in the 1D diffusion of drug in the central region. The corresponding axisymmetric geometry is also shown in Figure 1.13.

1.6 Geometry: setting the computational domain

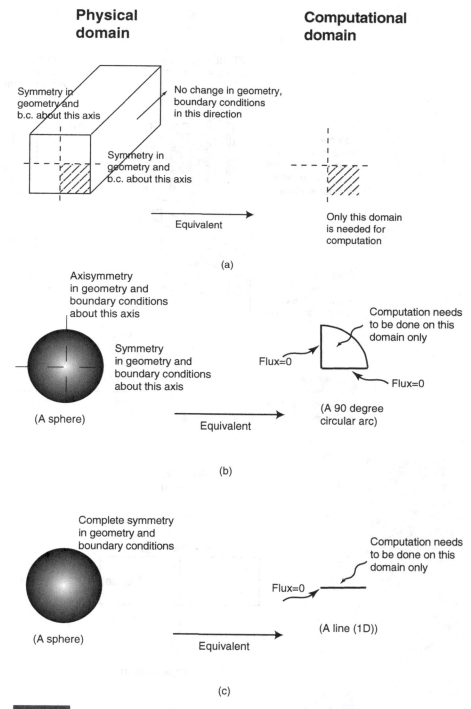

Figure 1.11

Possible reductions of various geometries by using: (a) symmetry; (b) symmetry and axisymmetry; (c) complete spherical symmetry.

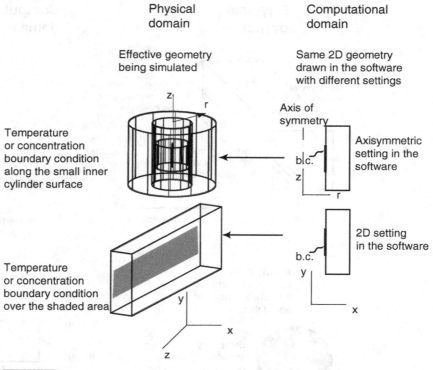

Figure 1.12

The 2D geometry on the right used for computation corresponds to two different 3D geometries, depending on whether the analysis is set to 2D or axisymmetry in the software.

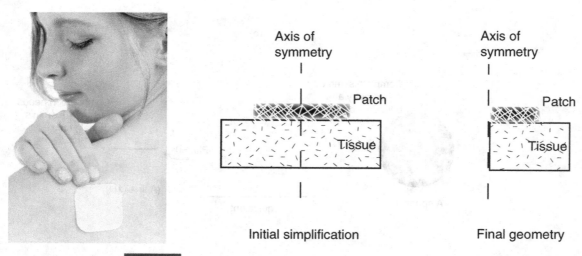

Figure 1.13

Example of a situation where we may be able to assume axisymmetry and thereby reduce the size of the computational domain. Although the original patch is square, unless we are interested near the edges, we can assume a circular patch of the same area and thereby treat it as axisymmetric.

Figure 1.14 shows a situation where axisymmetry cannot be assumed because of the non-axisymmetry of a number of factors. In the top figure, obviously, the geometry is non-axisymmetric. In the bottom figure, for the stent wire, the boundary conditions are different on the two sides (blood flow on one side and plaque on the other), making the problem not axisymmetric.

How to implement a 1D problem in 2D Although the software COMSOL can implement a truly 1D problem, most software is written for 2D or 3D problems. When we want to use the same software for a 1D problem, appropriate boundary conditions are used on the other dimension(s), typically a zero flux condition.

Box 1.1 Software implementation

Geometry can be created in COMSOL in two ways – drawn within COMSOL or imported. Examples of creating geometry within COMSOL can be seen in each of the 10 case studies in Chapter 6 (see Table 1.2 below). Geometry can also be imported from other sources, discussed in Section 2.7.2.

Table 1.2 Examples of choices among various geometries.

Geometry considerations	Examples in case studies
Dimensions	
1D	VII
1D implemented as 2D	I
2D	V
3D	VI
Axisymmetric	II, III, IV, VIII, IX, X
Number of regions (materials)	
One	II, III, VI, IX
Two or more	I, IV, V, VII, VIII, X
Solid and liquid	V, X
Size	
Infinite (1 or more dimensions)	I, II, IV, VII, VIII, IX

1.7 Governing equations

We need as many equations to describe the model as there are distinct physics. The three most distinct physics considered in this book are fluid flow, heat transfer and mass transfer. Depending on which of these physics is present in a particular

Axisymmetry cannot be assumed because the geometry is non-axisymmetric

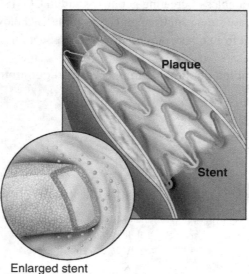

Axisymmetry cannot be assumed due to the boundary condition (on the stent wire) being different on the two sides (blood flow on one side and plaque on the other side)

Figure 1.14

Situations when axisymmetry cannot be assumed. From http://staff.washington.edu/leotta/research/vascular.html (top figure) and http://www.medartdesign.com/stock22.htm (bottom figure). Reproduced with permission.

situation, we will need the appropriate equation(s). These equations are applied over the entire computational domain, or selectively over its subdomains. If there is more than one physics in any subdomain, the number of equations to be applied must equal the number of physics in that subdomain.

As an example, the problem formulation illustration in Figure 1.2 involves three subdomains (gel, enamel and dentin) but each subdomain involves only the diffusion equation. We are ignoring any temperature differences in the entire computational domain. Making such an assumption requires good understanding of the physical process; for example, in a covered mouth, the teeth (Figure 1.6(a)) are all probably going to be at the same mouth temperature. We have also ignored the fluid flow equation, when we know there is fluid (saliva) flow around the teeth. We are assuming the saliva does not enter the region over which diffusion is taking place. Often, the only way we can say whether the assumptions we have made are reasonable is to compare results with those of a totally independent source such as an experiment.

1.7.1 Which governing equations?

Governing equations are statements of universal laws such as conservation of energy – they are derived in Chapter 7. An example of a set of governing equations for fluid flow, heat transfer and mass transfer in 3D in rectangular Cartesian coordinates is shown below. The symbols have their usual meanings and are shown in the List of symbols. Complete descriptions of the equations are provided in Chapter 7. Under each term in the equations below, the physical process leading to that term is noted. These terms are further discussed in the following section. The governing equations are always used in a simplified form – some considerations in simplifying the governing equation are discussed in the following section.

Conservation equation for total mass (continuity equation)

$$\frac{\partial \rho}{\partial t} + \frac{\partial}{\partial x}(\rho v_x) + \frac{\partial}{\partial y}(\rho v_y) + \frac{\partial}{\partial z}(\rho v_z) = 0 \tag{1.1}$$

Momentum conservation equations (fluid flow equations)

$$\rho \frac{\partial v_x}{\partial t} + \rho \left(v_x \frac{\partial v_x}{\partial x} + v_y \frac{\partial v_x}{\partial y} + v_z \frac{\partial v_x}{\partial z} \right) = -\frac{\partial p}{\partial x}$$
$$+ \mu \left(\frac{\partial^2 v_x}{\partial x^2} + \frac{\partial^2 v_x}{\partial y^2} + \frac{\partial^2 v_x}{\partial z^2} \right) + \rho g_x \tag{1.2}$$

$$\rho \frac{\partial v_y}{\partial t} + \rho \left(v_x \frac{\partial v_y}{\partial x} + v_y \frac{\partial v_y}{\partial y} + v_z \frac{\partial v_y}{\partial z} \right) = -\frac{\partial p}{\partial y}$$
$$+ \mu \left(\frac{\partial^2 v_y}{\partial x^2} + \frac{\partial^2 v_y}{\partial y^2} + \frac{\partial^2 v_y}{\partial z^2} \right) + \rho g_y \tag{1.3}$$

$$\underbrace{\rho \frac{\partial v_z}{\partial t}}_{\text{Transient}} + \underbrace{\rho \left(v_x \frac{\partial v_z}{\partial x} + v_y \frac{\partial v_z}{\partial y} + v_z \frac{\partial v_z}{\partial z} \right)}_{\text{Inertia}} = -\underbrace{\frac{\partial p}{\partial z}}_{\text{Pressure}}$$

$$+ \underbrace{\mu \left(\frac{\partial^2 v_z}{\partial x^2} + \frac{\partial^2 v_z}{\partial y^2} + \frac{\partial^2 v_z}{\partial z^2} \right)}_{\text{Viscous}} + \underbrace{\rho g_z}_{\text{Gravity}} \qquad (1.4)$$

Energy conservation equation (heat transfer equation)

$$\underbrace{\rho C_p \frac{\partial T}{\partial t}}_{\text{Transient}} + \underbrace{\rho C_p \left(v_x \frac{\partial T}{\partial x} + v_y \frac{\partial T}{\partial y} + v_z \frac{\partial T}{\partial z} \right)}_{\text{Convection}}$$

$$= \underbrace{k \left(\frac{\partial^2 T}{\partial x^2} + \frac{\partial^2 T}{\partial y^2} + \frac{\partial^2 T}{\partial z^2} \right)}_{\text{Diffusion}} + \underbrace{Q}_{\text{Generation}} \qquad (1.5)$$

Mass species conservation equation (mass transfer equation)

$$\underbrace{\frac{\partial c_A}{\partial t}}_{\text{Transient}} + \underbrace{\left(v_x \frac{\partial c_A}{\partial x} + v_y \frac{\partial c_A}{\partial y} + v_z \frac{\partial c_A}{\partial z} \right)}_{\text{Convection}}$$

$$= \underbrace{D_{AB} \left(\frac{\partial^2 c_A}{\partial x^2} + \frac{\partial^2 c_A}{\partial y^2} + \frac{\partial^2 c_A}{\partial z^2} \right)}_{\text{Diffusion}} + \underbrace{R_A}_{\text{Generation}} \qquad (1.6)$$

1.7.2 What terms remain in the governing equation?

The terms from the governing equations that we need to keep depend on the particular physics. The transient, convection, diffusion and generation terms just identified in the governing equations are kept or dropped depending on a number of factors, some of which are now discussed.

Transient Transient means changing with time, or unsteady. The transient term denotes rate of change of storage. Transient solutions can consume a lot of computer time and also be numerically challenging. Thus it is important to decide whether a transient solution is really necessary. The transient term *should be retained* when changes over time (during the time period of interest) are likely to be significant and are of interest. Sometimes, in a numerical solution, even though achieving steady state is of interest, we may have to start transient and run the program long enough for the process to reach steady state. The transient term *can be ignored* if we expect the process to reach a steady state for most of the time period of interest.

Convection The convection term in the heat or mass transfer equation represents the transport of energy or species, respectively, due to bulk flow. This term *should be retained* when the computational domain is a fluid, where movement (flow) is expected. Whether heat transport due to flow is stronger than that due to conduction can be checked by calculating the Peclet number ($Pe = Re \times Pr$). In the case of mass transfer, the Peclet number is calculated as $Pe = Re \times Sc$. The convection term *can obviously be ignored* in a solid region with no bulk flow through it, as it would be in a non-porous medium.

Diffusion This term represents the contribution to energy transport from conduction or diffusion. The term *is to be retained* when qualitative information suggests that diffusion is likely to be playing a role, which is most of the time. For example, if large temperature gradients are expected in the system (remember Fourier's law, $q'' = -k\partial T/\partial x$), conduction can be very significant. Similarly for concentration gradients. In some very special cases, this term *may be ignored*, such as: (1) uniform and rapid heat generation in a heating situation where the boundaries are insulating, so a temperature gradient does not develop. In mass transfer such a situation can be a uniform chemical reaction that generates or depletes species, with boundaries impermeable; (2) when diffusion is negligible, as for a large species such as a bacterium; (3) when the equation is used only as an analog and there is no real species to diffuse. This happens, for example, when we are modeling thermal burning of tissue, and the extent of the burn is treated as a species whose concentration accumulates with time (see Case study VIII in Chapter 6). See also the discussion on using zero diffusivity in Section 2.11.2.

Generation (or source) The generation term is also referred to as the source term. Depletion is the negative of generation, also called a sink term. For heat transfer, the generation term represents the contribution from volumetric generation of heat, converted from some other form of energy, as in electromagnetic, ultrasonic or other modes of heating. Metabolic heat generation is another type of heat generation, resulting from biochemical conversion. When such modes of heating or metabolic generation are present, this term *needs to be retained in the equation*. This term *can be ignored* when the amount of heat generation is small compared with energy transport due to conduction or bulk flow (convection). For example, metabolic heat generation may be small in some situations in comparison

> **Box 1.2 Software implementation and fundamentals**
>
> Implementation of governing equations in COMSOL is discussed in Chapter 2 starting in Section 2.8. Case studies showing examples of governing equations as they are obtained by simplifying various biomedical situations, including their computer implementation, are shown in the table below. Derivations of the governing equations are provided in Chapter 7 and background information for the source terms is in Chapter 8.
>
> Examples of various physics provided in the case studies in this book
>
> | Physical processes | Examples in case studies |
> |---|---|
> | **Fluid flow** | |
> | Steady | V |
> | Transient | VI, X |
> | **Heat transfer** | |
> | Transient | I, II |
> | Bioheat as source | VIII, IX, X |
> | Source exponentially decaying | X |
> | **Mass transfer** | |
> | Steady | V |
> | Transient | III |
> | Transient, with reaction | IV, VII, X |
> | Reaction only | VII, VIII |
> | **Coupled physics** | |
> | Heat transfer with Joule heating | IX |
> | Heat transfer with mass transfer | VIII, IX |
> | Mass transfer with fluid flow | V, X |
> | Mass transfer with multiple reactions | VII |
> | Fluid flow, heat transfer and mass transfer | X |

with other heating or cooling effects. The exact form of this term depends on the modes of heating and is discussed in detail in Section 8.2.

In mass transfer, the generation term represents the formation of a chemical species. A negative generation term (depletion term) results from the conversion of the chemical species into something else. For example, in studying oxygen transport, oxygen consumption in tissue can be important and therefore the generation term *needs to be retained*. This term *can be ignored* when the overall rates of reaction are much smaller than transport due to diffusion or bulk flow (convection). The exact form of this term depends on the type of reaction and is discussed in detail in Section 8.10.

Deciding on 2D/3D/axisymmetric How many terms are to be retained, i.e., whether a 2D or a 3D solution will apply, depends on the extent to which we can ignore variation in the other dimension(s). This has already been discussed in Section 1.6.3.

1.8 Boundary and initial conditions

Boundary conditions are statements describing how the process relates to its surroundings. Without the boundary conditions, the description (physical or mathematical) of the process is not complete. Discussion of boundary conditions can be divided into two parts: how many boundary conditions are needed, and what they are. The first of these is not our choice, but is decided by the governing equation and schematic that we choose. This is discussed first, followed by discussion of the type of boundary condition that is the most appropriate for a situation.

1.8.1 How many boundary conditions are needed?

For a numerical solution, all of the *external* boundaries of the computational domain, as defined in the software, would need boundary conditions for each of the primary variables (e.g., temperature in the heat equation) for which we are solving.

Thus, the problem formulation illustration in Figure 1.2 needs a different number of boundary conditions depending on the geometry chosen in Figure 1.6. The primary variable we are solving for is concentration, using the diffusion equation. For the 3D geometry, six boundary conditions are needed for the six surfaces, while for the 2D geometry, four boundary conditions are needed at the four edges (as mentioned in Figure 1.2) and, for the 1D geometry, two boundary conditions are needed at the points of the line.

At a surface, generally only one boundary condition can be specified. For example, if both boundary concentration and boundary flux are specified on the same surface, it is overspecified. Such overspecification of boundary conditions in a numerical software can give unpredictable (and wrong) results.

1.8.2 What kind of boundary condition?

Boundary conditions are described in detail in Chapter 7. A summary of the boundary conditions is shown in Table 1.3, including the case studies in this book where a particular boundary condition has been used in a complete problem. In practice the choice of boundary condition has to be done carefully. A constant flux boundary condition, for example, may not stay that way for a long time. Consider drug delivery via a patch. One can start by assuming a constant flux (mass per unit area per unit time) of drug enters the tissue out of the patch, but a constant flux cannot continue for long. As more drug comes out, the drug content of the reservoir decreases and the flux drops. The most realistic way to treat this is to consider the problem as shown in Figure 1.15(b) where the patch and the tissue are considered together; this is called a conjugate problem. Here the patch will lose drug while tissue will gain.

Box 1.3

Table 1.3 Choices in boundary conditions. The reader is urged to follow up on the case studies mentioned under each type of boundary condition, to see various examples of physical situations and how each situation was approximated by the particular boundary condition.

	Fluid flow	Heat transfer	Mass transfer
Value given	$u = u_s$ $v = v_s$ $p = p_s$	$T\|_{x=0} = T_s$	$c_A\|_{x=0} = c_{As}$
Examples in	Case study VI	Case studies I, VII	Case studies III, V
Flux given		$-k \frac{\partial T}{\partial x}\big\|_{x=0} = q_s''$	$-D_{AB} \frac{\partial c_A}{\partial x}\big\|_{x=0} = n_A$
Examples in	Case study VI	Case studies II, VII, IX, X	Case studies III, IV, V, VII
Convection		$-k \frac{\partial T}{\partial x}\big\|_{x=0}$ $= h(T\|_{x=0} - T_\infty)$	$-D_{AB} \frac{\partial c_A}{\partial x}\big\|_{x=0}$ $= h_m(c_A\|_{x=0} - c_{A\infty})$
Examples in		Case studies II, VII, X	

As another example, two of the possible boundary conditions that the problem formulation illustration in Figure 1.2 can have are: (1) constant flux of the H_2O_2 at

1.8 Boundary and initial conditions

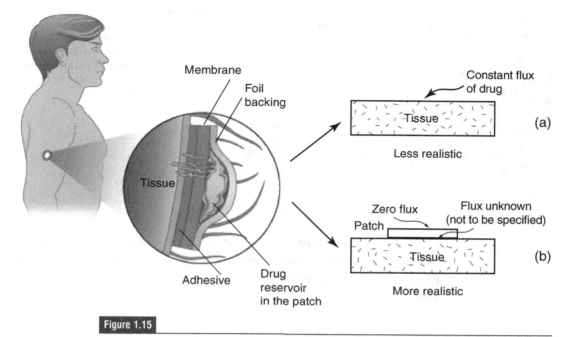

Figure 1.15

Realistic way of treating the boundary condition as a conjugate problem. For drug transport, the boundary condition at the skin-patch interface is treated more realistically in (b) by considering the patch and the skin together, called a conjugate problem. Consideration of a constant flux of drug into the tissue, as shown in (a), would be less realistic since the flux decreases in reality due to the patch having less and less of it.

the enamel surface (here we ignore the thickness of the gel containing H_2O_2); or (2) constant concentration of H_2O_2 (here also we ignore the gel thickness) on the enamel surface. From physical considerations, we need to see which one is more realistic. It turns out that a more realistic boundary condition in this case is neither of these two boundary conditions, but simply zero flux on the outside boundary of the gel while the concentrations in the gel, enamel and dentin tends to equilibrate, similar to what is suggested in Figure 1.15(b).

1.8.3 Boundary conditions changing with time

In real life, boundary conditions such as temperature or heat flux specified at a boundary can change with time. For example, if we are trying to model a process that depends on airflow during a respiratory cycle, as shown in Figure 1.16, velocity at the boundary becomes a function of time. Change in boundary conditions with time is routinely implemented in a numerical solution procedure. The variation with time is typically input into the software using discrete data points of boundary condition value and the corresponding time, or it can be provided as an equation

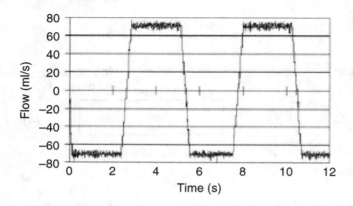

Figure 1.16

Flow rate of air during a typical human respiratory cycle as an example of a time-varying boundary condition. Positive pressure represents exhalation and negative pressure represents inhalation. Reprinted from Xu *et al.* (2006), with permission from Elsevier.

to a curve that fits the data, as discussed further in Section 2.9. As always, possible simplification should be considered. For example, when using the information in Figure 1.16, flow rate averaged over the breathing cycle may work just fine for some situations.

1.8.4 Boundary condition at an infinite region

For an infinite region problem, the directions that extend to infinity can have any one of the boundary conditions of constant value or zero flux specified. If the infinite region has been treated appropriately (see discussion in Section 1.6.2), either boundary condition is okay and in fact they will both be satisfied.

1.8.5 Boundary condition at an interface between two materials

One interesting and useful consequence of a numerical solution is that when there are two materials, as illustrated in Figure 1.17 for the towel and muscle tissue, the true boundary condition of flux continuity at the interface is automatically satisfied, by virtue of using the governing equation in both regions. In this example, at the interface between the towel and the muscle tissue, the heat flux is the same on either side of the interface, to satisfy energy conservation. This condition of same heat flux on either side is automatically satisfied through the use of the governing heat conduction equation and no additional setting of the boundary condition at the interface should be done. Stated another way, the interface is an internal boundary and no boundary conditions are needed at an internal boundary, as has been mentioned in Section 1.8.1. A complication can arise in mass transfer, however, due to partition coefficients at the interface, as discussed in Sections 7.13.4 and 9.17.

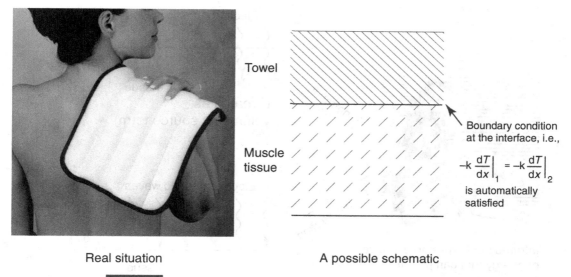

Real situation   A possible schematic

Figure 1.17

The boundary condition at an interface between two materials in an interior location is automatically satisfied. The picture on the left shows a reusable moist heat-pack, placed on the back, courtesy of Duro-Med Industries.

1.8.6 Flux boundary condition versus volumetric heat generation

In many heating processes such as microwave, ultrasonic, etc., the heating is treated as volumetric, through a heat generation term, discussed in Section 1.7.2. In some special cases of such heating, for example some instances of laser or infrared heating, the penetration of energy into the material is so small that the energy can essentially be considered to be absorbed at the surface. In these situations, the heating process is included through the boundary condition (flux is specified), as opposed to being included in the source term. Obviously, the heating process is to be included either as a source term or as a boundary condition, *but not both*, as illustrated in Figure 1.18.

1.8.7 Initial condition

When the process is changing with time, i.e., it is transient, we need to include the initial condition which is the condition at time $t = 0$. Unlike the type of problems solved in an undergraduate fundamentals class, where simplified analytical solutions are often the goal, in a numerical solution, the initial condition does not have to be constant over the entire domain, i.e., it can vary spatially. For example, the initial condition in a drug delivery problem may involve some previous presence of the drug that varies with position. In a heat transfer problem, initially, the material may not be at uniform temperature, but exhibit spatially varying temperatures.

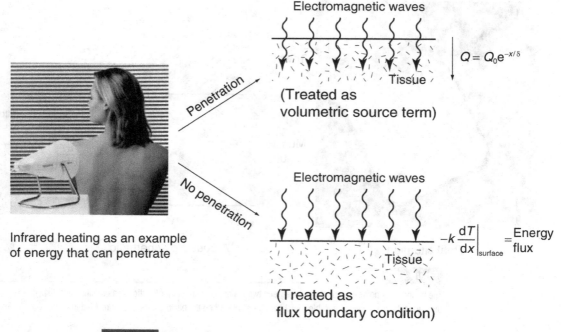

Figure 1.18

Schematic showing two possible ways (either as surface flux or as source term) of implementing radiative energy incident at a surface; depending on the penetration depth of the radiation into the material. Photograph from http://www.technilamp.co.za/Portals/140/images/IndustrialPhoto.jpg.

Box 1.4 Software implementation and fundamentals

Implementation of boundary conditions in COMSOL is discussed in Chapter 2 starting at Section 2.9. Fundamentals of boundary conditions are explained in Chapter 7 starting at Section 7.10. Implementation and details of initial conditions follow boundary conditions in the respective chapters.

1.9 Material properties

The governing equations in Section 1.7.1 that we are trying to solve have in them material properties (such as thermal conductivity in the heat equation). Solving these equations numerically will require data (i.e., numerical values) for these material properties. The material property data needed for each equation are listed in Table 1.4. Note that not all data are needed in all situations.

Table 1.4 Material properties needed for modeling. Reader is referred to the corresponding case studies (listed in Table 6.1) for examples of use. Variable properties, wherever used in case studies, are also noted. Chapter 9 discusses all properties in detail.

Process and properties	Example in case study
Heat transfer (conduction)	
Thermal conductivity	I, VIII-X, II (varies with temp.)
Density	I, II, VIII–X
Specific heat	I, VIII-X, II (varies with temp.)
Heat transfer (radiation)	
Surface absorptivity	
Penetration depth	
Mass transfer	
Mass diffusivity	III-V, VII, VIII, X
Partition coefficient	See Sections 7.13.5 and 9.17
Fluid flow	
Viscosity	V, VI, X
Density	V, VI, X

1.9.1 Questions to ask

In the problem formulation stage, while one can proceed assuming that material properties data will have to be found somehow, one can perhaps be a little more sensitive to efforts that are likely to be necessary in finding such data, and make some adjustments at this early stage. One can consider, for example: (1) are we likely to find the property data needed for the exact material that we need? If not, is there another material that would be similar and for which data are likely to be more readily available? In the case of a complex material, what simplification of the material can we get away with? (2) Can we estimate the property using empirical predictive equations? (3) How accurate must the data be? (4) When accurate property data are not available for a particular material, which is often the case, how do we get useful information from the simulation? These topics are addressed in greater detail in Chapter 9. It is a good idea to review relevant sections of Chapter 9 at this problem formulation stage.

1.9.2 Simplification is generally needed

Like other steps in problem formulation, finding properties also requires simplification. For example, most biomaterials are non-homogeneous. A specific example can be that thermal conductivity of a muscle tissue is different along the fiber and perpendicular to the fiber. Although software implementation of this is not difficult, implementation of some property non-homogeneity can be complex. The same can be said about property variation with temperature and composition. One needs to ask the question at this stage whether the additional complexities are critical to the physics being studied and, more importantly, to the goals for the simulation. Material in Chapter 9 can help in this process.

1.9.3 When accurate data are not available

For materials of interest in biomedical processes, a particularly common situation is the unavailability of property data for the exact material of interest. When specific property data are not available, perhaps the most useful route is to perform sensitivity analysis with respect to the unavailable data. Since the idea of computer-aided engineering (CAE) is to try many "what-if" scenarios anyway, it is only natural to examine the effect of property variations on the process. This way, we can see the effect of a range of properties on the process. If the outcome of the sensitivity analysis shows that the process is quite sensitive to property variations in the particular situation, more accurate data are needed for that particular property. Otherwise, the rough estimate of the property available (from tissues of similar composition, etc.) may be considered sufficient for design purposes. Sensitivity analysis is one of the most important tools in CAE and is not limited to use in the case of property variations; it can also be used in the case of other process parameter variations. This is also discussed in Chapter 5 (Section 5.4) and Chapter 9 (Section 9.4).

> **Box 1.5 Software implementation and fundamentals**
>
> Implementation of properties in COMSOL is discussed in Chapter 2 starting at Section 2.11. Examples of implementation of specific properties in the case studies is shown in Table 1.4 and the reader is referred to these case studies. Definitions of the properties and some data are provided in Chapter 9. Sensitivity analysis to accommodate uncertainty in data is discussed in Chapter 5.

1.10 Other input parameters

To complete the problem formulation, other input parameters in addition to the material property data noted above may be needed. Examples of such data can be heat or mass transfer coefficients at a surface, reaction rate constants, volumetric heating rates, etc. General discussion under material properties such as availability and accuracy of data is also applicable to such parameters. In some situations, non-availability of a parameter may require substantial changes in problem formulation. For example, referring to Figure 1.7, if heat or mass transfer coefficients are not available on the solid surface for case (1), we may require the formulation shown in case (3), which is substantially different and more complex. In the initial round of problem formulation, it is advisable to look hard for a parameter value if that value will allow a much simpler formulation, as would be the case for using heat and mass transfer coefficients in Figure 1.7.

> **Box 1.6 Software implementation and fundamentals**
>
> Implementation of parameters in COMSOL and the fundamentals behind them are scattered throughout the book. The reader is advised to consult the Index for the specific parameter.

1.11 Summary

Problem formulation is writing a physical problem in terms of its simplified mathematical equivalent (*i.e., creating a model*). Typically, a model consists of geometry (computational domain), governing equation, boundary conditions, properties and other parameters. We develop this model to be able to *simulate* the physical process on the computer. At every step in developing the model, for example in choosing a geometry, we need to balance between how realistic the model is and how computing intensive (time and effort) its simulation will be. This makes it critical to carefully simplify during every step of model building. The ultimate goal of the simulation, that is, what information we want to get out of it, guides us at each stage during the simplification.

The models developed by following the suggestions in this chapter can be implemented, that is, solved in the software COMSOL, following directions provided in Chapters 2, 3 and 4. Validation of the model and its sensitivity analysis can be done in accordance with Chapter 5. Fundamentals that help in deciding the steps in model development and its solution are provided in Chapters 7–10 and complete examples are provided in the Case studies in Chapter 6.

References

Androlowicz, J., Clark, I., Doerr, G., Netravali, N., Wynne, J. (2003). Hyperthermic Ablation of Hepatic Tumors by Inductive Heating of Ferromagnetic Alloy Implants. On the web at http://ecommons.library.cornell.edu/handle/1813/126

Anderson, J. C., Babb, A. L., and Hlastala, M. P. (2003). Modeling Soluble Gas Exchange in the Airways and Alveoli. *Annals of Biomedical Engineering*, **31**: 1402–1422.

Bahr, D., J. Tyler, P. Weinert and J. Egan. (2000). Heat and Moisture Transport in the Nasal Cavity. On the web at http://ecommons.library.cornell.edu/handle/1813/274

Bermudez, C., P. Davis, P. Gaborski, M. Hatfield and A. Vinegar (2004). Put your best teeth forward: A mass transfer study of Crest Whitestrips. On the web at http://ecommons.library.cornell.edu/handle/1813/129

Conaghey, O. M., Corish, J., Corrigan, O. I. (1998). The release of nicotine from a hydrogel containing ion exchange resins. *International Journal of Pharmaceutics*, **170**(2): 215–224.

Cao, H., Speidel, M. A., Tsai, J., Van Lysel, M. S., Vorperian, V. R., and Webster, J. G. (2002). FEM analysis of predicting electrode-myocardium contact from rf cardiac catheter ablation system impedance. *IEEE Transactions on Biomedical Engineering*, **49**, 520–525.

Dughiero, F. and Corazza, S. (2005). Numerical simulation of thermal disposition with induction heating used for oncological hyperthermic treatment. *Medical & Biological Engineering Computing*, **43**(1): 40–46.

Donnelly, B. J., Saliken, J. C., Ernst, D. S., Weber, B., Robinson, J. W., Brasher, P. M. A., Rose, M. and Rewcastle, J. (2005). Role of transrectal ultrasound guided salvage cryosurgery for recurrent prostate carcinoma after radiotherapy. *Prostate Cancer and Prostatic Diseases*, **8**(3): 235–242.

Gulsen, D., and Chauhan, A. (2004). Ophthalmic drug delivery through contact lenses. *Investigative Ophthalmology & Visual Science*, **45**(7): 2342–2347.

Ma, N., X. Gao, and X. X. Zhang (2004). Two-layer simulation model of laser-induced interstitial thermo-therapy. *Lasers in Medical Science*, **18**(4): 184–189.

Mongrain, R, R. Leask, J. Brunette, I. Faik, N. Bulman-Felemíng, T. Nguyen (2005). Numerical modeling of coronary drug eluting stents. *Studies in Health Technology and Informatics*, **113**: 443–58.

Niemz, M. H. (2002). *Laser-Tissue Interactions. Fundamentals and Applications*. Berlin, Springer-Verlag.

Poon, B. (2001). Drug delivery in the brain. On the web at http://ecommons.library.cornell.edu/handle/1813/263

Saltzman, W. M. and M. L. Radomsky (1991). Drugs released from polymers: Diffusion and elimination in brain tissue. *Chemical Engineering Science*, **46**(10): 2429–2444.

Tsuda, N., K. Kuroda and Y. Suzuki (1996). An inverse method to optimize heating conditions in RF-capacitive hyperthermia. *IEEE Transactions on Biomedical Engineering*, **43**(10): 1029–1037.

Tucker, R. D., C. Huidobro and T. Larson (2005). Ablation of stage T-1/T-2 prostate cancer with permanent interstitial temperature self-regulating rods. *Journal of Endourology*, September 1, 2005, **19**(7): 865–867.

Bunimovich, Y. Mintseris, J., Kim, B. and Mohan, V. (1999). Heating Effects of Dental Drilling. On the web at http://ecommons.library.cornell.edu/handle/1813/291

Xu, C., Sin, S. H., McDonough, J. M., Udupa, J. K., Guez, A., Arens, R., and Wootton, D. M. (2006). Computational fluid dynamics modeling of the upper airway of children with obstructive sleep apnea syndrome in steady flow. *Journal of Biomechanics*, **39**(11): 2043–2054.

1.12 Problems

1.12.1 Short questions

(1) Computer-aided engineering involves building a computer model out of a physical model. What are the main components of the computer model?

(2) Discuss the advantages and disadvantages of simulation-based design as opposed to prototype-based design.

1.12.2 Choice of domain size: heating tumors with ferromagnetic materials

Ferromagnetic materials heat until they reach their Curie point, the temperature at which they become non-magnetic and stop heating (Dughiero and Corazzo, 2005). This self-limitation can be used in selective heating of prostate tumors to destroy tissue (Tucker *et al.*, 2005). Consider a cylindrical rod, of the type shown in Figure 1.19, to be inserted in the prostate; heating is simulated for the required duration with the final temperature profile shown on the right. This is effectively an infinite domain problem. Comment on: (1) appropriateness of the size of the tissue domain chosen in the figure for computation, i.e., is it too small, too large or just right? (2) how do you know beforehand what size to choose? (3) explain the pros and cons of choosing too small or too large a size.

1.12.3 Choice of domain size: drug delivery in the brain

Figure 1.20 shows the placement of a polymer containing drug in the brain and the resulting drug concentration profile. Comment on the computational domain as to whether the size of the domain has been chosen appropriately. Also, should the domain size be chosen differently if the diffusivity or drug elimination rate changes? Explain your answer.

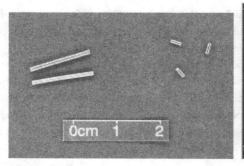 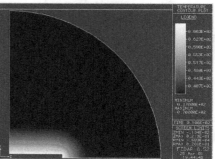

Figure 1.19

Cylindrical ferromagnetic implants (on the left) and simulated temperature contours after 10 seconds, using an axisymmetric geometry (Androlowicz et al., 2003).

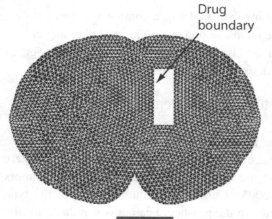 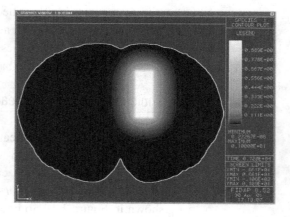

Figure 1.20

Cross-section of the brain showing mesh and drug placement (left figure). Cross-section showing drug concentration after two hours (right figure). From Poon (2001).

Questions on problem formulation: instructions

Below are a number of physical situations for which you can practise problem formulation, as discussed in this chapter. These situations are chosen mostly from past class projects from an undergraduate class (one semester duration) and are not intended to be particularly complex, as they would be in a research project of longer duration. Also, it is expected that the user will be obtaining additional anatomical, physiological and other details from various sources, as necessary, including from Chapters 7–9. For each situation below, do the following, *including providing reasons for your choice as appropriate*:

(1) *Provide the schematic of a simplified geometry which you would use to model the process (heat transfer, mass transfer, fluid flow). Mention whether it is 2D/3D/axisymmetric. Provide approximate size (dimensions, which would be needed in solving the problem numerically) of the computational domain based on your best understanding of the problem.*

(2) *Provide a list of assumptions that lead to the schematic and the rest of the problem formulation.*

(3) *What governing equation(s) would you need to solve? Be specific, retaining only the terms (transient, reaction, etc.) needed for the given scenario and providing reasons for the terms dropped. You also need to show the respective regions over which a particular equation is being solved.*

(4) *What boundary conditions are needed? Show conditions for all variables and at all boundaries that would be needed to solve this problem using numerical software. Preferably show this information on the schematic above.*

(5) *Provide all the initial conditions that may be necessary and the regions over which they are needed, preferably on the schematic.*

(6) *List all material properties that you will need to solve this problem.*

(7) *List any other parameter(s), such as a heat transfer coefficient, that would be needed to solve this problem.*

(8) *Pick two of the assumptions that you have made and show how these assumptions can be relaxed, i.e., how your model can be made more general.*

1.12.4 Drug delivery in the brain

Consider drug delivery from a polymer wafer into brain tissue shown in Figure 1.21. Formulate the problem of drug diffusion from the polymer matrix and its elimination in the tissue (see Salzman and Radomsky, 1991).

1.12.5 Heat generation from tooth drilling

A tooth is made up of three distinct layers: enamel, dentin and pulp (see Figure 1.22). Drilling of the tooth becomes necessary at times to remove decayed tooth matter. The objective of this student design (Bunimovich *et al.*, 1999) project from the past was to be able to lessen the pain caused by the frictional heat of the drilling process acting on the nerves in the pulpal layer. The faster the tooth is drilled the more frictional heat is produced; however, faster drills allow dentists to manipulate the drill more easily, and finish the drilling faster, which may cause less heat to be dissipated into the tooth. The drilling speed that would best suit the process would be one that limits the heat-induced pain, while also making it

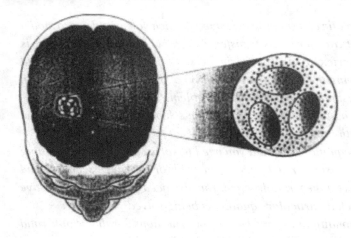

Figure 1.21
Schematic showing cylindrical drug wafers placed in brain tissue.

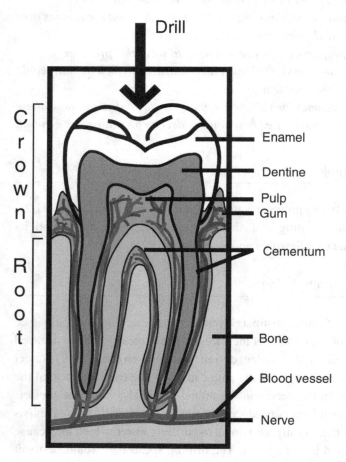

Figure 1.22
Cross-section of a tooth. Image from http://commons.wikimedia.org/wiki/Image:Tooth_Section.svg.

possible to drill fast enough to reduce the duration of heating. Provide the problem formulation to model the heating process as the drill enters the tooth.

1.12.6 Drug delivery from a stent

Balloon angioplasty and coronary stent (Figure 1.23) deployment are powerful techniques in the treatment of advanced coronary artery disease (Mongrain *et al.*, 2005). Recent advances have led to the development of drug-eluting stents designed to release a drug polymer that inhibits restenosis, a bodily defense mechanism characterized by tissue ingrowth and reblockage of the artery. Assume the stent material, shown in Figure 1.23, is impregnated with an anti-cancer or immunosuppressive drug that diffuses through the surrounding tissues over a critical period of time. In this study, you would like to model diffusion of the drug Rapamycin. Provide a problem formulation for this purpose. The stent geometry can be simplified to concentric rings and it is sufficient to model diffusion from one of these rings.

1.12.7 Laser-interstitial thermal therapy (LITT)

Laser interstitial thermal therapy (LITT) is a well-established surgical method (Ma and Zhang, 2004) used in the treatment of tumors (see Figure 1.24; the tumor is indicated by the arrows). During the procedure laser radiation is transmitted down a quartz optical fiber to an applicator. The applicator diameter is 1.0 mm with an applicator length of 4 cm (see figure on the right). The laser light is diffused along the entire length of the applicator to produce a homogeneous radiation out of the applicator, in the radial direction. Laser power (total laser energy released) is 25 W. To prevent the formation of gas bubbles the applicator temperature should not exceed 100 °C. Provide a problem formulation to analyze the extent of tissue damage (as related to the temperature history) of regions surrounding the LITT probe in a liver. For the laser heating (penetration depth in tissue is 1 mm), mention (1) what form of equation you will use; (2) what would be the value at the applicator surface?

1.12.8 Nitrogen elimination in alveoli

Nitrogen elimination is of particular importance to scuba divers, as excess nitrogen gas in the bloodstream can create bubbles which can cause many medical complications including the commonly known "bends." Nitrogen elimination from the bloodstream occurs through diffusion inside the blood, the tissue layer of the alveolar wall (and the capillary wall) and the air in the alveolus. Our interest is in modeling the rate of nitrogen transport from an aleveolus that is surrounded by blood capillaries (Anderson *et al.*, 2003). Figure 1.25 shows the geometry and one possible schematic of the transport process.

42 Problem formulation

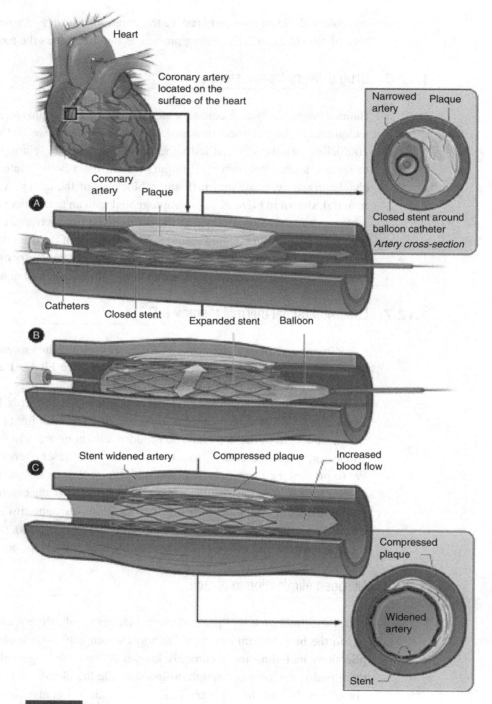

Figure 1.23

Stent placement in a coronary artery. Figure reproduced from NHLBI – part of the National Institute of Health and the US Department of Health and Human Services (http://www.nhlbi.nih.gov/).

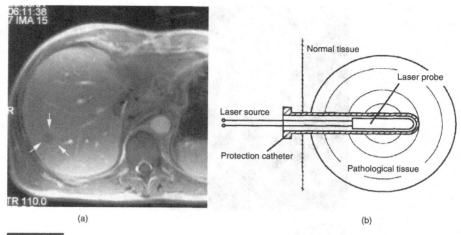

Figure 1.24

(a) A liver tumor (within the arrows) and (b) a schematic of the LITT process, Ma *et al.* (2004). With kind permission from Springer Science + Business Media.

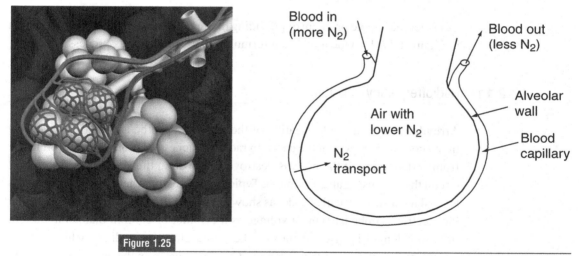

Figure 1.25

Alveoli on the left and an enlarged alveolus on the right.

1.12.9 Drug release from a nicotine patch

Nicotine patches are used as a transdermal drug delivery method to help subjects quit smoking. An illustration of such a patch is shown in Figure 1.15. Those trying to quit smoking are encouraged to place the patch on the upper arm, as illustrated in Figure 1.26. We would like to simulate the diffusion of nicotine from the patch into the bloodstream (Conaghey *et al.*, 1998).

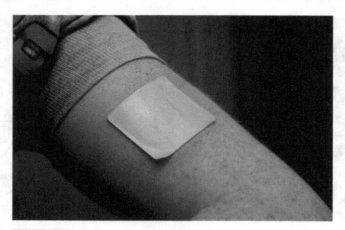

Figure 1.26

Picture showing placement of a nicotine patch. From http://en.wikipedia.org/wiki/Nicotine_patch

1.12.10 Therapeutic heating

Consider the simple problem of providing therapeutic heat onto the body, as shown in Figure 1.17. Provide a problem formulation to simulate this process.

1.12.11 Radiofrequency ablation

Arrhythmia is an irregular beating of the heart that can be caused by improperly timed contractions within the heart. In radiofrequency ablation (RFA), heart tissue from certain parts of the heart is destroyed by heating thus interrupting the short circuit that is causing an arrhythmia. Typically, the region where this is performed is around the atrioventricular node, as shown in Figure 1.27. For simplicity, consider the electrical probe carrying a voltage as a cylinder of 2.6 mm diameter, 7.5 mm of whose length is inserted into the heart muscle (myocardium, which can be considered much thicker than 7.5 mm). Current flows from the probe to the ground that is attached to the heart muscle at another location. We would like to simulate the temperature rise in the heart tissue from electrical heating.

1.12.12 Radiofrequency heating to destroy a tumor

Figure 1.28, from the work of Tsuda *et al.* (1996), shows a very simplified view of a section of the lower abdomen with a tumor. The tumor will be heated by placing electrodes with varying voltages around the stomach. The bolus, placed between

1.12 Problems

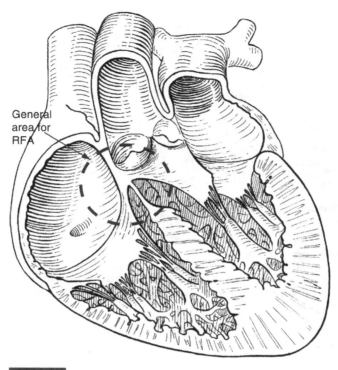

Figure 1.27

Cross-section of the heart showing heart muscle and the region (dashed circle) around the atrioventricular node.

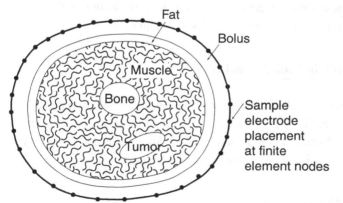

Figure 1.28

Cross-section of the abdomen showing tumor and possible placement of electrodes.

the stomach and the electrodes, is a water bag which serves to reduce localized high temperatures. About 30 electrodes will be placed around the periphery, where voltages can be applied independently. We would like to model the temperature rise for some assumed placement of voltages around the stomach.

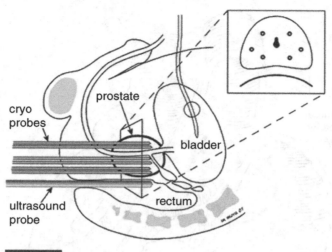

Figure 1.29

Illustration of cryosurgery showing placement of several cryoprobes within the prostate gland. In this problem, only two cryprobes are to be placed at short distance from each other. Reprinted by permission from Macmillan Publishers Ltd: Donnelly *et al.* (2005).

1.12.13 Cryosurgery

In cryosurgery, very low temperatures are used to freeze and thereby destroy abnormal tissue. Consider cryosurgery of a prostate gland using multiple probes, as illustrated in Figure 1.29. In our application, we would be placing only two cryoprobes at a short distance from each other. We would like to obtain the tissue temperature around the probes as a function of position and time.

1.12.14 Drug release from a therapeutic contact lens

The use of a therapeutic contact lens can be a possible method for treating glaucoma (Gulsen and Chauhan, 2004). We would like to model the delivery of the drug timolol maleate from a hydrogel contact lens into the aqueous humor of the eye. Provide a problem formulation to simulate the drug transport.

1.12.15 Oxygen transport in alveoli

Figure 1.25 shows how a respiratory tube ends in a tiny spherical alveolus (hollow air sac), surrounded by a capillary network. Air flows in and out of the sac as shown. In the enlarged view of an alveolus, oxygen diffuses from the air in the alveolus into the capillary, while carbon dioxide diffuses from the blood within the capillary into the alveolus. Provide the problem formulation to simulate the rate of oxygen transport from the air into the blood. List two of the major uncertainties in the assumption you made in arriving at the problem formulation and, for each uncertainty, state the changes (in any of schematic, ge, bc, properties, etc.) you would make to improve the model.

1.12.16 Refractive laser vision surgery

One of the recent techniques associated with correcting vision loss due to disorders is refractive laser vision surgery (Niemz, 2002). This procedure involves the ablation (tissue destruction) of the cornea using laser heating to correct vision loss. Assume the central region of the cornea is to be ablated by a certain amount using a laser beam whose spot size is much smaller than the cornea. We would like to simulate the region of ablation as a function of time and the power of the laser beam. For anatomical details, see Figure 1.30 and Figure 9.2 in Section 9.5.

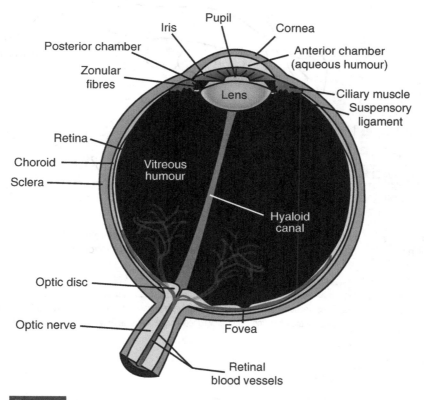

Figure 1.30

Section of an eye. A therapeutic contact lens will be placed on the cornea. From http://en.wikipedia.org/wiki/File:Schematic_diagram_of_the_human_eye_en.svg.

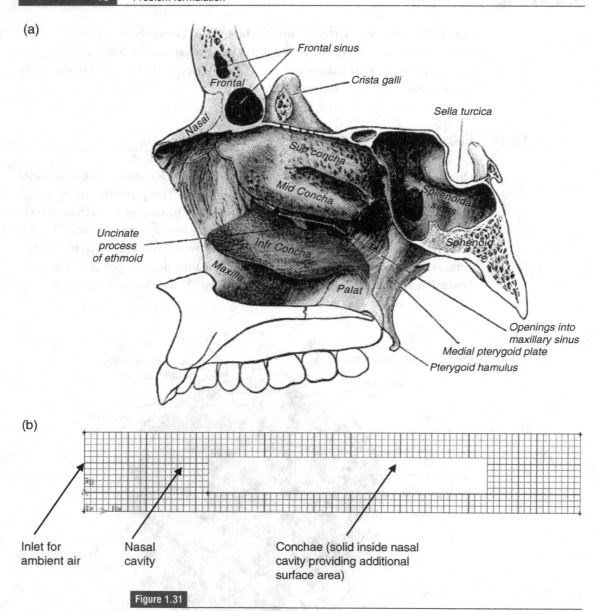

Figure 1.31

(a) Cross-section of the nose and (b) a highly simplified schematic of the nasal cavity perpendicular to the section in (a). Nose figure taken from http://en.wikipedia.org/wiki/Image:Gray170.png and schematic from Bahr *et al.* (2000).

1.12.17 Conditioning of air in the nose

When we breathe, our nasal passages take in the air and condition it before sending it to the lungs. If the air is dry, the nasal cavity humidifies it as the air blows over the interior surface. If the air is too cold, the same surfaces also warm it up. Figure 1.31 shows a cross-section of the nose and a schematic of a highly simplified nasal cavity as a 2D problem (solid in the middle is an effective representation of entire conchae). Complete the rest of the problem formulation.

2 Software implementation 1
What to solve (preprocessing)

"I really hate this damned machine. It never does quite what I want. But only what I tell it."

In the previous chapter, the physical problem was transformed into a mathematical model consisting of the governing equations and boundary conditions. We solve these equations using a computational software, also known as computer-aided engineering (CAE) software. Implementation of the model, i.e., solution of the equations in a computational software requires that we tell the software (1) what equations to solve; and (2) how to solve them. The former is the subject of this chapter while "how to solve" is the topic of the next chapter. Telling the software what equations to solve is critical since the computer will solve the exact problem it is *told* to solve, not necessarily what we *intend* to.

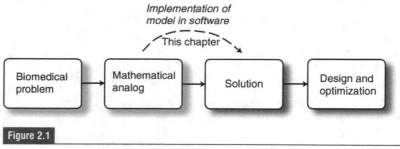

Figure 2.1
Implementation in computational software of the problems formulated in Chapter 1 is the subject of this chapter.

2.1 Choosing a software

Several factors need to be considered in choosing a software for a particular problem or type of problems. The very first question, of course, is, can the software solve the necessary physics? Often, it is better to assume a more complex version of the problem at hand as this may reflect the future need. The capabilities of a software with respect to handling a complex physics may not be straightforward and one has to dig deeper into the documentation available for the software, and beyond the sales literature. If the software cannot solve the necessary physics in its exact form, how difficult is it to customize it for the intended purpose? Ease of use or the user interface is perhaps the next most important assessment to make. An easy to use interface can save a tremendous amount of time, particularly in the early stages of learning. Another important item is the cost (of the software and the required hardware), although, for university use, access seems to be easy to come by from department, college, university or even national facilities such as the National Center for Supercomputing Applications (NCSA), located at the University of Illinois at Urbana-Champaign.

Situations can arise when the available software will not solve the most comprehensive versions of the equations needed, but time and resources will not permit moving to the most appropriate software right away. Some initial insight into the model needs to be obtained before making further investment. The obvious choice in such circumstances is to simplify the problem so that it can be solved using the available software, even if with limited utility.

2.2 Software is not to be used as a blackbox

Before we even begin to discuss the details of implementing the equations in a software, it is critical to remind ourselves that a computational software can easily provide results that are colorful, but garbage. Although working with computational software is conceptually analogous to pressing buttons on a calculator, the type of computation that we are talking about here is not nearly as well behaved as calculating a sine function, for example. The user is solely responsible for making sure the results are accurate. For example, before running the software, one should be able to describe the problem and have sufficient understanding of the physics to be able to recognize when a valid solution is obtained. In other words, one should have a qualitative picture of what to expect before using the software.

2.3 Organization of a typical CAE software: preprocessing, processing and postprocessing

There is a large amount of computer-aided engineering (CAE) software available today. The group of software programs that solve fluid flow, heat transfer and mass transfer is generally referred to as computational fluid dynamic (CFD) software. Computational software exists for other types of physics, such as mechanics and electromagnetics (not surprisingly called computational mechanics and computational electromagnetics). A website that lists a variety of CFD software is *http://www.cfd-online.com/Wiki/Codes*.

The organization of a typical CAE software is generally divided into three parts – preprocessing, processing and postprocessing. These steps are illustrated in Figure 2.2 and are now discussed individually.

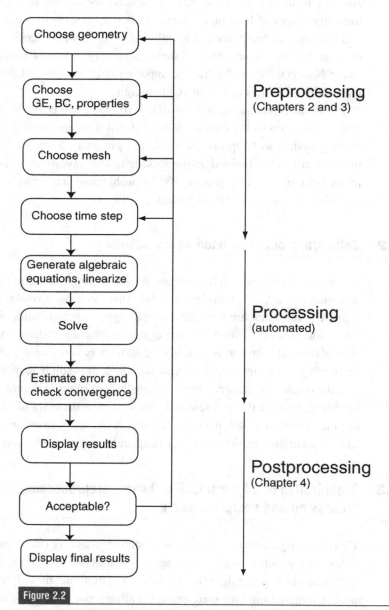

Figure 2.2

Flow chart for model development and solution, showing the steps of preprocessing, processing and postprocessing.

- **Preprocessing** This step comprises input of the required data in the CAE software. The required data comprises of (a) what to solve and (b) how to solve.

  The items needed to specify "what to solve" typically involve the following
  - **Geometry:** in this step, the computational domain for the problem is specified (drawn) in the software.
  - **Governing equations:** the set of mathematical equations that describe the physical problem are chosen in the software in this step.
  - **Boundary conditions:** the appropriate boundary conditions corresponding to each governing equation being solved are specified.
  - **Initial conditions:** the initial conditions for the problem are set in this step.
  - **Properties:** the material properties (such as thermal conductivity, density, etc.) needed for the problem are specified.

  The items needed to specify "how to solve" typically involve solution strategies including the following
  - **Meshing:** in this step, the computational domain is divided into small simple shapes (known as elements) to help solve the governing equations numerically.
  - **Time steps** (for a transient problem): if the problem is time-dependent, the time over which the problem needs to be solved and the time step increment to be used by the solver are specified.
  - **Approach for solving algebraic equations:** to solve the set of algebraic equations obtained from the original partial differential equations in the processing (or solution) stage, there are a number of different solvers available. In this step, an appropriate solver which is expected to work for the problem is chosen.
  - **Tolerances:** in order to control the error in the calculated solution, tolerances are set in the CAE software in this step.
- **Processing** This step is typically automated and performed by the CAE software based on the input provided in the previous preprocessing step. The governing partial differential equations are transformed into a system of algebraic equations and the unknown values (such as temperature, concentration, velocity, etc.) are determined.
- **Postprocessing** This step involves visualizing the solution obtained at the previous step.

The CAE software **COMSOL** will be used in this text for all these steps. In this chapter, we will look at implementing the items related to "what to solve" under the preprocessing step in COMSOL. The implementation of items related to "how to solve" the problem under the preprocessing step will be discussed in the next chapter (Chapter 3). Chapters 4 and 5 include processing and postprocessing in COMSOL.

2.4 Some general guidelines to preprocessing

General guidelines to choosing the geometry and dimensions are discussed in Chapter 1. Governing equations are mentioned in Chapter 1 with guidelines on how to drop any individual term in these equations. The governing equations are also derived in Chapter 7, showing the origin of each of the terms. The same for boundary and initial conditions, i.e., guidelines are provided in Chapter 1 (Section 1.8) and basics are provided in Chapter 7 (Section 7.10). Origins of the source terms and their mathematical formulations are discussed in Chapter 8. Properties and where to get them are discussed at length in Chapter 9.

Units of input data need to be consistent and the user is responsible for ensuring the consistency. Most CAE softwares do not check for units. Figure 2.3 shows one consistent set of units. In this example, the geometry is drawn using the units as m. Then, the velocities are entered in m/s, time in s, temperature in °C, thermal conductivity in W/m°C, density in kg/m^3, specific heat capacity in J/kg°C and the heat source term in W/m^3 in order to maintain the consistency of the units. Likewise,

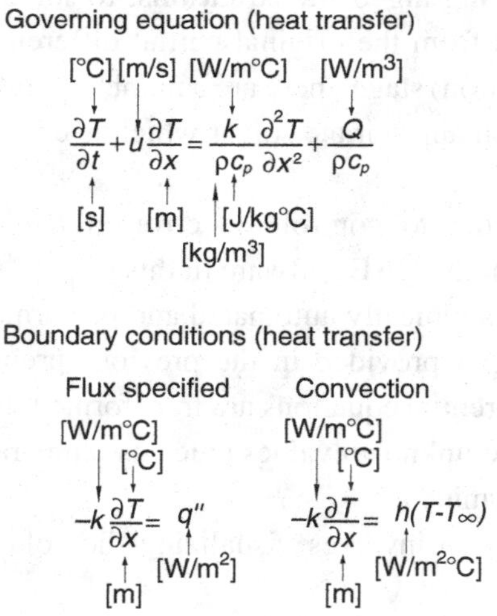

Governing equation (heat transfer)

$$\frac{\partial T}{\partial t} + u\frac{\partial T}{\partial x} = \frac{k}{\rho c_p}\frac{\partial^2 T}{\partial x^2} + \frac{Q}{\rho c_p}$$

with units [°C], [m/s], [W/m°C], [W/m^3], [s], [m], [J/kg°C], [kg/m^3].

Boundary conditions (heat transfer)

Flux specified: $-k\frac{\partial T}{\partial x} = q''$ with [W/m°C], [°C], [W/m^2], [m].

Convection: $-k\frac{\partial T}{\partial x} = h(T-T_\infty)$ with [W/m°C], [°C], [W/m^2°C], [m].

Figure 2.3 Use of a consistent set of units for the variables, with the governing equation and boundary conditions for heat transfer as an example.

for the boundary conditions. It must be noted that temperatures in COMSOL are in K (Kelvin) by default. Sometimes we prefer to solve the governing equations in a non-dimensional form. Non-dimensionalization of the governing equations is discussed in Section 7.7.

2.5 Introduction to preprocessing in a computational software (COMSOL)

COMSOL is a general purpose computer program that solves the governing equations and boundary conditions using a numerical method called the finite-element method (FEM). For a discussion on FEM, see Section 10.6. The software can simulate many different physics by being able to solve the corresponding equations. Such physics includes, but is not limited to, heat transfer, mass transfer, fluid flow, electromagnetics and solid mechanics. Such "multi-physics" capabilities allow the user to incorporate various kinds of physics (such as electromagnetics and heat transfer) for solving a particular problem. One-dimensional, two-dimensional, axisymmetric and three-dimensional steady state or transient simulations in complex geometries are possible in COMSOL. In the next few sections of this chapter, we will look at the various capabilities in COMSOL related to our interest in heat transfer, mass transfer and fluid flow and the implementation of different preprocessing steps discussed earlier.

Figure 2.4 shows the initial window of COMSOL that opens up when the program is run (either from the Windows Desktop or the Start menu in Windows). The user needs to select the appropriate geometry and the analysis type for the problem. For example, geometry types such as 1D, 2D, axisymmetric or 3D and analysis types such as heat transfer or convection-diffusion are set for the particular problem. Once the geometry and analysis type are selected, the main COMSOL window (shown in Figure 2.5) opens up. As shown in the figure, the main COMSOL window is divided into three parts: (1) the menu options on top which are used for performing all the tasks inside COMSOL, such as saving files, defining constants, drawing geometry, meshing, specifying solver settings, postprocessing, etc; (2) the model tree on the left which shows the selected analysis types; (3) the drawing area which is used to draw the geometry, select different geometry components and display the results after solution. For detailed overview of the COMSOL user interface, please refer to the COMSOL documentation that comes with the product the detail of which can be found under Help in the main menu. We will now look at how to select the geometry and analysis type in COMSOL.

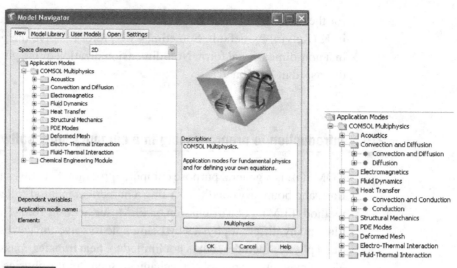

Figure 2.4

Overview of COMSOL. The left figure shows the MODEL NAVIGATOR window which is the initial startup screen of COMSOL. The **space dimensions** (geometry type) and **application mode** (analysis type) for the problem are specified here. The underlying menu items for the common application modes that will be used in this text are shown on the right.

2.6 Geometry and analysis type

2.6.1 Geometry

Methods to obtain the geometry can be divided into two broad classes: (1) drawing of the geometry inside the analysis software or, in some cases, in another software included with the analysis software; or (2) obtaining it from a computer-aided design (CAD) program where it is drawn or automatically obtained. In this book, the discussion on geometry creation will be primarily limited to the first type, i.e., the geometry created inside COMSOL itself. A brief discussion on obtaining geometry from a CAD program is included in Section 2.7.2. To draw the geometry in the main COMSOL window (Figure 2.5), firstly, the geometry type needs to be selected. The different geometry options are shown in Figure 2.6.

Box 2.1 Software implementation: geometry type

The geometry type in COMSOL is selected using the path: COMSOL Multiphysics (icon on Desktop or from Start Menu) >> Model Navigator >> Space Dimensions

---

2.6 Geometry and analysis type

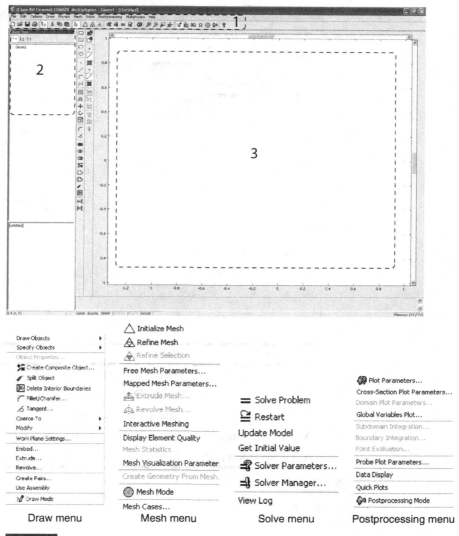

Figure 2.5

The main COMSOL window shown on the top opens up once the geometry and analysis type (shown in Figure 2.4) for the problem are selected. The different options available under the most commonly used menu items are also shown. Items in the Draw menu are used for drawing the geometry. The Mesh menu is used for dividing the geometry into small elements. The Solve menu is used for specifying the solver setting for the problem and to solve it subsequently. The Postprocessing menu is used for visualizing the results obtained from the simulations.

Software implementation 1

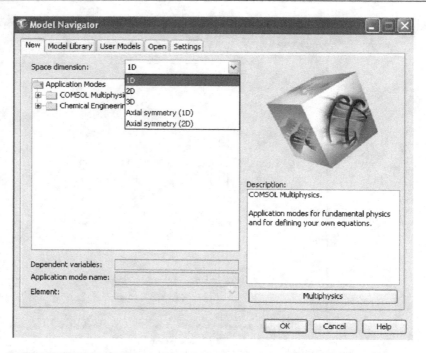

Problem	Path
1D	Space Dimensions >>1D
2D	Space Dimensions >>2D
1D	Space Dimensions >>3D
Axial Symmetry(1D)	Space Dimensions>> Axial Symmetry (1D)
Axial Symmetry(2D)	Space Dimensions>> Axial Symmetry (1D)

Figure 2.6

Geometry options in COMSOL.

2.6.2 Analysis

We will be dealing with fluid flow, and heat and mass transfer problems in this text. COMSOL has additional options for other types of physics such as electromagnetics and ultrasonics, discussed in Chapter 8, whose implementation will not be covered here. The different options under analysis types are shown in Figure 2.7.

2.6 Geometry and analysis type

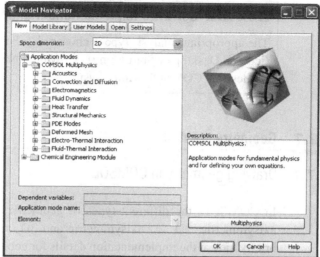

Mass transfer (COMSOL Multiphysics >> Convection and Diffusion)	**Heat transfer** (COMSOL Multiphysics >> Heat Transfer)	**Fluid flow** (COMSOL Multiphysics >> Fluid Dynamics)

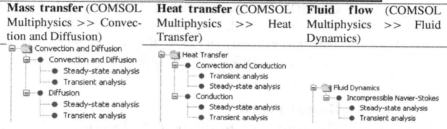

Select **Multiphysics** and add two or more different analysis types to solve coupled momentum, heat and species transfer problems as shown below:

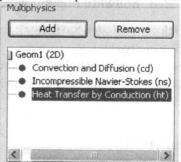

Figure 2.7

Analysis options in COMSOL. Different physics can be selected such as heat transfer, mass transfer or fluid flow. Different physics may be combined using the multiphysics options as shown.

> **Box 2.2 Software implementation: analysis type**
>
> The appropriate analysis type is chosen as follows: COMSOL Multiphysics (icon on Desktop or Start Menu) >> Model Navigator >> Application Modes >> COMSOL Multiphysics

2.7 Geometry creation

2.7.1 Drawing geometry in COMSOL

In COMSOL, the different types of geometries that are possible are: 1D, 1D axisymmetric, 2D, 2D axisymmetric and 3D, as shown in Figure 2.6. We will briefly look at the implementation details for geometry creation in COMSOL. The COMSOL user's guide includes a detailed discussion of geometry creation capabilities in COMSOL. Additionally each of the ten Case studies in Chapter 6 uses a different geometry and shows their implementation in COMSOL. For a quick reference, see Table 1.2 for the geometries used in the different Case studies.

Geometry in 1D

The geometry in 1D is a line or a collection of lines. Points may be created to specify a boundary condition, or to divide the computational domain into subdomains.

> **Box 2.3 Software implementation**
>
> (1) Select (in **Model Navigator**): Space dimension >> 1D or Axial Symmetry (1D)
> (2) Select (in **Main COMSOL window**): Draw >> Draw Object or Specify Object >> Point or Line. See Figure 2.8 for details.

Geometry in 2D

Geometrical shapes such as polygons (rectangle, square, triangle, etc.), circles and ellipses, or their combinations can be drawn in COMSOL. The required geometry can be drawn using either a bottom-up approach in which the 2D structure is created starting from lines and points, or a top-down approach in which the 2D geometries are created directly. For example to create a rectangle, we may first create four different lines and join them (bottom-up) or specify the width and height of the rectangle to create it directly (top-down).

2.7 Geometry creation

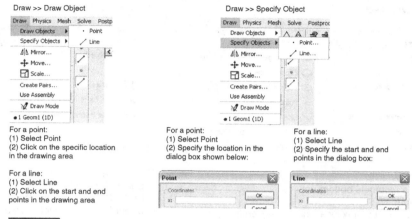

Figure 2.8

Geometry options in COMSOL for 1D problems. The steps needed to draw a point or line in COMSOL using two different methods are shown.

Box 2.4 Software implementation

(1) Select (in **Model Navigator**): Space dimension >> 2D or Axial Symmetry (2D)

(2) Select (in **Main COMSOL window**): Draw >> Draw Object or Specify Object >> Select rectangle, square, circle, etc. See Figure 2.9 for details.

Geometry in 3D

In 3D, different shapes such as block, cone, cylinder, sphere, etc. can be drawn in COMSOL. Again a bottom-up approach can be used, starting from 2D geometries and using extrusion or revolution tools, or a top-down approach can be used by directly choosing the 3D shape.

Box 2.5 Software implementation

(1) Select (in **Model Navigator**): Space dimension >> 3D

(2) Select (in **Main COMSOL window**): Draw >> Select block, cylinder, cone, etc. See Figure 2.10 for details.

Other geometry creation tools in COMSOL

Figure 2.11 shows the tools and options in COMSOL for performing actions such as setting grids, displaying geometry properties, adding and subtracting geometries,

Draw >> Draw Object

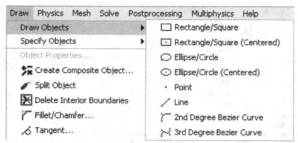

Example:

(1) To draw a rectangle bottom-up:
Select Line
Click on the first vertex location, subsequently click on the second, third and fourth vertices, and finally right click on the first vertex again to create the rectangle

(2) To draw a rectangle top-down:
Select Rectangle
Click on the first vertex location in the drawing window and then click on the diagonally opposite vertex location

Draw >> Specify Object

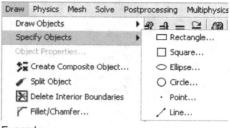

Example:

To draw a rectangle top-down:
Select Rectangle
Under Size, specify the Width and Height
Under Position, select Base as Corner or Center and specify the x and y coordinates of the corner or the center

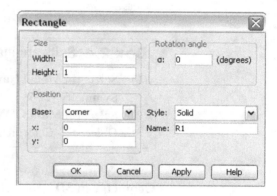

Figure 2.9

Geometry options in COMSOL for 2D problems. Steps for drawing a rectangle in COMSOL using top-down and bottom-up approaches are shown. Other geometric shapes can be drawn similarly.

dividing geometries into different parts, etc. These items provide additional options to the user for creating complex geometries or modifying geometries created using the operations discussed in the previous sections.

> **Box 2.6 Software implementation**
>
> All these tools can be found in the **Main COMSOL window** under **Options** or **Draw**, as shown in Figure 2.11.

2.7 Geometry creation

Draw options
(in the main menu)

Examples :
(1) To draw a block bottom-up starting from a rectangle:
  Click on Work Plane Settings and select the plane (x-y, y-z or z-x)

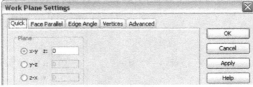

Draw the required Rectangle in the selected 2D plane
Click on Extrude, select the rectangle and specify the length of
the block in the third direction under Distance to create the block

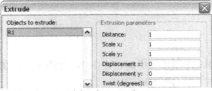

(2) To draw a block top-down:
  Select Block
  Under Length, specify the x, y and z dimensions
  Change other parameters such as Base, Axis base point, etc. as
  needed

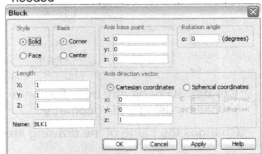

Figure 2.10

Geometry options in COMSOL for 3D problems. Steps for drawing a 3D rectangular block in COMSOL using top-down and bottom-up approaches are shown.

2.7.2 Obtaining exact geometry from a CAD program

A CAD (computer-aided design) software is typically used to build the geometry (especially complex ones) on the computer. Examples of such software include AutoCAD, CADAM, CADKEY, and Pro/Engineer. These softwares can export the geometry in a number of formats which an analysis program (such as COM-SOL) can read. Such formats include Parasolid, ACIS, STEP, STL, IGES, Native

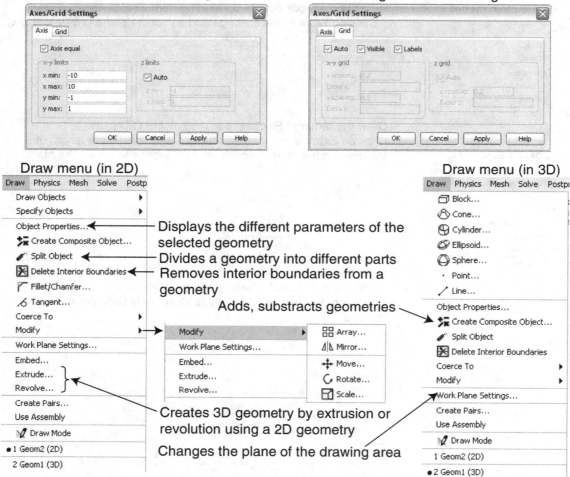

Figure 2.11

Brief overview of tools in COMSOL for creating and modifying geometries. The figure shows how to set grids, display attributes of a selected geometry, divide a geometry into different parts, remove interior boundaries from a geometry, add or subtract geometries, create a 3D shape by extrusion or revolution and change the plane of the drawing area. Refer to COMSOL user's guide for details.

CATIA V4 and Pro/E. Sometimes these files are imported into another software that specializes in mesh generation (see example in the following section). Another example of mesh generation software is GAMBIT which can import many of the above formats.

Obtaining complex geometries is a difficult job and automated ways of obtaining them are constantly being sought. There are several competing 3D scanning technologies. One such technology, of particular interest for biomedical applications, is 3D geometry generation using data from CT scan machines. Magnetic Resonance Imaging (MRI) can also be used to obtain images that can be reconstructed to form a 3D geometry. Another method to obtain exterior geometry is to use laser scanning.

As an example, Figure 2.12 shows the upper airway of a horse in the software GAMBIT. This geometry was obtained by stitching the 2D CT scanned images of the horse in a geometry reconstruction software to obtain a 3D geometry in STL format. After making some necessary changes in the geometry like smoothing, creating flow boundaries and removing unwanted parts using geometry manipulation software, it was imported into GAMBIT and meshed. The sequence of steps used to obtain the geometry is shown in Figure 2.12.

Another example is shown in Figure 2.13. In this case, laser scanning is performed by a VITUS/smart 3D Body Scanner (Human Solutions, Troy, MI). Figure 2.13 shows the setup for such scanning and the 3D laser scan. The scanner provides a wireframe representation of the surface with a triangulated surface.

2.8 Governing equations

Once the geometry is decided, the appropriate equations representing the physics that will be solved over the geometry (domain) are decided. The governing equations are introduced in Chapter 1 starting from Section 1.7, and their details and derivations are provided in Chapter 7. Depending on the problem, the appropriate terms in the energy, species or momentum (Navier–Stokes) equations need to be selected in the software. Figure 2.14 shows the governing equations for the different physics and their COMSOL implementation. All the equations are shown in 1D for convenience. For terms not needed in the model, the coefficients of the respective terms are entered as zero. Refer to the next section for details.

2.8.1 Convection, transient and source terms

As shown in Figure 2.14, three of the terms in the governing equations are convection, transient and source. These may be turned on or off depending on the problem. This is done by selecting the appropriate Analysis Type in the Model Navigator window and entering the coefficient of the respective terms as zero, as shown in Figure 2.15. The equations below are shown for heat transfer. These terms can be turned on/off in the momentum and species equations in the same way.

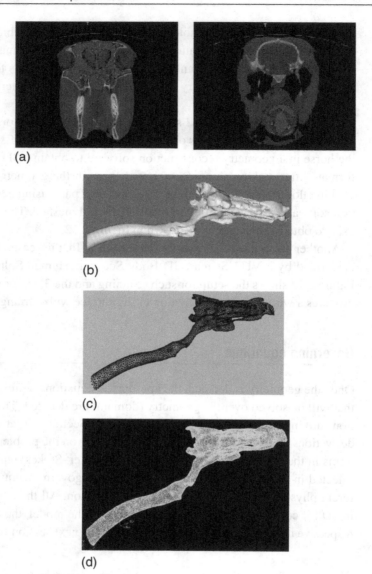

Figure 2.12

Obtaining geometry of a horse upper airway for airflow analysis: (a) CT scanned images; (b) 3D geometry reconstructed from CT scanned images, in the software MIMICS; (c) smoothed 3D geometry with inflow and outflow surfaces and unwanted parts removed in the software MAGICS; (d) final geometry with the 3D mesh in software GAMBIT (see Rakesh et al., (2008), for more details).

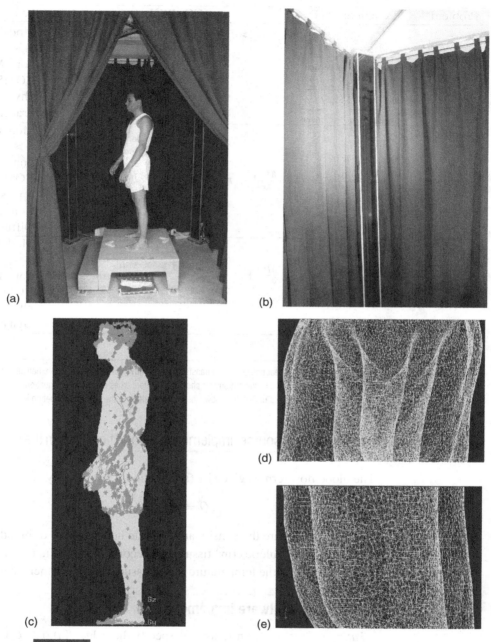

Figure 2.13

3D laser scan of human body (a), (b) setup for the laser scan by VITUS/smart 3D body scanner; (c) wireframe geometry obtained after the scan as seen in GAMBIT; (d) detailed view of the chest; (e) detailed view of the legs (courtesy of Cornell Body Scan Research Group, Dept. of Textiles and Apparel, Ashdown, Loker and Schoenfelder).

Problem	Equations (1–D)	COMSOL	Path

Energy	$\underbrace{\frac{\partial T}{\partial t}}_{\text{transient}} + \underbrace{u\frac{\partial T}{\partial x}}_{\text{convection}} = \underbrace{\frac{k}{\rho c_p}\frac{\partial^2 T}{\partial x^2}}_{\text{conduction}} + \underbrace{\frac{Q}{\rho c_p}}_{\text{source}}$	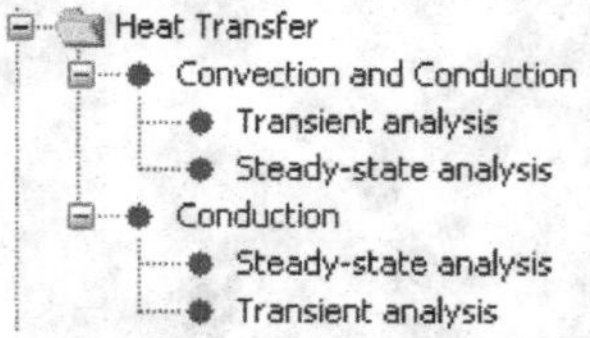 Heat Transfer
		— Convection and Conduction
		  — Transient analysis
		  — Steady-state analysis
		— Conduction
		  — Steady-state analysis
		  — Transient analysis
			Model Navigator >> Application Modes >> COMSOL Multiphysics >> Heat Transfer

Species	$\frac{\partial c_A}{\partial t} + u\frac{\partial c_A}{\partial x} = D_{AB}\frac{\partial^2 c_A}{\partial x^2} + r_A$	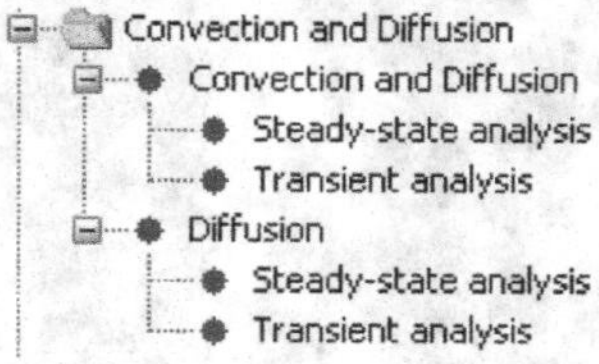 Convection and Diffusion
		— Convection and Diffusion
		  — Steady-state analysis
		  — Transient analysis
		— Diffusion
		  — Steady-state analysis
		  — Transient analysis
			Model Navigator >> Application Modes >> COMSOL Multiphysics >> Convection and Diffusion

Momentum	$\frac{\partial u}{\partial t} + u\frac{\partial u}{\partial x} = -\frac{1}{\rho}\frac{\partial P}{\partial x} + \frac{\mu}{\rho}\frac{\partial^2 u}{\partial x^2} + X$	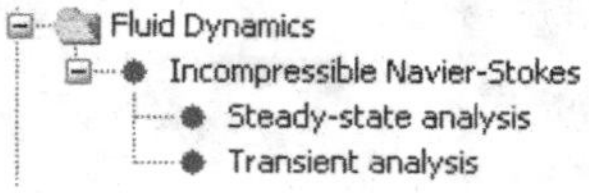 Fluid Dynamics
		— Incompressible Navier-Stokes
		  — Steady-state analysis
		  — Transient analysis
			Model Navigator >> Application Modes >> COMSOL Multiphysics >> Fluid Dynamics

Figure 2.14

Governing equations for heat transfer, mass transfer and fluid flow along with their implementation in COMSOL. The column under COMSOL shows the options for choosing different types of governing equations in COMSOL graphical user interface (GUI) and that under Path shows the path for selecting the particular equation in the GUI.

2.8.2 Example of heat source: implementing blood flow term in the bioheat equation

The blood flow term is given by (see Section 7.5 for theory):

$$Q = \rho_b c_b \dot{V}_b^v (T_a - T)$$

where ρ_b and c_b are the density and specific heat capacity of blood and $\dot{V}_b^v$ is the flow rate, in (m^3 blood)/(m^3 tissue) per second. T_a is the temperature of arterial blood, while T is the temperature of the tissue. This implemented as follows:

> **Box 2.7 Software implementation**
>
> *Path*: Options >> Constants >> Specify the value of rho_b, c_b, V_b and T_a (as shown in case of Reactions in the next section)
> Then, the blood flow term is implemented as the source term, Q.
> *Path*: Physics >> Subdomain Settings >> Physics >> Type rho_b*c_b * V_b*(T_a - T) in the Heat Source, Q, field (as shown in Figure 2.16)

2.8 Governing equations

Problem	Equations (1–D)	COMSOL implementation	Path
Steady-state conduction	$0 = \underbrace{\dfrac{k}{\rho c_p}\dfrac{\partial^2 T}{\partial x^2}}_{\text{conduction}} + \underbrace{\dfrac{Q}{\rho c_p}}_{\text{source}}$	Heat Transfer → Convection and Conduction (Transient analysis / Steady-state analysis); Conduction → **Steady-state analysis** / Transient analysis	Model Navigator >> Application Modes >> COMSOL Multiphysics >> Heat Transfer
Transient conduction	$\underbrace{\dfrac{\partial T}{\partial t}}_{\text{transient}} = \underbrace{\dfrac{k}{\rho c_p}\dfrac{\partial^2 T}{\partial x^2}}_{\text{conduction}} + \underbrace{\dfrac{Q}{\rho c_p}}_{\text{source}}$	Heat Transfer → Convection and Conduction (Transient analysis / Steady-state analysis); Conduction → Steady-state analysis / **Transient analysis**	Model Navigator >> Application Modes >> COMSOL Multiphysics >> Heat Transfer
Convection	$\underbrace{\dfrac{\partial T}{\partial t}}_{\text{transient}} + \underbrace{u\dfrac{\partial T}{\partial x}}_{\text{convection}} = \underbrace{\dfrac{k}{\rho c_p}\dfrac{\partial^2 T}{\partial x^2}}_{\text{conduction}} + \underbrace{\dfrac{Q}{\rho c_p}}_{\text{source}}$	Heat Transfer → Convection and Conduction; u [u] m/s x-velocity; v [v] m/s y-velocity (Check the variable names)	Model Navigator >> Heat Transfer Physics >> Subdomain Settings
No convection	$\underbrace{\dfrac{\partial T}{\partial t}}_{\text{transient}} = \underbrace{\dfrac{k}{\rho c_p}\dfrac{\partial^2 T}{\partial x^2}}_{\text{conduction}} + \underbrace{\dfrac{Q}{\rho c_p}}_{\text{source}}$	u [0] m/s x-velocity; v [0] m/s y-velocity	Physics >> Subdomain Settings
Source	$\underbrace{\dfrac{\partial T}{\partial t}}_{\text{transient}} + \underbrace{u\dfrac{\partial T}{\partial x}}_{\text{convection}} = \underbrace{\dfrac{k}{\rho c_p}\dfrac{\partial^2 T}{\partial x^2}}_{\text{conduction}} + \underbrace{\dfrac{Q}{\rho c_p}}_{\text{source}}$	Q [Value] W/m³ Heat source	Physics >> Subdomain Settings
No Source	$\underbrace{\dfrac{\partial T}{\partial t}}_{\text{transient}} + \underbrace{u\dfrac{\partial T}{\partial x}}_{\text{convection}} = \underbrace{\dfrac{k}{\rho c_p}\dfrac{\partial^2 T}{\partial x^2}}_{\text{conduction}}$	Q [0] W/m³ Heat source	Physics >> Subdomain Settings

Figure 2.15

Selecting governing equation with appropriate terms in COMSOL. Shown is the example for heat transfer. Similarly, the appropriate terms can be selected for a mass transfer or fluid flow problem.

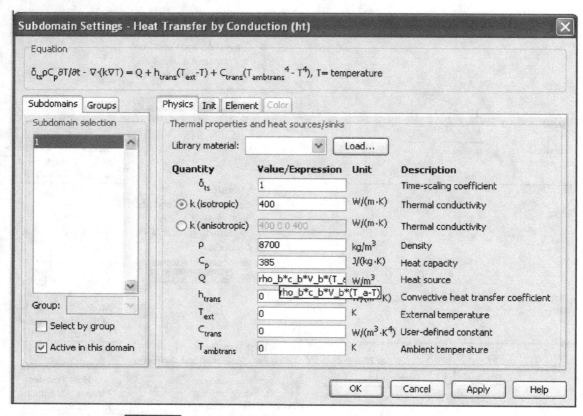

Figure 2.16

Implementation of the bioheat equation in COMSOL. The expression for the blood flow term is entered as a source term, Q, in the software.

2.8.3 Example of mass source: implementing reactions

> **Box 2.8 Software implementation**
>
> *Path*: Physics >> Subdomain Settings >> Physics >> Type the form of the reaction equation in the Reaction Rate field (as shown in Figure 2.17).

For example, a first-order reaction with one species and no dependence on temperature can be implemented by entering $kr*c$ in the **Reaction Rate** field, **R** as shown in Figure 2.17. The value of reaction rate constant, kr, can then be defined as a **Constants** menu item as shown in Figure 2.18.

2.8 Governing equations

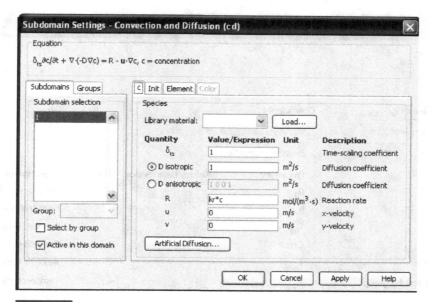

Figure 2.17

Implementation of reactions in COMSOL. In this example, the reaction term is of the form, $r_A = k''c_A$, a first-order reaction with k'' as the reaction rate constant.

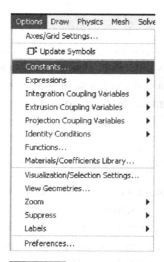 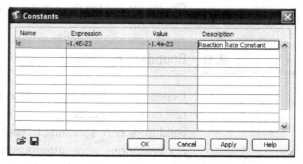

Figure 2.18

Defining the reaction rate constant in the **Constants** window in COMSOL.

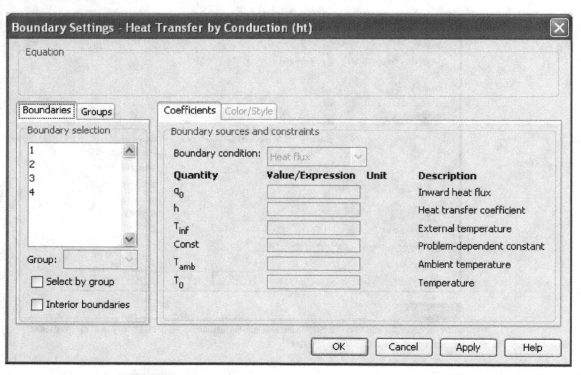

Figure 2.19

Boundary conditions window in COMSOL.

Similarly, temperature dependence can be specified by entering the expression: $kr*c*exp(Ea/(Rg*T))$. The values of reaction rate constant (kr), activation energy for the reaction (Ea) and the universal gas constant (Rg) can be specified in the **Constants** field in the same way under the **Constants** window.

2.9 Boundary conditions

Depending upon the problem at hand, i.e., whether solving the energy, momentum or mass balance equation (or a combination of these), we need the boundary conditions corresponding to each governing equation used. The types of boundary conditions were introduced in Section 1.8 and their details are provided in Chapter 7 starting from Section 7.10. The different types of boundary conditions needed for energy, momentum and mass balance equations with their implementation in COMSOL are described in Figure 2.20.

2.9 Boundary conditions

Boundary condition	COMSOL implementation	Path
Constant temperature $T = \text{constant}$	T_0 [VALUE] K Temperature	Physics >> Boundary Settings >> Boundary condition >> Temperature >>Enter Temperature
Constant flux $-k\frac{dT}{dx} = \text{constant}$	q_0 [VALUE] W/m² Inward heat flux	Physics >> Boundary Settings >> Boundary condition >> Heat Flux >> Enter Flux value
Convection $-k\frac{dT}{dx} = h(T - T_\infty)$	h [VALUE] W/(m²·K) Heat transfer coefficient T_{inf} [VALUE] K External temperature	Physics >> Boundary Settings >> Boundary condition >> Heat Flux >> Enter heat transfer coefficient and Ref Temp.
Constant species concentration $c_A = \text{constant}$	c_0 [VALUE] mol/m³ Concentration	Physics >> Boundary Settings >> Boundary condition >> Concentration >> Enter Species Conc.
Constant species flux $-D_{AB}\frac{dc_A}{dx} = \text{constant}$	N_0 [VALUE] mol/(m²·s) Inward flux	Physics >> Boundary Settings >> Boundary condition >> Flux >> Enter Species Flux
Species convection $-D_{AB}\frac{dc_A}{dx} = h_m(c_A - c_{A\infty})$	N_0 [h_m*(c-c_inf)] mol/(m²·s) Inward flux Define mass transfer coefficient, h_m, and Ref Conc., c_inf, in the Constants field (under Options)	Physics >> Boundary Settings >> Boundary condition >> Flux >> Enter convective species transfer equation
Constant velocity $u = \text{constant}$	u_0 [VALUE] m/s x-velocity v_0 [VALUE] m/s y-velocity	Physics >> Boundary Settings >> Boundary condition >> Inflow/Outflow velocity >> Enter Velocity
No slip (at the wall) $u = 0$	Boundary condition: No slip	Physics >> Boundary Settings >> Boundary condition >> No Slip
Constant pressure $P = \text{constant}$	p_0 [VALUE] Pa Pressure	Physics >> Boundary Settings >> Boundary condition >> Pressure >> Enter Pressure

Figure 2.20

Setting up various boundary conditions in COMSOL. Refer to screen in Figure 2.19 for the complete window.

2.9.1 General implementation

Path: Physics >> Boundary Settings

2.9.2 Special cases: time-varying boundary conditions

Temperature, species concentration, flux (heat and species), velocity and pressure values can also be defined as time-varying quantities. These variables will be denoted as "dependent variables" in the subsequent discussion.

COMSOL provides us with two options to specify time-varying boundary conditions:

- The dependent variable can be defined as a function of time by specifying a set of data points (dependent variable versus time), as shown in detail in Table 2.1. In the example in Table 2.1, heat flux has been defined as a function of time. Other dependent quantities can be defined similarly.
- The second option in COMSOL is to specify the algebraic expression representing the dependent variable as a function of time; discussed in detail in Table 2.2.

2.9.3 Example: specifying a parabolic inlet velocity profile

In some cases, we may want to model a situation where the inlet velocity profile can be specified as a parabola. For example:

$$u = 10 - 40y^2$$

We will implement this equation in COMSOL as shown in the example below:

Step 1: Activate the momentum equation
Path: **Model Navigator** >> **Application Modes** >> **COMSOL Multiphysics** >> **Fluid Dynamics** (Select the convective term option by specifying the coefficient, as described earlier in Figure 2.15, if you want the solver to solve for convection in the heat transfer or species transfer equations)

2.9 Boundary conditions

Table 2.1 Setting up a time-varying boundary condition: using data points.

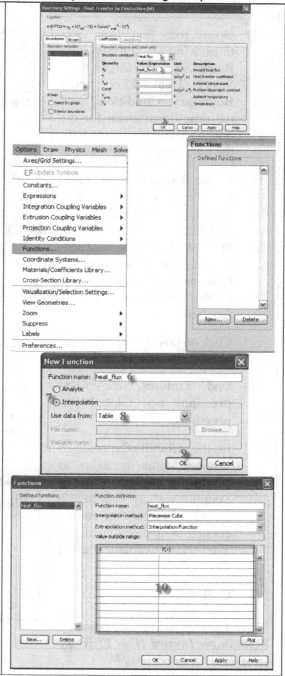

(1) Open the Boundary Settings window: **Physics** >> **Boundary Settings** >> Select **Heat Flux** under **Boundary Condition** from the drop-down menu for the boundary you want to specify the time-varying heat flux for.

(2) In the **Inward Heat Flux, q_0**, field, enter heat_flux(t). heat_flux(t) tells COMSOL that heat_flux (to be specified in Step 4 later) is a function of time. heat_flux is any arbitrary user-defined name chosen for the function.

(3) Click on **OK**. We will now define the function, heat_flux(t).

(4) Under **Options**, click on **Functions**...

(5) Click on the **New** button.

(6) Next to Function name: type in heat_flux.

(7) Check **Interpolation**. By checking this, we are directing the solver to interpolate values of heat flux between different times.

(8) Select **Table** next to the Use data from box.

(9) Click **OK**.

(10) Enter the time and the corresponding heat flux values in the text boxes. The first column (under x) shows the values of time and f(x) represents the corresponding heat flux values. Click on **OK**.

Table 2.2 Setting up a time-varying boundary condition: using a function.

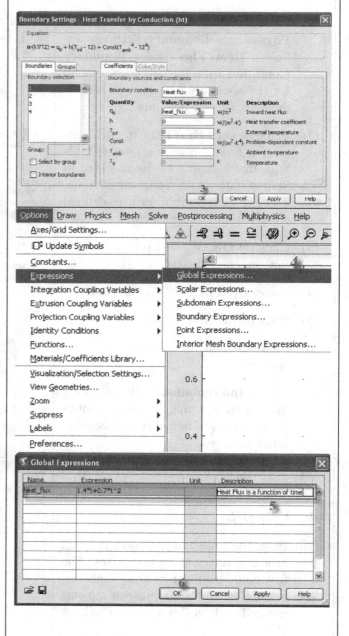

(1) Open the Boundary Settings window: **Physics** >> **Boundary Settings** >> Select **Heat Flux** under **Boundary Condition** from the drop-down menu for the boundary you want to specify the time-varying heat flux for.

(2) In the **Inward Heat Flux**, q_0, field, enter heat_flux. heat_flux tells COMSOL that heat_flux (to be specified in Step 4 later) is a function of some other variables. heat_flux is any arbitrary user-defined name chosen for the function.

(3) Click on **OK**. We will now define the function, heat_flux.

(4) Under **Options**, click on **Expressions** >> **Global Expressions.**

(5) In the **Global Expressions** window, define heat_flux as a function of time. In this example we say heat_flux = $1.4t+0.7t^2$. heat_flux can be defined as a function of any independent variables using this method.

(6) Click **OK**.

*Note: The expression can be directly input in the **Inward Heat Flux**, q_0, field as in the next example.*

Step 2: Specify the parabolic inlet velocity boundary condition

(1) Open the Boundary Settings window: **Physics >> Boundary Settings >>** Select **Inflow/Outflow Velocity** under **Boundary Conditions** from the drop-down menu for the boundary you want to specify the parabolic velocity profile for.
(2) Specify the parabolic velocity expression in the appropriate (x or y) velocity field. Here we specify the x-velocity as a parabolic velocity profile.
(3) Click on **OK**.

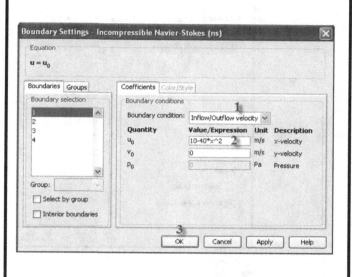

2.10 Initial conditions

Initial temperature, concentration, velocity or pressure for a time-dependent problem can be set in COMSOL by using the appropriate window and entering the values in the respective fields, as shown in Figures 2.21–2.23.

> **Box 2.9 Software implementation**
>
> *Path*: Physics >> Subdomain Settings >> Init Tab

2.11 Material properties

Material properties such as thermal conductivity, density, specific heat, diffusivity and viscosity need to be specified in the computational software depending upon the particular problem. For a heat transfer problem, the material properties that

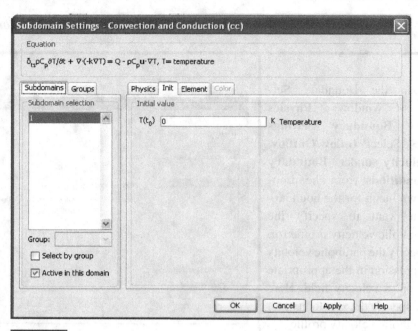

Figure 2.21

Initial Condition window in COMSOL for a heat transfer problem. The initial temperature needs to be set in the appropriate field.

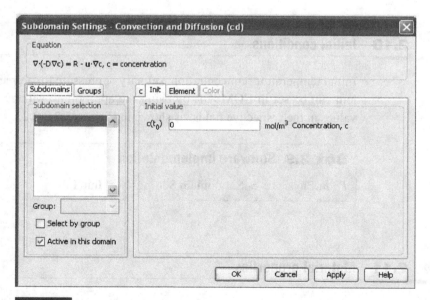

Figure 2.22

Initial Condition window in COMSOL for a mass transfer problem. The initial concentration is entered in the appropriate field.

2.11 Material properties

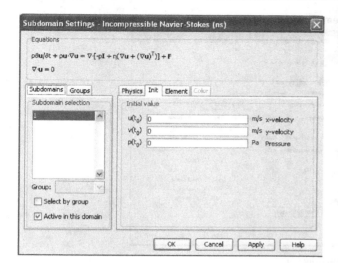

Figure 2.23

Initial Condition window in COMSOL for a fluid flow problem. The initial velocities and pressure are entered in the appropriate fields.

are needed are thermal conductivity (k), density (ρ) and specific heat (c_p). For a mass transfer problem, diffusivity (D) is needed and for a fluid flow problem, density (ρ) and viscosity (μ) are needed. Table 1.4 in Section 1.9 provides a list of the material properties and their implementation in the different Case studies in Chapter 6. Figure 2.24 shows the material properties window in COMSOL for the three types of problems. For material properties that are constant, the constant values are entered directly into the respective fields, as shown in Figure 2.24. Variable properties are discussed in the next section.

Box 2.10 Software implementation

Path: Physics >> Subdomain Settings >> Physics Tab

2.11.1 Variable properties

Property values can be functions of temperature, time, space coordinates or any other variable. This is defined the same way as we specified heat flux as a function of time in the *time-varying boundary condition* example discussed earlier in Section 2.9.2. One example of variable property values is covered in Case study II (Chapter 6), where thermal conductivity and specific heat are specified as a function of temperature by using a set of data points. Another example is shown in Figure 2.25, where diffusivity is specified as a function of distance for a mass transfer problem.

(a) **Heat transfer**: Inputs – thermal conductivity (k), density (ρ), specific heat (c_p)

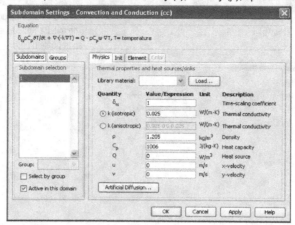

(b) **Mass transfer**: Input – diffusivity (D)

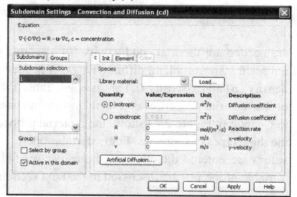

(c) **Fluid flow**: Input – density (ρ), viscosity (μ)

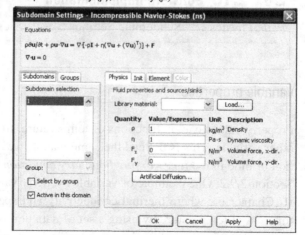

Figure 2.24

Material properties panels in COMSOL for heat transfer, mass transfer and fluid flow problems.

2.11 Material properties

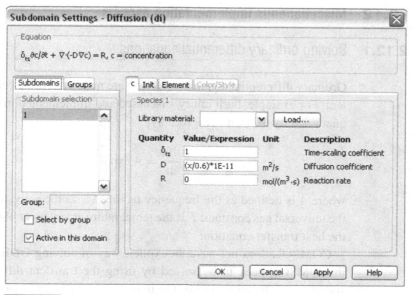

Figure 2.25

Specifying variable properties in COMSOL. In this case, the diffusivity is a function of space and is given by $D = x/0.6 \times 10^{-11}$ m^2/s. Similarly, the material properties can be a function of time or any other variable.

Box 2.11 Software implementation

Path: Physics >> Subdomain Settings >> c Tab

2.11.2 Zero diffusivity

In problems where there is no movement of the mass species inside the domain due to diffusion, the diffusivity needs to be set to zero in the computational software. However, in some software (e.g., FIDAP), a diffusivity value of zero (dropping of the diffusion term) is not acceptable. Here the trick is to use a diffusivity value that is small enough so that the problem does not change. It is not known a priori what value of diffusivity is small enough for a problem, therefore, it requires one to try progressively smaller diffusivity values until the diffusivity value has no effect on the solution. In the software COMSOL, zero diffusivity is allowed. COMSOL handles zero diffusivity by using specialized stabilization techniques for solution.

2.12 Miscellaneous implementation aspects

2.12.1 Solving ordinary differential equations

Ordinary differential equations (ODEs) are needed to solve certain types of problem. For example, burn injury, discussed and implemented in Case study VIII, is quantified by the term, Ω, given by:

$$\frac{d\Omega}{dt} = A \exp\left[-\frac{\Delta E}{RT}\right]$$

where A is defined as the frequency factor, ΔE as the activation energy and R is the universal gas constant; T is the temperature at any point obtained from solving the heat transfer equation.

COMSOL does not have an explicit way of solving ODEs. Therefore ODEs in COMSOL need to be solved by using the transient diffusion equation and specifying diffusivity as zero. The mass source term can be defined accordingly to incorporate any other terms that are needed. For example in this case, the following formulation can be used:

$$\frac{\partial c_A}{\partial t} = D_{AB}\frac{\partial^2 c_A}{\partial x^2} + r_A$$

with $D_{AB} = 0$ and $r_A = A \exp[-E/RT]$. As you can see, this case is similar to the one discussed in Section 2.11.2 for implementing immobile mass species. Since COMSOL can handle zero diffusivity values as discussed in that section, an ODE of this type can be solved easily.

2.12.2 Using logical expressions

Sometimes logical expressions are needed to define certain quantities. For example, take the case of a drug delivery problem where the drug is supplied to the surface of the skin through a patch. The patch is applied in a cyclic way such that it is applied to the body for one day, taken off the next day and then placed back on the third day. Assuming that we can use a constant flux boundary condition on the skin surface to model the diffusion of drug from the patch to the skin, we need to implement the cyclic application and removal in the model. We can incorporate this by using a logical expression for the flux at the skin surface in COMSOL (also shown in Figure 2.26):

2.12 Miscellaneous implementation aspects

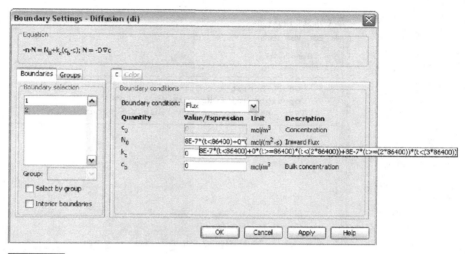

Figure 2.26

Specifying logical expressions in COMSOL. In this case a boundary condition has been specified as a logical expression.

$$\text{Flux} = 8E - 7 * (t < 86400)$$
$$+ 0 * (t >= 86400) * (t < (2 * 86400))$$
$$+ 8E - 7 * (t >= (2 * 86400)) * (t < (3 * 86400))$$

where t is the time in s and 8×10^{-7} (in g/m^{2}s) is the flux supplied by the patch when it is applied to the skin. The expressions in parentheses are 1, if they are true and 0, if false. For example, if the time (during the solution process) is 3600 s then $(t < 86\,400)$ is true and hence, its value is 1. On the other hand, the expressions $(t >= 86\,400)$, $(t < (2 * 86\,400))$, $(t >= (2 * 86\,400))$ and $(t < (3 * 86\,400))$ are false and therefore equal to zero. Therefore, for the first day (up to 86 400s), when $(t < 86\,400)$ is true:

$$Flux = 8E - 7$$

Similarly, if $(t >= 86\,400)$ and $(t < (2 * 86\,400))$:

$$Flux = 0$$

Finally, if $(t >= (2 * 86\,400))$ and $(t < (3 * 86\,400))$:

$$Flux = 8E - 7$$

Similarly material properties, initial conditions and source terms can be defined using logical expressions.

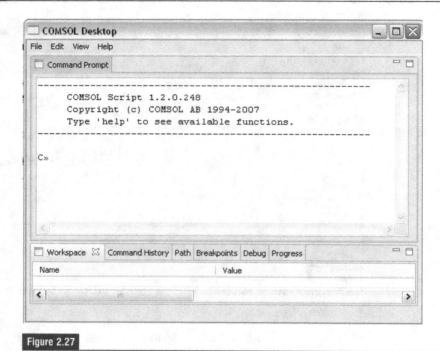

Figure 2.27

The COMSOL Script window. Commands may be typed in the command prompt for performing different tasks such as creating geometry, meshing etc.

2.12.3 Modeling in COMSOL Script

Instead of using the COMSOL graphical user interface (GUI), which was the topic in this chapter, COMSOL programming language can be used in COMSOL Script (shown in Figure 2.27) to build and solve models. Models can be developed in COMSOL Script by two alternative means:

- **Scripting from scratch** To build the model, commands to perform different operations such as geometry creation, meshing, specifying governing equations, boundary conditions, initial conditions and material properties, and selecting the solver can be entered directly in a text file. The file can then be run from the COMSOL Script Command line to solve the model. This process is similar to writing codes in a programming language such as MATLAB or C. Please refer to the COMSOL Script user's guide, for the commands needed to perform the different operations for building and solving the model.
- **Create text file using COMSOL GUI** The text file with the commands to perform the different operations discussed above can also be created automatically using the GUI. In this case, the model is built in the COMSOL GUI using

the process discussed in this chapter, and Chapters 3 and 4 on software implementation. The file can then be saved in .m format instead of .mph format that is generally used to save the COMSOL files. The .m file is a text file that can be opened in Notepad, WordPad or any other text editor such as MATLAB editor or COMSOL Script editor and it contains all the commands for the model built in the GUI. Changes can then be made in the .m file by typing additional commands to perform advanced analysis. An example of using the GUI to create a .m file and then use it for subsequent computations is shown in Case study III (Chapter 6).

In this book, we will primarily focus on using the COMSOL GUI for building and solving computational models. The subsequent chapters on software implementation (Chapters 3–4) are similar to this chapter in discussing the details about implementation in context of the GUI. The reader is referred to the COMSOL Script user's guide for further details about building and solving models using COMSOL Script.

References

Anonymous. (2009). Additional resources on computer implementation, from the work of students at Cornell University. On the web at
http://instruct1.cit.cornell.edu/courses/bee453/ >> Software Resources.

Rakesh, V., Datta, A. K., Ducharme, N. G. and Pease, A. P. (2008). Simulation of turbulent airflow using a CT based upper airway model of a racehorse. *Journal of Biomechanical Engineering*-Transactions of the ASME **130**(3):031011-1-031011-13.

2.13 Problems

2.13.1 Short questions

(1) What are the three steps in a computer-aided engineering software (such as COMSOL)? Hint: One of the steps is *processing*.

(2) Show a set of consistent units that you will use in a software for the rest of the variables in the following equation, where c is measured in μg/cc.

$$\frac{\partial c}{\partial t} + u\frac{\partial c}{\partial x} = D\frac{\partial^2 c}{\partial x^2} + r_A \tag{1}$$

(3) Show a consistent set of units that you will use in a software for the variables in solving fluid flow (Navier–Stokes equations, only the x component is reproduced here),

$$\rho\left[\frac{\partial u}{\partial t} + u\frac{\partial u}{\partial x} + v\frac{\partial u}{\partial y} + w\frac{\partial u}{\partial z}\right] = \rho g_x + \mu\left[\frac{\partial^2 u}{\partial x^2} + \frac{\partial^2 u}{\partial y^2} + \frac{\partial^2 u}{\partial t^2}\right] - \frac{\partial p}{\partial x} \quad (2)$$

2.13.2 Questions on software implementation

For each problem formulation question in Chapter 1 (Section 1.12.3), state how the following items will be implemented by selecting from the various choices available in COMSOL:

(1) Geometry type.
(2) Analysis type.
(3) Governing equation.
(4) Boundary conditions.
(5) Initial conditions.
(6) Material properties.

3 Software implementation 2 How to solve (preprocessing)

In the previous chapter, the mathematical model consisting of the geometry, governing equations and boundary conditions was implemented in a software. That is, the software now knows the computational domain and which equations to solve. We still have to tell the software how to solve the equations (from a set of parameter choices built into the software), which is the subject of this chapter.

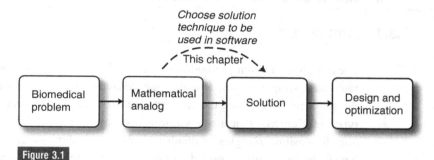

Figure 3.1
Instructing the software with detailed solution techniques to solve the model (set of equations) defined in Chapter 2 is the subject of this chapter.

3.1 Which numerical method to use

Today there is a choice of many numerical methods to solve governing equations – for example, the finite difference method, the finite volume method and the finite element method. Selection of a method is typically via the software chosen since any given software is based on one of these methods. Thus, once the software is chosen, the basic numerical method is fixed. For example, COMSOL uses the finite element method to solve governing equations. Details of the finite element method are provided in Chapter 10. Selection of the software is typically based on many considerations including whether it can solve the types of problems one is interested in, ease of use and cost.

3.2 Items needed in specifying the solution methodology

The items that need to be supplied to the computational software for it to solve the governing equation in the computational domain (Section 2.3) are:

- Mesh.
- Time steps (for a transient problem).
- Approach for solving algebraic equations.
- Tolerances for how accurate a solution is needed.

In this chapter, we will look at what these items mean and how they are implemented in COMSOL.

3.3 How to discretize the domain: mesh

3.3.1 Elements and mesh

For the finite element method, a mesh is a discretization of a geometric domain into small, simple shapes. The small discretized regions are known as elements and the set of points defining the elements are known as nodes. Examples of different types of elements that are possible are shown in Figure 3.2 – line elements in one dimension, triangles or quadrilaterals in two dimensions and tetrahedra, or hexahedra or prisms in three dimensions. An example of a computational domain consisting of a mesh having two-dimensional quadrilateral elements is shown in Figure 3.3.

3.3.2 Structured (mapped) versus unstructured (free) mesh

A structured or mapped mesh can be recognized by all interior nodes of the mesh having an equal number of adjacent elements. The mesh generated by a structured mesh generator is typically all quadrilateral (in 2D) or hexahedral (in 3D). Structured mesh generators employ unique iterative smoothing algorithms which attempt to align elements with boundaries or physical domains. For difficult boundaries, structured mesh generation requires the domain to be split up into simple blocks which are then meshed automatically.

Unstructured or free mesh generation, on the other hand, allows any number of elements to meet at a single node. The most common unstructured meshes are triangular (in 2D) and tetrahedral (in 3D), although quadrilateral and hexahedral

3.3 How to discretize the domain: mesh

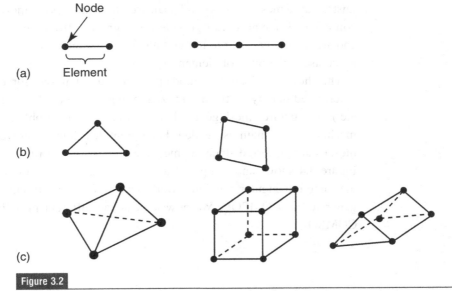

Figure 3.2

Types of finite elements. (a) One-dimensional line elements; (b) two-dimensional elements: three-node triangle, four-node quadrilateral; (c) three-dimensional elements: four-node tetrahedron, eight-node hexahedron, six-node prism (wedge).

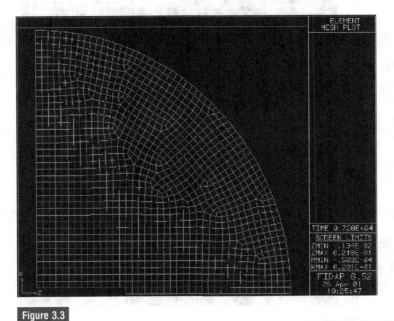

Figure 3.3

A mesh formed using two-dimensional four-noded quadrilateral elements.

unstructured meshes are possible. Unstructured meshes are more flexible in fitting complicated domains, can provide rapid grading from small to large elements and are relatively easier to refine (increase the number of elements) and de-refine (decrease the number of elements).

The choice between structured and unstructured meshes for a simple domain is determined mainly by the discretization method (finite difference, finite element etc.). A structured mesh generally helps to reduce the problem size by efficiently meshing the domain as needed. For a complex domain, however, unstructured meshes are preferred since the meshing process can be fully automatic and fast. Figure 3.4 shows an example of a structured and unstructured mesh created for the same rectangular geometry. Both structured and unstructured meshes can be generated in COMSOL. We now look at the different meshing capabilities of COMSOL.

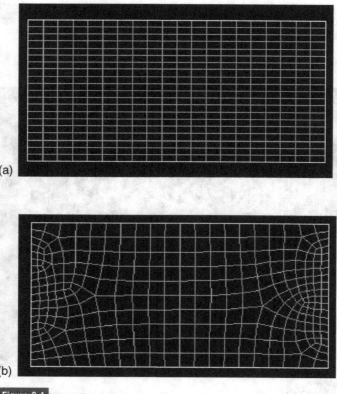

Figure 3.4

Examples of (a) a structured mesh; (b) an unstructured mesh.

3.3.3 Meshing in COMSOL: 1D geometry

In 1D problems, the computational domain is a line; meshing it leads to the creation of 1D line elements. The different features of the mesh that can be controlled in COMSOL are the element size (and thereby the total number of elements) and the element growth size which represents the maximum rate at which the element size can grow from a region with small elements to a region with larger elements. Meshing in 1D always leads to a structured mesh. The 1D meshing options in COMSOL are shown in Figure 3.5. An example of a 1D mesh is shown in Figure 3.6.

> **Box 3.1 Software implementation**
>
> *Path*: Mesh >> Free Mesh Parameters

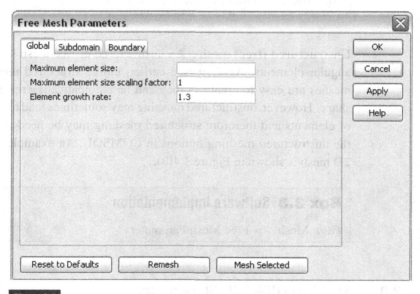

Figure 3.5

Mesh options in COMSOL for 1D problems. The size of the line elements can be controlled by specifying the maximum element size. Element growth rate is set to create a mesh with varying element size. If there is more than one subdomain in the overall computational domain, different maximum element sizes and element growth rates can be specified for each subdomain using the **Subdomain** tab.

Figure 3.6

Example of a 1D mesh with ten 1D line elements (eleven nodes).

3.3.4 Meshing in COMSOL: 2D geometry

For 2D problems, both structured or unstructured mesh can be created in COMSOL.

Structured (mapped) mesh Structured mesh in 2D consists of quadrilateral elements. The size of the these quadrilateral elements can be controlled by selecting a predefined mesh size or meshing the 1D boundaries first and then meshing the 2D domain. For domains with complex shapes, the boundaries have to be divided and meshed manually in order to create a structured mesh. The various mapped meshing options in COMSOL are shown in Figure 3.7. An example of a structured 2D mesh is shown in Figure 3.4(a).

> **Box 3.2 Software implementation**
>
> *Path*: Mesh >> Mapped Mesh Parameters

Unstructured (free) mesh An unstructured or free mesh in 2D consists of triangular elements. As discussed earlier, unlike structured meshes, unstructured meshes are easy to create as they can be automatically created for any domain shape. However, unstructured meshing may sometimes lead to very large numbers of elements and therefore structured meshing may be needed. Figure 3.8 shows the unstructured meshing options in COMSOL. An example of an unstructured 2D mesh is shown in Figure 3.4(b).

> **Box 3.3 Software implementation**
>
> *Path*: Mesh >> Free Mesh Parameters

3.3.5 Meshing in COMSOL: 3D geometry

For a 3D geometry, three different types of mesh are possible: (a) fully structured mesh; (b) fully unstructured mesh; (c) combination of unstructured and structured mesh. All these meshing possibilities are shown in Figure 3.9.

Structured mesh A structured mesh in 3D consists of hexahedral or prism-shaped elements, as shown in Figure 3.9. To create a fully structured mesh in COMSOL for a 3D problem, first a 2D boundary needs to be meshed using structured elements (quadrilaterals), as detailed in the previous section.

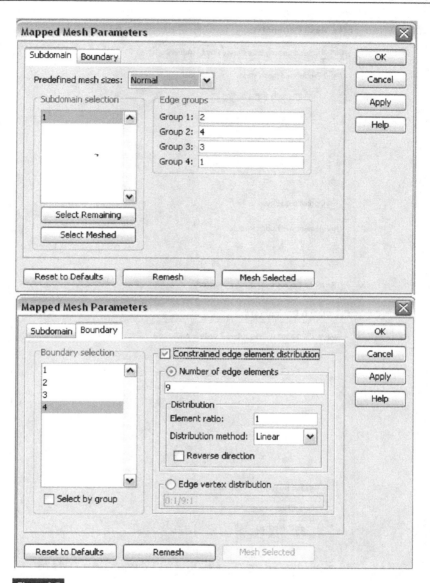

Figure 3.7

Mesh options in COMSOL to create structured (mapped) mesh for 2D problems. The top figure shows the options under the **Subdomain** tab. Here **Predefined mesh sizes** can be set to create different levels of mesh density starting from **Extremely coarse** to **Extremely fine**. Under the **Boundary** tab, as shown on the bottom figure, the number of elements on the different boundaries of the computation can be set first and then the domain can be meshed. Also the size ratio between the last and first element along a boundary can be set under the **Element ratio** field. These options are used to create more elements in specific regions of the domain.

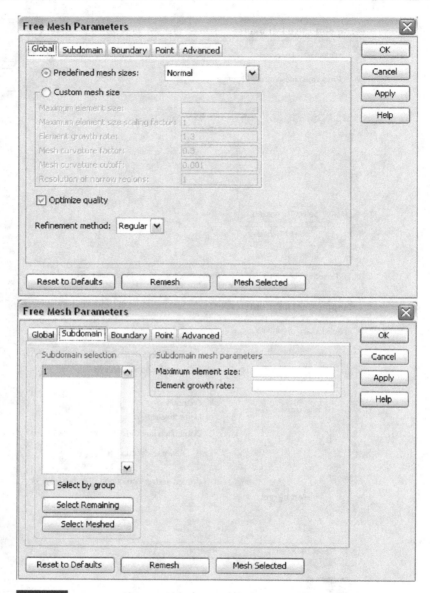

Figure 3.8

Mesh options in COMSOL to create an unstructured (free) mesh for 2D problems. Here again, as in structured meshing, under the **Global** tab, shown on the top, **Predefined mesh sizes** can be set. Under the **Subdomain** tab, shown on the bottom figure, the **Maximum element size** and **Element growth rate** can be set appropriately for different subdomains to control the mesh sizes. Similarly, under the **Boundary** tab (not shown in the figure) the element sizes can be controlled.

3.3 How to discretize the domain: mesh

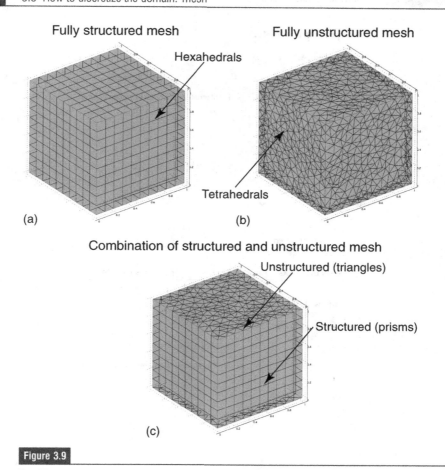

Figure 3.9

The different types of meshes along with the element shapes that are possible for a 3D geometry.

Subsequently the 2D mesh in the boundary is extended by sweeping in the third direction to mesh the full 3D domain. The options for extending a 2D mesh to create a a fully structured mesh in 3D in COMSOL is shown in Figure 3.10. The number of elements and the element size ratio in the sweep direction can be set to control the mesh size.

> **Box 3.4 Software implementation**
>
> Mesh a 2D boundary first using structured meshing (see previous section on 2D structured meshing for details):
> *Path*: Mesh >> Mapped Mesh Parameters
> Subsequently, mesh the 3D domain:
> *Path*: Mesh >> Swept Mesh Parameters

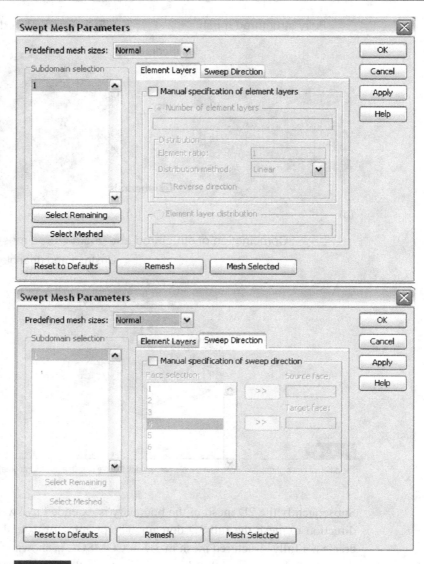

Figure 3.10

Mesh options in COMSOL to sweep a 2D mesh in the third direction to create a structured mesh for 3D problems. Note here that a 2D mesh must first be created on a boundary of the 3D domain, as shown in Figure 3.7. Under the **Element Layers** tab shown at the top, **Predefined mesh sizes** can be selected. Also, the number of element layers in the sweep direction and the element size ratio between the last and first element can be specified by selecting **Manual specification of element layers**. Under the **Sweep Direction** tab, shown at the bottom, the direction of sweep can be manually specified. Although sweep direction is determined automatically, manual specification of sweep direction may be necessary for complex geometries.

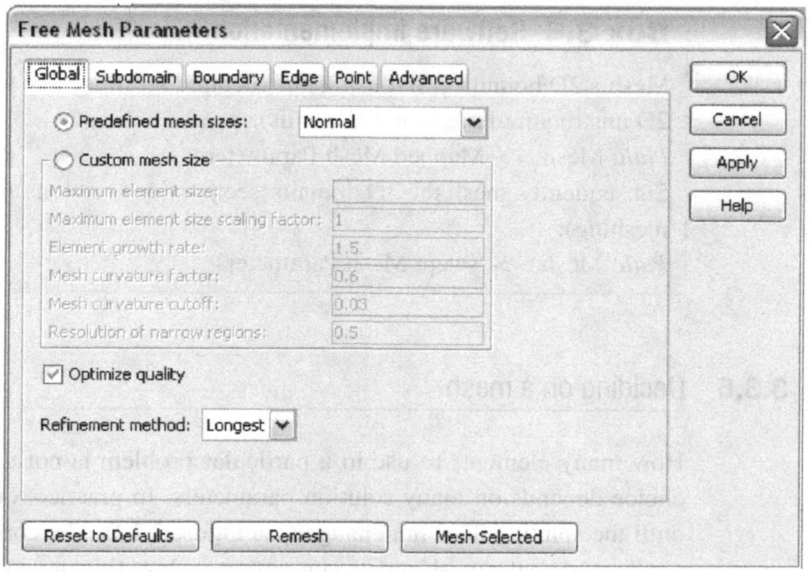

3.3 How to discretize the domain: mesh

Figure 3.11

Mesh options in COMSOL to create an unstructured mesh for 3D problems. Under the **Global** tab, **Predefined mesh sizes** can be set. Under the **Subdomain** and **Boundary** tabs (not shown in the figure), the **Maximum element size** and **Element growth rate** can be set for different subdomains and boundaries to control the mesh sizes as needed.

Unstructured mesh An unstructured mesh in a 3D geometry consists of tetrahedral elements, as shown in Figure 3.9. An unstructured mesh can be created directly for a 3D problem by selecting the **Free Mesh Parameters** option as shown in Figure 3.11.

> **Box 3.5 Software implementation**
>
> *Path*: Mesh >> Free Mesh Parameters

Combination of structured and unstructured mesh Instead of creating a structured mesh in a boundary and then sweeping it to map the 3D geometry (as discussed above) to create a fully structured mesh, the boundary can be meshed using unstructured elements (triangles) and then this triangular mesh can be swept to create the 3D mesh. In this case the 3D elements that are formed are prismatic in shape and are structured in the direction of the sweep. In the direction perpendicular to the direction of the sweep, the mesh is unstructured. An example of such a mesh is shown in Figure 3.9.

> **Box 3.6 Software implementation**
>
> Mesh a 2D boundary first using unstructured meshing (see previous section on 2D unstructured meshing for details):
> *Path*: Mesh >> Mapped Mesh Parameters
> Subsequently, mesh the 3D domain (see previous section on 3D structured meshing):
> *Path*: Mesh >> Swept Mesh Parameters

3.3.6 Deciding on a mesh

How many elements to use in a particular problem is not always obvious. The choice depends on many solution parameters. In practice, one refines the mesh until the solution does not change (see Figure 3.12). The correct solution should not depend on the mesh and it is important to show this step in order to confirm the validity of the results. This process is known as mesh convergence analysis and is discussed in detail in Section 5.3.

The idea is to use as few elements as possible, but to use sufficient. Computation can easily become intensive. The amount of time needed to form the element matrices increases linearly with number of elements and can take 20–80% of the total time. The rest of the time is taken in solving the equations and depends on the solution method used. Figure 3.13 shows how the computational time increases with the increase in the number of elements in the mesh for a drug diffusion example (Case study III in Chapter 6).

Increased mesh density, therefore, can reduce error but increases computing time. A uniform mesh may require a very large number of nodes to solve some problems. To reduce the total number of elements, a non-uniform mesh is used, with more elements in the region where spatial variations are more significant.

3.4 How to choose a time step

For numerical solution of a time-dependent (transient) problem, the time over which the problem needs to be solved can only be approached in small increments. For such problems, the time step increments (Δt) that the solver takes to solve the governing equations is critical. If the time step size is too large, the solution may be unrealistic or unstable (see discussion in Section 10.13 for theory and details). Small time steps generally provide more accurate results, but need more

3.4 How to choose a time step

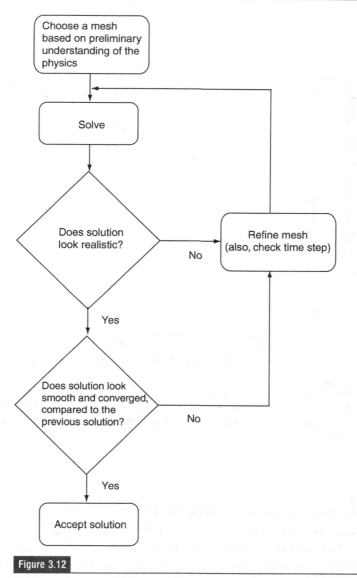

Figure 3.12

Typical procedure for mesh refinement, to obtain an acceptable mesh.

computation. Therefore the choice of time step increment in the solver is critical and once the mesh is decided (see previous section), the next task is to select an appropriate time step. Time steps can be fixed or variable. Fixed time stepping refers to the case when the time steps (Δt) taken by the solver are the same from the beginning to the end of the solution process. In the case of variable time

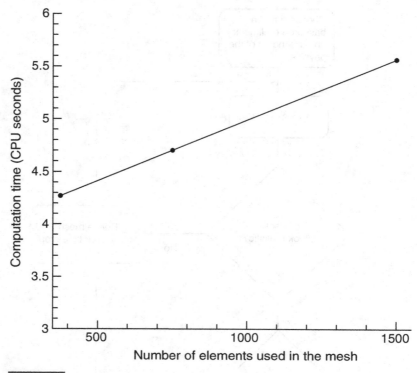

Figure 3.13

As the number of elements in the mesh increases, computation time also increases. The somewhat linear increase shown for this problem is not always the case and computation time can increase drastically with increase in the number of elements.

stepping, the solver automatically changes the time steps by calculating errors at each increment and modifying the time step accordingly.

Deciding on the time step is far from trivial. It has been a topic for many researchers and many research articles have been written about it. When the solution is changing rapidly with time, smaller time steps are necessary. Conversely, when variables are not changing rapidly with time, large time steps may be acceptable. In a transient problem, it is generally a good idea to use variable Δt such that small time steps are used when the solution is changing rapidly and large time steps are used when the solution is not changing rapidly. Sometimes, very small time steps need to be forced, particularly initially. In practice, the time step is chosen primarily by trial and error. Figure 3.14 shows a schematic of the typical iterative process for a transient problem.

3.4 How to choose a time step

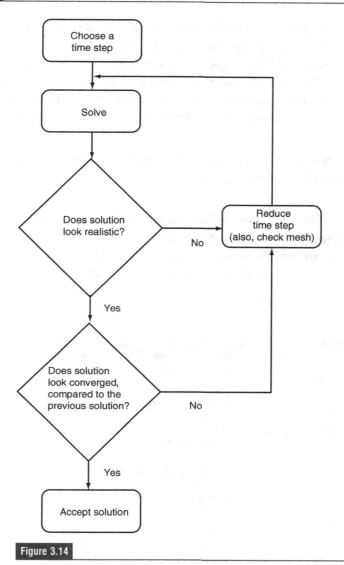

Figure 3.14

Typical procedure for choosing the time step in a transient calculation.

3.4.1 Implementation of time step in COMSOL

There are three options to specify time increments in COMSOL:

(1) **Default** By default, COMSOL uses variable time steps. It decides on the variable time step size automatically depending on the problem using the DASPK (Brown *et al.*, 1994) time stepping algorithm.

(2) **Fixed Time Steps** The problem can be made to run with fixed time steps in COMSOL. These settings are described below.
(3) **Variable Time Steps** Parameters in variable time step integration can also be controlled by the user as shown below. For variable time step integration, the time increments for successive steps are determined by the error control via the tolerances set by the user (discussed in Section 3.5.2).

The parameters used to control the time step size in the **Solver Parameters** panel in COMSOL (shown in Figure 3.15):

- **Initial Time Step:** is the time step size entered by the user which the solver uses for the first time increment.
- **Maximum Time Step:** is the time step size entered by the user to limit the time increment taken by the solver at any step.

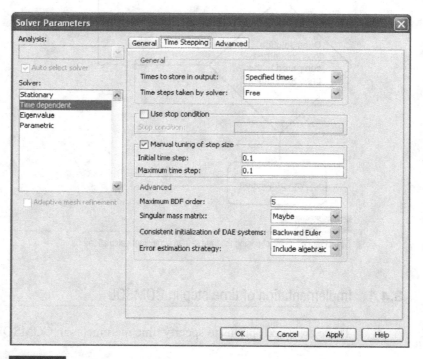

Figure 3.15

Time stepping options in COMSOL.

> **Box 3.7 Software implementation**
>
> *Path*: Solve >> Solver Parameters >> Time Stepping Tab >> *Check* Manual tuning of step size:
>
> **Default** Uncheck *Manual tuning of step size*.
>
> **Fixed time steps** Set the *Initial Time Step* equal to the *Maximum Time Step* to use fixed time stepping for the problem.
>
> **Variable time steps** Set the *Initial Time Step* and *Maximum Time Step* differently to use variable time stepping for the problem.

3.4.2 Specifying end times and saving data in COMSOL

To save the solution to the time-dependent model at specific times, the output times are entered in the **Times** field under **Time Stepping**, as shown in Figure 3.16. The notation 0:0.1:1 tells the solver to save the solution starting at time t = 0 s, with steps of 0.1 s up to the end time t = 1s. Here, 1 is the end time of the simulation.

> **Box 3.8 Software implementation**
>
> *Path*: Solve >> Solver Parameters >> General tab

3.5 How to choose a solver to solve the system of linear equations

Using the finite element method, the governing partial differential equations are converted to a linear system of equations. Therefore a linear system of equations is solved eventually by the computational software to obtain the solution. Two different methods exist for solving the linear set of equations – **direct** or **iterative**. The direct method is the default solution method in COMSOL and uses Gaussian elimination. The advantage of the direct method is that a solution is always obtained after a finite number of operations. However, for large simulations (more than about 100 000 degrees of freedom), the computer memory and CPU time required by a direct solver becomes very large. Iterative methods obtain a solution starting from an initial guess and converging to the exact solution in a finite number of iterations. The advantage of an iterative method is that it may reduce storage and CPU time requirements. However, it may also lead to slow or

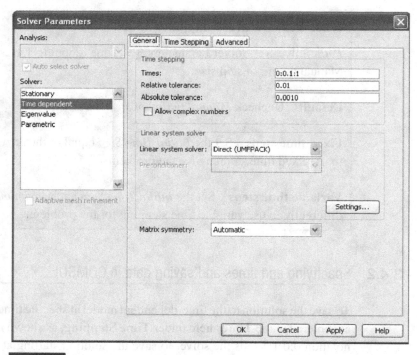

Figure 3.16

Specifying output times (specific times in a transient calculation for which the solution is saved).

irregular convergence, depending on the particular problem. Therefore, according to which solver is chosen, memory requirements and speed of solution will vary. It is hard to give general guidelines. For small sets of equations, it is not critical to choose a solver.

The reader is directed to Section 10.6 for detailed theory behind the finite element method, solution of the linear system of equations, and discussion about direct and iterative solvers.

3.5.1 Solver selection in COMSOL

The different direct and iterative linear system solvers available in COMSOL are:

- **Direct solvers**
  (1) Direct (UMFPACK)
  (2) Direct (SPOOLES)
  (3) Direct (PARDISO)
  (4) Direct Cholesky (TAUCS)

3.5 How to choose a solver to solve the system of linear equations

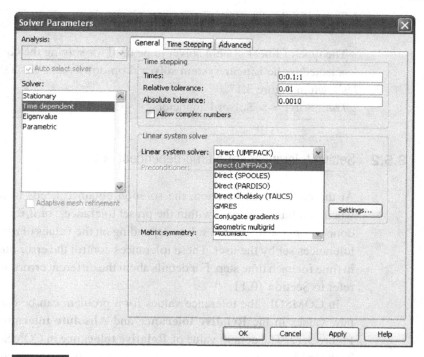

Figure 3.17
Choosing a solver in COMSOL to solve the linear system of equations from the finite element formulation of governing equation and boundary condition.

- **Iterative solvers**
  (1) GMRES
  (2) Conjugate gradients
  (3) Geometric multigrid

We will discuss the two most common solvers used for solving the types of problems covered in this text. One is a direct solver and the other one is iterative.

(1) **Direct (UMFPACK)** UMFPACK is the default linear system solver. The direct solvers UMFPACK is preferable for 1D and 2D models as well as 3D models with few degrees of freedom.
(2) **GMRES** GMRES linear system solver uses the restarted generalized minimum residual method, which is an iterative method for solution of general linear systems of equations. This solver should be used for models with many degrees of freedom (more than 100 000) for which the direct solver needs too much memory.

> **Box 3.9 Software implementation**
>
> The type of linear system solver to be used for solving the equations can be selected in the **Linear system solver** drop-down menu in COMSOL, as shown in Figure 3.17.
> *Path*: Solve >> Solver Parameters >> General tab

3.5.2 Setting tolerances for a time-dependent problem

At the end of each iteration, the solution obtained needs to be checked to see whether it has converged within the preset tolerances, or if it is diverging. This is done automatically by the solver depending on the values of *absolute* and *relative* tolerances set by the user. These tolerances control the error due to discretization in time for each time step. For details about the different errors and their definition, refer to Section 10.11.

In COMSOL, the tolerance values for a problem can be changed by entering new values in the **Relative tolerance** and **Absolute tolerance** fields shown in Figure 3.17. The default value of **Relative tolerance** in COMSOL is 0.01. It may be desirable in some cases to increase or decrease this value to control the accuracy or performance of the solver. Similarly the absolute tolerance, which has a default value of 0.001, needs to be adjusted for a particular problem depending on the absolute values of the solution parameters. For example, for a heat transfer problem with expected temperatures in the range 273–373 K, the value of absolute tolerance is very small compared to temperatures obtained in the simulations ($0.001 << 273$ K), therefore the absolute tolerance may need to be increased for faster solution of the problem.

3.6 Problems

3.6.1 Guidelines for mesh and time step

What are the general guidelines for (1) selecting the mesh (number of nodes and their placements); (2) selecting the time step in a transient problem?

3.6.2 Need for non-uniform mesh

We prefer to use a non-uniform mesh in the problem of therapeutic heating (See Section 1.12.10) for which a schematic is shown in Figure 1.17. Sketch how you would like to mesh the domain, giving reasons for your choice.

3.6.3 Choice of mesh

For each problem formulation question in Chapter 1 (starting from Section 1.12.4), determine what kind of mesh is needed from these choices: structured or unstructured, uniform or non-uniform. If a non-uniform mesh is required, show the regions with higher density of mesh elements. Explain the reasons for your choices.

3.6.4 Choice of mesh: Case studies VI, VIII, IX and X

Consider the software implementation of Case studies VI, VIII, IX and X (Chapter 6). Specify the type of mesh used for each case study: structured / unstructured, uniform / non-uniform. Comment whether the meshing strategies are appropriate. Explain why or why not.

3.6.5 Choice of time steps: Case studies I, II, III and IV

Complete Case studies I, II, III and IV (Chapter 6). Demonstrate that the results are independent of the time steps chosen for each case study. Refer to the discussion in Section 3.4 and flowchart shown in Figure 3.14 to perform the analysis.

4 Software implementation 3 Visualizing and manipulating solution (postprocessing)

In the previous two chapters, the mathematical model, consisting of the governing equations and boundary conditions, was implemented in a software, together with instructing the software how to solve the equations. These steps together are called *preprocessing*, as discussed earlier. After these steps, the computer simply solves the equations – a step called *processing*, which is completely automated. Once a "solution" is available, we can visualize and analyze it to make sense of it. Remember computational results are not a solution to our problem until we have validated the results (the topic of the next chapter). This chapter simply presents ways of visualizing and manipulating the results from computation using a software, and its relation to other chapters is shown in Figure 4.1. This stage is also called *postprocessing*. Postprocessing is the further processing of raw data produced by the computation. A simple postprocessing step could be the display, as a graph, of the data produced by computation, for example, a plot of concentration calculated at a point as a function of time.

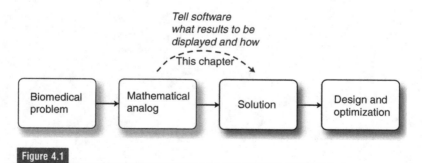

Figure 4.1

Visualization and manipulation of the solution are the subjects of this chapter.

Standard quantities to be displayed or standard computations to be made are built into a typical software. If this is all that is needed, one simply learns the respective commands from the software interface. Sometimes and especially for a newcomer to modeling, it may not be obvious as to which raw or derived information is most relevant in relating back to the physical problem. Thus, in this chapter, we first give examples of the different types of information that are useful in various biomedical situations. The rest of the chapter shows how to obtain the information in one particular software, namely COMSOL. We skip details when the software makes it obvious what parameter is required to be entered.

4.1 Useful information in a biomedical context

Information at a location Temperature or concentration at a location, at any time or as a function of time, is the simplest type of information that may be necessary.

Spatial average Sometimes detailed variation is not that important and an average value over a region is all that is needed. For example, in therapeutic heating, average temperature for a region may be of interest. In drug delivery, average concentration over a region may be the quantity we are interested in.

Spatial variation The simplest information on spatial variation is to locate the point or region for maximum or minimum values. For example, in cryosurgery, we may be interested in the boundary of the region over which the temperature is less than $-45\ °C$, which would be the region where tissue destruction by freezing occurs.

A more comprehensive way to provide spatial variation over a region for a continuous variable such as temperature or concentration is to have contours. For a quantitative representation of variations or uniformity in a region, a statistical measure such as standard deviation can be used.

Time variation Time variation for a single location can be provided as a function of time as a line graph. Time variation for an entire region as a function of time can be provided as a series of contour plots or, more effectively, as a movie. Although mostly qualitative, a movie is very effective in obtaining a visual picture of how the quantities change with time.

One may be interested in how much drug has entered through a particular boundary as a function of time. This means the time-varying flux has to be integrated over time.

Secondary quantities Examples of a secondary quantity are heat or mass flux at a location on a surface. This is useful, for example, when a certain amount of drug per unit area and per unit time has to enter a surface. Time variation of flux is also important in drug delivery as, in many cases, an attempt is made to obtain a more or less constant rate of release over time.

Another example of a secondary quantity is mass balance at a location. For example, often in drug delivery we would like to know how much drug has left the drug capsule. This can be found by knowing how much drug is remaining in the capsule and subtracting that from the initial amount of drug in the capsule.

Arbitrary functions We often require customized information which depends on the variables calculated. For example, thermal conductivity may be some function of temperature (e.g., $k = 0.4692 + 0.001\,161T$) and we would like to plot the variation of thermal conductivity in the domain. We can define functions of concentration or temperature according to our needs.

Examples of various postprocessing options (computation and display of the above quantities) from the Case studies in Chapter 6 are listed in Table 4.1. The reader is referred to the appropriate pages of these Case studies.

We will now look at the detailed steps of how to obtain the plots of these quantities in COMSOL. Table 4.2 lists the most commonly used variables in COMSOL. The symbols shown in the table can be used to access the solution variables in order to define expressions and functions during postprocessing (as well as preprocessing).

Table 4.1 Postprocessing used in the different Case studies. The reader is referred to Section 6.1 for more details of these Case studies.

Postprocessing	Example in Case study
Information at a location	I, II, III, IV, IX
Spatial average	X
Spatial variation	I, II, III, V, VI, VII, VIII, IX, X
Time variation	I, II, III, IV, VII, IX
Secondary quantities	V
Arbitrary functions	II (objective function), IV (amount absorbed)

Table 4.2 Commonly used variables in COMSOL.

Variable	Symbol
Time	t
Location	x, y, z (for 3D problem); x, y (for 2D); r, z (for axisymmetric)
Temperature	T
Concentration	c, c2, c3 etc.
Velocity	u, v, w

4.2 Obtaining data at a particular location

The value of a computed variable at any particular location in the domain may be needed during postprocessing. The point of interest may either be a vertex that was created during the geometry creation process, or any other arbitrary location inside the geometry.

> **Box 4.1**
>
> To obtain data at a vertex, use the **Point Evaluation** option under the **Postprocessing** menu:
> Path: **Postprocessing >> Point Evaluation**

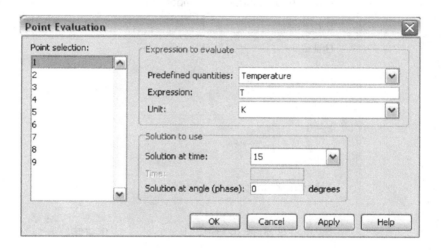

Inputs

(1) Select the appropriate point under **Point selection**.
(2) Select the variable for which the data is needed from the **Predefined quantities** menu, or specify an expression using the computed variables in the **Expression** field. For example, to get the temperature at the point, select **Temperature** from **Predefined quantities**. For obtaining some function of temperature at the selected point such as $0.4692 + 0.001\,161 T$, set $0.4692 + 0.001\,161 * T$ as **Expression**. Similarly, different mathematical functions can be used using any combination of variables that are solved in the problem.

Software implementation 3

(3) If the problem is time-dependent, select the time at which the data are needed from the **Solution at time** menu.

> **Box 4.2**
>
> To obtain data at any arbitrary location, use the **Data Display** option under the **Postprocessing** menu:
>
> Path: **Postprocessing** >> **Data Display** >> **Subdomain**

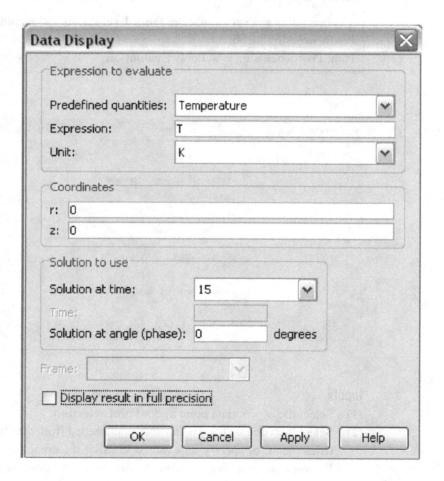

Inputs

(1) Select the variable for which the data is needed from the **Predefined quantities** menu or, specify an expression using the computed variables in the **Expression** field.

(2) Set the coordinates of the desired point under **Coordinates**. For a 2D problem specify x and y coordinates, for an axisymmetric problem specify r and z and for a 3D problem set x, y and z.
(3) If the problem is time-dependent, select the time at which the data are needed from the **Solution at time** menu.

The value of the selected variable is shown in the message log (which is the area below the graphics panel in the main COMSOL window).

Example

Using the "Burn injury in blood-perfused skin" Case study (Chapter 6), we will now obtain the degree of tissue injury at the location (0.002, 0.004) at 15 s.

(1) Select **Postprocessing >> Data Display >> Subdomain** to open the **Data Display** dialog.
(2) Select Concentration under **Predefined quantities**.
(3) Set **r** as 0.002 and **z** as 0.004 under **Coordinates**.
(4) Select 15 as the **Solution at time** under **Solution to use**. Click **OK**.

The value of degree of tissue injury is given as 806.8 at the specified location, which corresponds to second to third degree burn (refer to the Case study for details).

4.3 Plotting transient data at one or more points, line or surface as a function of time

> **Box 4.3**
>
> To plot a variable as a function of time (for a time-dependent problem) at a vertex, use the **Domain Plot Parameters** option under the **Postprocessing** menu:
> Path: **Postprocessing >> Domain Plot Parameters >> Point** tab

Software implementation 3

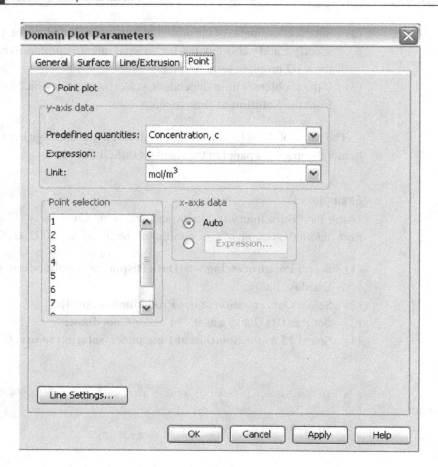

Inputs
(1) Select the variable for which the data is needed from the **Predefined quantities** menu, or specify an expression using the computed variables in the **Expression** field.
(2) Select the appropriate point under **Point selection**.
(3) In the **General** tab, select the times to be used to plot the variable. By default, all output times are selected.

Box 4.4

To plot a variable as a function of time (for a time-dependent problem) at any arbitrary location inside the geometry, use the **Cross-Section Plot Parameters** option under the **Postprocessing** menu:
Path: **Postprocessing** >> **Cross-Section Plot Parameters** >> **Point** tab

4.3 Plotting transient data at one or more points, line or surface as a function of time

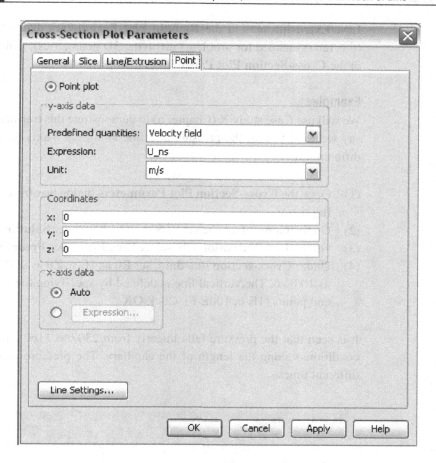

Inputs

(1) Select the variable for which the data is needed from the **Predefined quantities** menu, or specify an expression using the computed variables in the **Expression** field.

(2) Set the coordinates of the desired point under **Coordinates**. For a 2D problem specify x and y coordinates, for an axisymmetric problem specify r and z and for a 3D problem set x, y and z.

(3) In the **General** tab, select the times to be used to plot the variable. By default, all output times are selected.

To plot the transient data at more than one point on the same graph, check **Keep current plot** under the **General** tab and set and plot data at the different points by the above procedure.

Similarly, any variable can be plotted along a specified line (in the case of 1D/2D/3D problems) or surface (in 2D/ 3D problems) using the **Surface**,

Line/Extrusion tabs in the **Domain Plot Parameters** dialog box or the **Slice** (this tab is enabled for problems involving 3D geometries), **Line/Extrusion** tabs in the **Cross-Section Plot Parameters** dialog box.

Examples

We will use Case study X (Chapter 6) to demonstrate this transient line plot. Let's say we want to plot the pressure at a line along the z-axis in the capillary at 3 different times.

(1) Open the **Cross-Section Plot Parameters** dialog box from the **Postprocessing** menu.
(2) On the **General** tab, select 0, 600 and 1200 under **Solutions to use**.
(3) On the **Line/Extrusion** tab, select **Pressure** under **Predefined quantities**.
(4) Under **Cross-section line data**, set **R0** as 1E-6, **Z0** as 0, **R1** as 1E-6 and **Z1** as 100E-6. The vertical line is defined by specifying the start (1E-6, 0) and end points (1E-6, 100E-6). Click **OK**.

It is seen that the pressure falls linearly from 2307 to 1160 (specified boundary conditions) along the length of the capillary. The pressure drop is the same at different times.

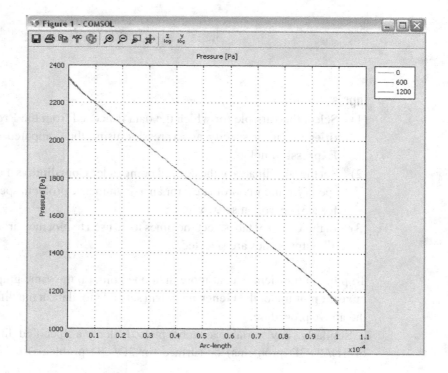

Another example of a transient data plot at a vertex can be seen in Case study IX. Examples of transient data plots at arbitrary points can be seen in Case studies I, II and III. An example of a line plot at different times can be seen in Case study VII. An example of a line plot at steady state (for a steady-state problem) can be seen in Case study V.

4.4 Obtaining surface/contour plots (in 2D problems) for observing variation within a region

To plot a variable over a subdomain (in 2D problems) in order to look at the spatial variation, use the **Contour** or **Surface** plot option in the **Plot Parameters** dialog box. Contour plots in COMSOL are plots that show the variation of a particular variable in a subdomain by lines of constant magnitude. Surface plots (in COMSOL) are color-filled plots that show the variation using a continuous color display, instead of discrete lines representing specific values. Both types of plot (contour and surface) display essentially the same information and should be used as needed. As a result, in some places in the text, contour plots have been used interchangeably with surface plots. Both types of plot are described here.

4.4.1 Contour plot

> **Box 4.5**
>
> To generate a contour plot use the **Contour** tab in the **Plot Parameters** dialog box:
>
> Path: **Postprocessing** >> **Plot Parameters** >> **Contour** tab (in 2D problems)

Inputs (see figure next page)
(1) In the **General** tab, check **Contour** as the **Plot type**; uncheck all other types of plot.
(2) In the **Contour** tab, select the desired variable for postprocessing from the **Predefined quantities** menu, or specify an expression using the computed variables in the **Expression** field.
(3) Set the number of contour levels under **Contour Levels** in the **Number of levels field**. This will divide the data automatically into the number of levels that are specified. You can also plot the data at specific levels. Use the **Vector with isolevels** field and specify the levels in this case.

Software implementation 3

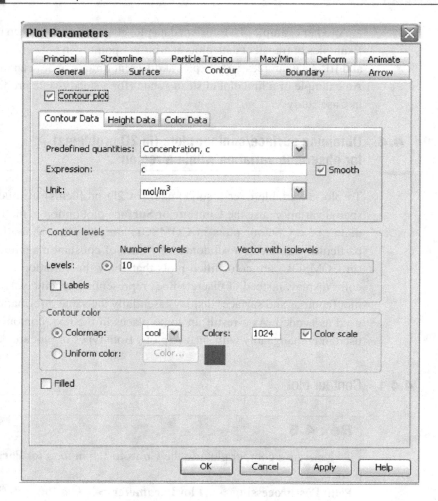

4.4.2 Surface plot

> **Box 4.6**
>
> To generate a surface plot use the **Surface** tab in the **Plot Parameters** dialog box:
> Path: **Postprocessing** >> **Plot Parameters** >> **Surface** tab (in 2D problems)

Inputs (see figure next page)

(1) In the **General** tab, check **Surface** as the **Plot type**; uncheck all other types of plot.

4.4 Obtaining surface/contour plots (in 2D problems)

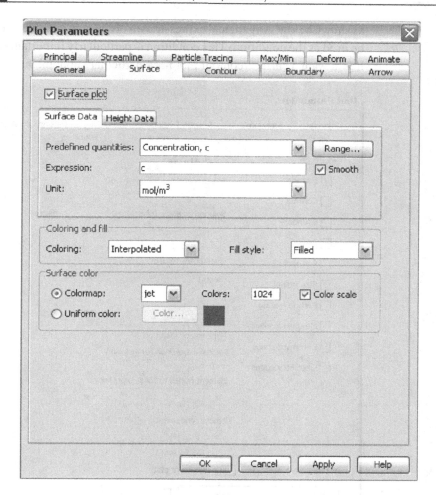

(2) In the **Surface** tab, select the desired variable for post-processing from the **Predefined quantities** menu, or specify an expression using the computed variables in the **Expression** field.

(3) The plots will be created using a range of values automatically calculated by the software. To change this range select **Range** and specify the minimum and maximum values.

4.4.3 Additional options for surface and contour plots

For both contour and surface plots, additional conditions can be specified so that the output satisfies these constraints. For example, if the domain for the problem is 1 m and you want to look at the variation only for distances in the x direction up to 0.1 m. This can be specified in the **General** tab in the **Plot Parameters** dialog box, as shown below. Any logical expression involving any variable in the problem may be used as the condition. This condition is particularly important in 3D geometries if you want a surface/contour plot for a particular subdomain.

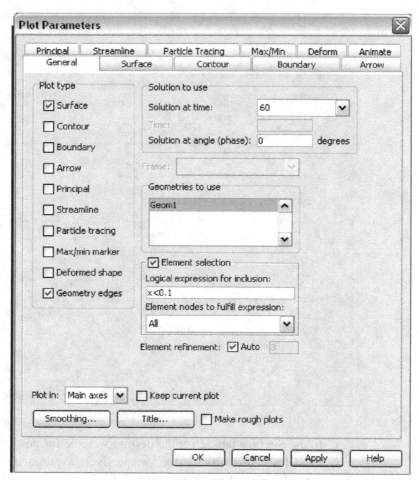

Path: **Postprocessing** >> **Plot Parameters** >> **General** tab

Input

Check the **Element selection** box and specify the logical expression for use in plotting the surface/contour plot. These settings are used in addition to the surface/contour plot settings discussed earlier.

Examples

We will now use the case study on "Thermal ablation of hepatic tumors" (Case study 1, Chapter 6) to obtain a surface plot of temperature and to plot the region that has temperature greater than 50 °C (323 K).

4.5 Obtaining a surface plot in a 3D problem

(1) Open the **Plot Parameters** dialog box from the **Postprocessing** menu.
(2) On the **Surface Page**, select **Concentration, c** in the **Predefined quantities** area.
(3) On the **General** page select the **Surface** check box for a surface plot.
(4) Check the **Element selection** box and type T>323 in the **Logical expression for inclusion**. Click **OK**.

An alternative method to obtain the same plot would be to change **Range** of temperatures to be plotted in the **Surface** tab.

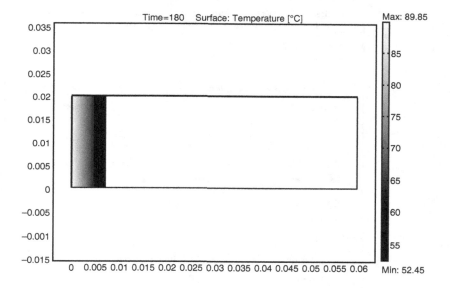

It is seen that only the region of the tumor close to the probe is heated to temperatures greater than 50 °C and areas of tissue farther away from the probe are not heated. This confirms that the procedure is effective.

Additional examples of surface and contour plots can be seen in Case studies II, III, V, VIII and X.

4.5 Obtaining a surface plot in a 3D problem

In 3D problems, spatial variations can be plotted using a surface plot for the entire geometry (or a part of it), or taking cross-sections at different parts of the geometry and obtaining surface plots at those sections.

4.5.1 For the entire geometry

> **Box 4.7**
>
> To generate a surface plot for the entire geometry, use the **Subdomain** Tab in the **Plot Parameters** dialog box:
>
> Path: **Postprocessing** >> **Plot Parameters** >> **Subdomain** tab (in 3D problems)

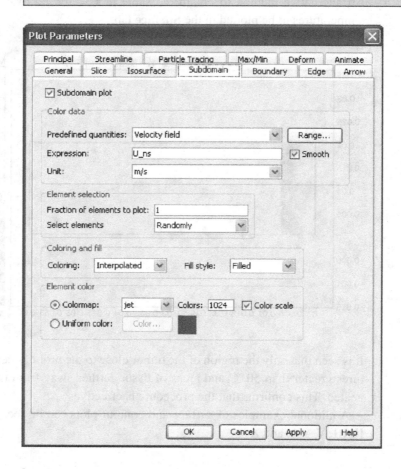

Inputs

(1) In the **General** Tab, check **Subdomain** as the **Plot type**; uncheck all other types of plot.

(2) In the **Subdomain** tab, select the desired variable for postprocessing from the **Predefined quantities** menu, or specify an expression using the computed variables in the **Expression** field.

4.5 Obtaining a surface plot in a 3D problem

(3) The plots will be created using a range of values automatically calculated by the software. To change this range select **Range** and specify the minimum and maximum values.

(4) To plot a particular region of the geometry: In the **General** tab, check the **Element selection** box and type the condition in the **Logical expression for inclusion** (usage described in detail in the previous section).

4.5.2 For different sections of the geometry

> **Box 4.8**
>
> To generate surface plots at different sections of the geometry, use the **Slice** Tab in the **Plot Parameters** dialog box:
> Path: **Postprocessing** >> **Plot Parameters** >> **Slice** tab (in 3D problems)

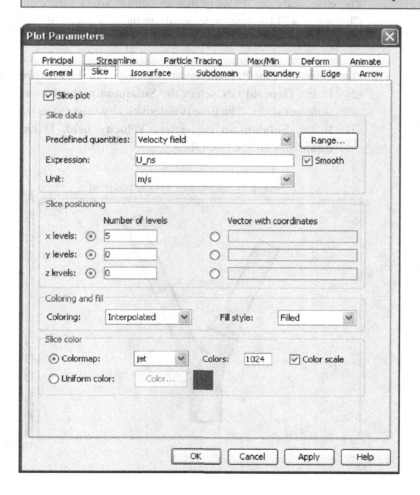

Inputs

(1) In the **General** tab, check **Slice** as the **Plot type**; uncheck all other types of plot.
(2) In the **Slice** tab, select the desired variable for postprocessing from the **Predefined quantities** menu, or specify an expression using the computed variables in the **Expression** field.
(3) Set the number of slices (cross-sections) under **Slice positioning** in the **Number of levels** field. You can set the slices in the x, y or z direction by filling the appropriate field. This will divide the geometry automatically into the number of slices that are specified. You can also plot the data at a specific slice. Use the **Vector with coordinates** field and specify the vector representing the plane of the desired slice in this case.

Examples

The case study on "Flow in human carotid artery bifurcation" (Case study VI, Chapter 6) will be used to demonstrate the process of obtaining a surface plot in the geometry at the end time.

(1) Open the **Plot Parameters** dialog box from the **Postprocessing** menu.
(2) In the **General** tab, select the **Subdomain** check box for surface plots at different slices. The time is automatically selected as 0.75 (end time).
(3) In the **Subdomain** tab, select **Velocity field, U_ns** in the **Predefined quantities** area. Click **OK**.

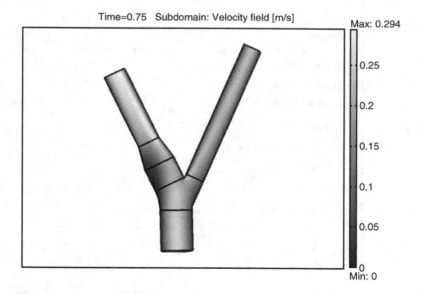

We can confirm from the figure that no slip boundary condition is satisfied at the wall. It is also observed that near the bifurcation the flow stagnates at the outer side of the internal carotid artery. Another example, obtaining a slice plot in a 3D problem, can be found in the same Case study.

4.6 Obtaining average values at a particular time or as a function of time

The average value of a variable in the entire region or a part of it and its variation with time (for a time-dependent problem) can be obtained in COMSOL. However, there is no direct method to obtain average values in COMSOL and hence, we must use the integration tool in the software to determine the average values, as described below.

4.6.1 Average values at any particular time

> **Box 4.9**
>
> To obtain average values of a variable at any particular time, use the **Subdomain Integration** dialog box from the **Postprocessing** menu:
> Path: **Postprocessing** >> **Subdomain Integration**

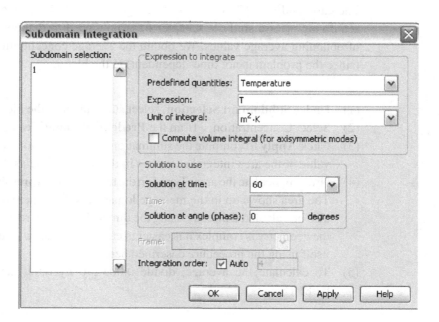

Procedure

(1) Under **Subdomain Selection**, select the appropriate region (subdomain) for which the average value needs to be obtained.

(2) Select the desired variable for obtaining the average from the **Predefined quantities** menu, or specify an expression using the variables in the **Expression** field.

(3) Select **Compute volume integral (for axisymmetric modes)**, if the problem is axisymmetric.

(4) Select the time at which the average value needs to be computed from the **Solution at time** menu.

(5) Click **Apply** to compute the area/volume weighted summation of temperature ($\sum T_i \Delta A_i$ or $\sum T_i \Delta V_i$). The value of the integral shows up in the message log (below the graphics window in COMSOL).

(6) Now to calculate the area (or volume if the problem is axisymmetric or 3D): Type **1** in the text box for **Expression** and click **OK**. Again, the appropriate quantity (area for 2D, volume for axisymmetric/3D problems) shows up in the message log.

(7) To calculate the average, divide the value obtained in (5) (area/volume integral) by (6) (area/volume) manually.

Example

The case study on "Elimination of nitrogen from the blood stream during deep sea diving" (Case study V, Chapter 6) is now used to demonstrate the process of obtaining average values. We obtain the average concentration at steady state (since the problem is not time-dependent) in the blood.

(1) Under **Subdomain Selection**, select subdomain 1 (the blood flow region).

(2) Select Concentration, c from the **Predefined quantities** menu.

(3) Click **Apply** to compute the summation of concentration (area integral). The value of the area integral shows up in the message log as 0.003 246.

(4) Now to calculate the area: Type **1** in the text box for **Expression** and click **OK**. The area shows up in the message log as 0.04. Note here that this value can also be calculated directly using the dimensions of the rectangle. However, if the geometry is complex it is easier to calculate the area/volume in COMSOL itself using the procedure described here.

(5) To calculate the average, divide 0.003 246 by 0.04 manually, to obtain 0.081 15.

4.6.2 Average values as a function of time

> **Box 4.10**
>
> To obtain average values of a variable as a function of time, use **Integration Coupling Variables** under **Options** as shown below:

For 2D, 3D problems

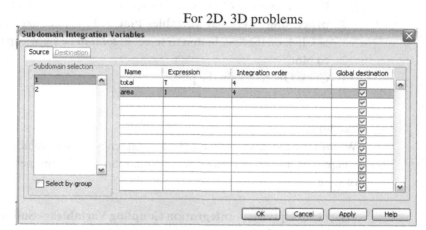

For axisymmetric problems

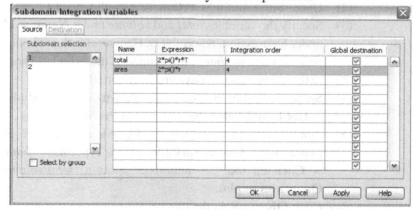

(1) Go to **Options** > **Integration Coupling Variables** > **Subdomain Variables**. **Subdomain Integration Variables Window** opens up.

(2) In the **Subdomain Integration Variables** window, select the appropriate region (subdomain) for which the average value needs to be obtained.

(3) Specify two user defined variables:
   (a) the first calculates the area/volume integral of the desired variable in the subdomain, e.g., type *total* under **Name** and T under **Expression** if you

want to calculate the average temperature (type c if you need average concentration, and similarly other variables);

(b) the second calculates the area/volume of the subdomain, e.g., type **area** under **Name** and 1 under **Expression** as shown below. *Note: For an axisymmetric problem type, 2*pi()*r*T and 2*pi()*r*1 to obtain the volume integral and volume, respectively.*

(4) Set up the rest of the problem and solve it.
(5) Plot the average of the selected variable as a function of time. Go to **Post-processing** > **Global Variables Plot** > Type *total/area* under **Expression** > Click on the ">" button to select it; *total/area* shows up under **Quantities to plot**.
(6) Click **OK** to obtain the plot.

Examples

We will use the case study on "Radiofrequency cardiac ablation" (Case study IX, Chapter 6), to demonstrate how to obtain the average temperatures as a function of time.

(1) Go to **Options** > **Integration Coupling Variables** > **Subdomain Variables**. **Subdomain Integration Variables Window** opens up.
(2) Under **Subdomain Selection**, select subdomain 1.
(3) Specify the variables *total* and *volume* under **Name** as 2*pi*r*T and 2*pi*r under **Expression**, respectively. Here *total* is the volume integral of the temperatures and *volume* is the volume of the cylinder. These have been defined in this form as it is an axisymmetric problem.
(4) Set up the rest of the problem as shown in the case study and solve it: **Solve** > **Get Initial Value**. **Solve** > **Solve Problem**.
(5) Plot the average of the selected variable as a function of time. Go to **Post-processing** > **Global Variables Plot** > Type *total/volume* under **Expression** > Click on the ">" button to select it; *total/volume* shows up under **Quantities to plot**.
(6) Click **OK** to obtain the plot.

The average temperature plot (see following page) shows that the temperature of the entire domain does not rise significantly.

Another example of plotting average values as a function of time can be seen in Case study X, where average tumor temperature is plotted as a function of time and the data is exported in a text format. Using the procedure described above, variables such as total amount lost or absorbed in a particular region can be determined for the problem. For an example showing such a plot, see Case study IV.

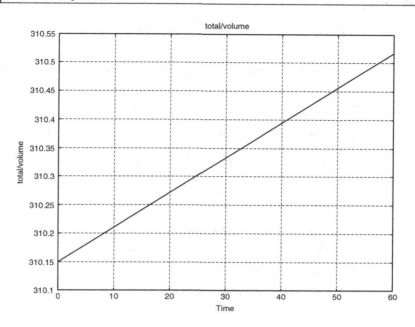

4.7 Obtaining arbitrary functions of computed variables

Arbitrary functions involving computed variables can be obtained by specifying the expression in the **Expression** field (instead of choosing the variable in the **Predefined quantities** menu) in all the postprocessing methods discussed above. For example, as in Section 4.2, a function of T such as $0.4692 + 0.001\,161T$ can be plotted by setting $0.4692 + 0.001\,161 * T$ in the **Expression** field in any postprocessing dialog box to look at the variation of the function of T instead of T (temperature).

4.8 Creating animations

> **Box 4.11**
>
> Animations/movies provide an excellent way of demonstrating results during a presentation. They can be created in COMSOL using the **Animate** tab in the **Plot Parameters** dialog box:
> Path: **Postprocessing** > **Plot Parameters** > **Animate** tab

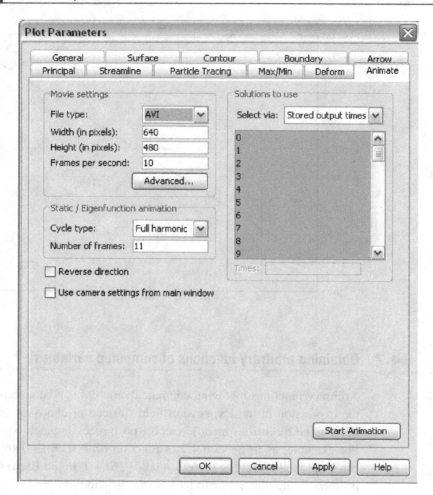

Inputs

(1) In the **General** tab, select the type of plot (surface, contour, slice etc.) that is needed for the movie.

(2) Go to the appropriate plot tab and change the settings as desired. For example, if you need a movie of the surface plot at different times, check **Surface** in the **General** tab and then go to the **Surface** tab and select the variable, ranges etc.

(3) In the **Animate** tab, select the times at which you want to save the results for the movie: select **Solution to use** > **Stored output time** > Hold **Ctrl** key and select the different times. (Remember by default all times are selected. You do not want to use that setting as the computer may need a lot of memory to make the movie file.)

4.9 Dedicated plotting and postprocessing software

Software such as Microsoft Excel, MATLAB, Tecplot and Ensight can be used for additional postprocessing of the data obtained from the analysis in COMSOL. For example, transient data at a point can be exported from COMSOL, as shown in the Case study X (Chapter 6). Once exported it can be used to do calculations inside another software program such as a spreadsheet application (e.g., Excel) or MATLAB.

To export data in text format for any plot:

(1) In the General tab of any postprocessing dialog box (**Plot Parameters**, **Cross Section Plot Parameters** or **Domain Plot Parameters**), select **Plot in** as **New Figure**. This setting will create the desired plot in a new window.
(2) Display the appropriate plot (using the different settings discussed in the chapter). The plot will show up in a new window.
(3) Click on the **Save Current Plot in ASCII file** button in the **Figure** Window (fourth button on the top denoted by **ASC=**).
(4) Specify the file name and save the file in a desired folder. This file is a text file and can be used for further calculations.

The most common need for exporting data is for a transient plot at a point or for a plot along any line. Details of postprocessing by exporting data can be seen in Case study X.

Tecplot and Ensight are two of the advanced tools that offer 2D/3D visualization, analysis and plotting capabilities. To use these software packages, the data must, however, be converted to a format suitable for import into them. More details about these packages can be found from their own individual websites.

4.10 Relating to the goals of the simulation: guidelines for postprocessing

The postprocessing techniques discussed earlier can create a set of beautiful plots of the problem. However, the whole point of doing a simulation is not to get a large collection of plots, rather it is to get concise and meaningful answers relevant to the goal of solving the problem. It is, therefore, essential to create and analyze only those plots that are relevant to your problem. Before choosing any type of plot for postprocessing, you should consider the following questions and the associated guidelines:

(1) **When is a transient line plot for a point important?** A transient plot for a point may be necessary if the point is a location of interest, that is, it represents a location where you want to know how the variable varies as time progresses. For example, in the case study "Radiofrequency cardiac ablation" we want to heat the tissue using an electrode to a particular temperature, but not beyond a certain other temperature. In this case, we know that the highest temperatures will be at places where the electrode touches the tissue. We therefore obtain a line plot representing temperatures at a point on the surface of the tissue as a function of heating time, to make sure that the temperatures are in the acceptable range. It would, however, not be that important to show a transient temperature plot for a point far away from the electrode as we almost expect that the temperatures would not vary significantly in those regions.

(2) **When do you need to show spatial variation?** A surface/contour plot may be important for problems where one wants to observe how the variable of interest varies in the domain at any given time. For example, in Case study II, we want to determine if the wart region reaches the desired low temperature for destruction after the cryosurgery procedure, and at the same time that the normal tissue is not cooled to very low temperatures. Calculation of average temperatures (discussed next) may also be useful in some cases. However, when it is critical that some solution variable does not fall below a minimum limit or exceed a maximum limit, then the spatial variation plot is critical.

(3) **Would an average be useful for a particular analysis?** Averages are useful if we are more interested in things happening in the overall domain rather than the spatial variation. For example, in a drug diffusion problem (similar to Case study II), we are interested in determining the total amount of drug delivered to the blood stream or the average concentration of the drug delivered as a function of time. A spatial variation in such cases may not provide critical information.

(4) **Would calculation of fluxes be useful? Would a flux plot be more relevant than the primary variable (i.e., temperature or concentration)?** Sometimes calculation of fluxes may be more important than the primary variable. For example, if we want to calculate the heat or mass transfer coefficient at a particular boundary using the computed solution, we need to determine the heat or mass flux at that boundary.

(5) **When do you present the data in tables instead of figures (plots)?** Results may sometimes be needed to be presented in tabular form. This is generally the case when we want to report discrete values such as data at different locations or at different times and average values of different subdomains, or after particular time intervals. Plots in such cases may not be very specific and useful information may become hidden.

4.11 Analysis of data obtained from postprocessing

Interpretation of results is as important as the actual analysis itself. First and foremost the computed data should be validated. This is discussed in detail in Chapter 5. Once the results are validated, we should try to reason out scientifically why the data is behaving the way it is. We should try to answer questions such as:

(1) **Why do the temperature/concentration/pressure vary with time the way they do?** Specifically look at the material properties, boundary conditions and source terms (their values and dependence on time) being used in the problem to figure out the behavior of the results.

(2) **Why do the temperature/concentration/pressure vary with space the way they do?** Again look at the input parameters to answer this question. Check if they are varying with the solution variable. Look at their values in different regions of the geometry in trying to determine the reason for the observed results. Look at the geometry and determine if the shape of the geometry is leading to the observed results.

(3) **How do we interpret the surface/contour plots?** The first thing that we use these plots for is to visually check that the problem we have solved is indeed the problem we are trying to solve. For example, we can visually check whether a no-flux condition is being satisfied at a particular boundary. This process is known as qualitative check and is discussed in detail in Chapter 5. We can also visually check if the material properties, initial conditions, other boundary conditions and time stepping are being properly implemented as desired. Once we confirm that the problem is correctly implemented in the software, we can try to learn from the solution.

4.12 Presenting the simulation results to others

A written report provides a formal way for the modeler to describe what has been accomplished and to offer an analysis or interpretation of what it means. Within an organization, reports are often the primary way to let colleagues know what one has been doing. Supervisors may judge the work based on the quality of one's reports. An example of detailed instructions for reporting projects performed in an undergraduate class can be seen in the Appendix to this chapter.

4.13 Problems

4.13.1 Postprocessing: Case study I

Complete Case study I (Chapter 6) and perform the following postprocessing operations.

(1) Temperature versus time plot at a point for different boundary conditions:
   (i) Plot the temperature at the point (0.0075, 0.01) as a function of time.
   (ii) Save the data in text format.
   (iii) Change the boundary condition. The probe temperature has been set at a temperature of 90 °C. Change the temperature to 70 °C.
   (iv) Solve the problem again.
   (v) Save the temperature versus time data at the same point for the new boundary condition in text format.
   (vi) Make a single plot using the data obtained in steps (ii) and (v) using Microsoft Excel or any other plotting software.

(2) Create a movie of the process. Select 6 different times: 30, 60, 90, 120, 150 and 180 to create the movie. By default all times are selected. You do not want to use that setting as the computer will need a lot of memory to make the movie file.

4.13.2 Postprocessing: Case study II

Complete Case study II (Chapter 6) and perform the following postprocessing operations.

(1) Plot temperature contours in the domain at five different levels with labels: 100, 150, 200, 250 and 300 K.
(2) Obtain a surface plot of thermal conductivity in the domain after 15 s.
(3) Calculate the average temperatures in the wart and the normal tissue regions after 15 s. Remember the problem is axisymmetric.
(4) Plot average temperature in the wart region as a function of time.

4.13.3 Postprocessing: Case study III

Complete Case study II (Chapter 6) and perform the following postprocessing operations.

(1) Plot the drug flux to the blood stream at the end of each day on the same graph for the week-long application of the patch modeled in the case study.
(2) Calculate the total amount of drug supplied by the patch to the tissue in a week. Remember the problem is axisymmetric.
(3) Calculate the amount of drug in the tissue after one week.
(4) Using (2) and (3) above, calculate the amount of drug supplied to the blood stream.

4.13.4 Postprocessing: Case study IV

Complete Case study IV (Chapter 6) and perform the following postprocessing operations.

(1) Plot concentration versus time at the mid-point (0.3375, 0.1125) of the aqueous humor.
(2) Obtain the surface plot for the aqueous humor region only at time 45 000 s (12.5 h).
(3) Calculate and plot the amount of drug lost in the tear film and aqueous humor regions due to the sink term, as a function of time.

4.13.5 Postprocessing: Case study VII

Complete Case study VII (Chapter 6). Calculate and plot the average antibody, antigen and complex concentrations in the tumor as functions of time on the same graph. Remember that the problem is spherically symmetric.

4.13.6 Postprocessing: Case study X

Complete Case study X (Chapter 6). Create a movie that shows the change in radius of the blood capillary and the transport of oxygen in the capillary and tumor regions as a function of time.

4.14 Appendix

This section presents an example of detailed instructions for reporting projects in an undergraduate class at Cornell University. Project reports following the instructions below can be seen on the website *http://courses.cit.cornell.edu/bee4530/*

Instructions for project reports

For this class you will write a comprehensive report on your final design project. The report will generally present:

- what you did
- how you did it
- what resulted
- what those results mean

The format of a report varies with organizations, but the following elements are typical in a report and we will follow this format. The report should not exceed 25 pages, in total. (*Note: The order in which you write the report will be different from the order in which the following sections (elements) appear. You will probably write the Objectives first, Appendix A and B next, followed by results, etc. Only when you are finished with the entire report are you in a position to write the introduction and finally the Executive Summary.*)

[1] Title and executive summary (1 paragraph)

These appear on a separate page. Toward the top, you present the title of your report, your name, your course number and the date. The executive summary is typically addressed to a busy manager who wants a quick understanding or a reminder of your work and its significance. This summary should be able to stand on its own, as a quick snapshot of the entire report. In one paragraph, it summarizes briefly, in order, the main points of your report. The summary also maintains the purpose, context and consequences of your project, and it stresses the important data and conclusions.

[2] Introduction and design objectives (2–3 paragraphs)

You inform the reader of the purpose and scope of the report and set the stage for what follows. It should include:

- what the project is about
- background and importance of the problem
- problem schematic

- design objectives
- preview of the organization of the report itself

[3] Results and discussion

This section presents your major findings – it describes how you reached each of these. It includes and discusses relevant figures (integrating them into the text or placing them in an appendix).

- Describe the process qualitatively using the computational results obtained
- Describe sensitivity of the process to parameter variations
    - include contour, vector and line plots, as appropriate.
    - summarize important trends; discuss these in relation to appropriate graphs and tables. Include only those graphs and tables that help to clarify. Additional graphs and tables may be placed in the appendix. Discuss every graphic (however briefly) in the text.
    - use the present tense for generalizations that always apply; employ the past tense to describe what you did or what your study revealed.
- Address these questions (refer to Section 5.2 for details):
    - how do your results compare with those expected?
    - how might you explain any unexpected results?

[4] Conclusions and design recommendations

Interpret your results in the context of the questions you set out to address in the Introduction and design objectives phase, mentioned above. This section must include:

- design recommendations
- discussion on realistic constraints. Such realistic constraints in design may include economic, environmental, sustainability, manufacturability, ethical, health and safety, social and political. Please discuss at least two of these constraints in relation to your project.

[5] Appendices

Appendix A Mathematical statement of the problem
- geometry or schematic (unless already shown in [2])
- exact form of all (velocity/temperature/concentration) governing equations

- exact form of the initial and boundary conditions used
- all input parameters in a table (or plot) in consistent units

Appendix B Solution strategy
- what solver was used to solve the algebraic equations
- how was the time stepping done (for transient problem only)
- what tolerance (relative and absolute) was used – explain this using one of the solution variables
- a plot of the element mesh. What type of element was used?
- convergence of the solution and mesh refinement (important)

Appendix C If you have not included important (and relevant) visuals, data or derivations in the Results section, then you should place them here (be sure each figure and table is appropriately numbered and titled).

Appendix D References

References (articles, books, web pages, etc.) should take the form:
Author(s). Date. Title. Publisher, Place of publication
Example:
Datta, A.K. and V. Rakesh. 2009. An Introduction to Modeling of Transport Processes: Applications to Biomedical Systems. Cambridge University Press, Cambridge, UK.

5 Validation, sensitivity analysis, optimization and debugging

"Regardless of the sources, the engineer who uses the software is held responsible for the results." (Cook, 1995)

Unfortunately, the nature of the numerical solutions discussed in this book is such that what appears to be a solution (sometimes based on colorful plots) may not be a solution at all. We need to *verify the model*: an unverified model can simply be, well, garbage! Even the state-of-the-art in modeling software has very few built-in checks for the validity of a solution. The software takes no responsibility whatsoever in providing a valid solution; *the user is completely responsible*.

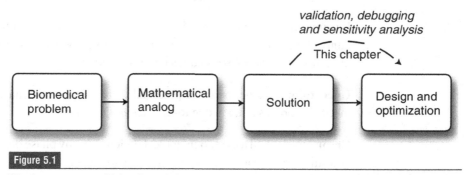

Figure 5.1

This chapter concerns validation of the solution obtained and its dependence on solution parameters (sensitivity analysis).

Thus, the question is, how do we trust the results we have obtained? In a research project, we will most likely need to compare the results from our model with experimental data. We are sometimes fortunate enough to find relevant experimental data reported in the literature, so we need not always develop our own elaborate experimental setup, which in itself might constitute a substantial research project. For a class project, we are unlikely to find the time and resources needed to develop our own experiments. In this case, we will have to depend on published experimental results. If we do not find published experimental results, how do we trust our results? To address this issue, in this chapter we will try to provide some measures that will give us confidence in our simulation results, as illustrated in the schematic of Figure 5.1.

When we do have a working solution, we are often interested in how the solution changes as its many parameters, such as properties or geometry, change. This is termed 'sensitivity analysis'. In industrial design of biomedical equipment, when developing a novel procedure in a clinical setting or in a research scenario, such sensitivity analysis is often the primary benefit of modeling. This chapter therefore also includes some discussion of sensitivity analysis and simple optimization. Finally, when simulation is not working, there are some guidelines on what to try. But these are just guidelines. We will talk about some of these at the end of the chapter.

5.1 Types of errors and error reduction

This section (adapted from Slater, 2003) provides a classification of uncertainties and errors that cause simulation results to differ from their true or exact values.

5.1.1 Defining uncertainty and error

It is important to understand the distinction that we will maintain between uncertainty and error. Here we follow the definitions provided by the American Institute of Aeronautics and Astronautics (AIAA) guidelines:

Uncertainty is defined as a potential deficiency in any phase or activity of the modeling process that is due to the *lack of knowledge*.

Error is defined as a recognizable deficiency in any phase or activity of modeling and simulation that is *not due to lack of knowledge*.

The key word in the definition of *uncertainty* is "potential", which indicates that deficiencies may or may not exist. *Lack of knowledge* has primarily to do with lack of knowledge about the physical processes that go into building a model. Sensitivity and uncertainty analyses, discussed later in the chapter, can be used to better determine uncertainty.

The definition of error implies that the deficiency is identifiable upon examination. Errors can also be classified as *acknowledged* or *unacknowledged*.

Acknowledged errors can be identified and possibly removed through established procedures (examples include round-off error and discretization error). Otherwise they can remain in the code, estimated and listed.

Unacknowledged errors cannot be identified through established procedures, so they may persist within the code of simulation (examples include computer programming errors or usage errors).

One can also differentiate between *local* and *global* errors. Local errors occur at a grid point or cell, whereas global errors affect the entire domain. We are interested here in global errors in a solution that account for local errors at each grid point, but are more than just the sum of the local errors. Local errors are transported, advected and diffused throughout the grid.

The definition of error presented here is different from what an experimentalist may use, which is "the difference between the measured value and the exact value." Experimentalists usually define uncertainty as "the estimate of error." These definitions are inadequate for computational simulations because the exact value is typically not known. Further, these definitions link error with uncertainty. The definitions provided in the above paragraphs are more definite because they differentiate error and uncertainty according to what is known.

5.1.2 Classification of errors

Various types of errors can be classified as below. These errors relate to both the software and how it is used.

Physical approximation error Physical modeling errors are due to uncertainty in the formulation of a model and deliberate simplifications of the model. These errors deal with continuum models only. Converting a model to discrete form for the code is discussed later in connection with discretization errors. Errors in the modeling of fluids or solids problems are concerned with the choice of the governing equations that are solved and models for the fluid or solid properties. Further, the issue of providing a well-posed problem can contribute to modeling errors.

Sources of uncertainty in physical models have been listed as: (1) the phenomenon is not thoroughly understood; (2) parameters used in the model are known but with some degree of uncertainty; (3) appropriate models are simplified, thus introducing uncertainty; and (4) an experimental confirmation of the models is not possible or is incomplete (Mehta, 1998). Even when a physical process is known to a high level of accuracy, a simplified model may be used within the CFD code for the convenience of a more efficient computation.

For example, in Figure 5.2, several assumptions are made such as: (1) no boiling or related latent heat; (2) constant thickness of steam layer; (3) skin layer has

 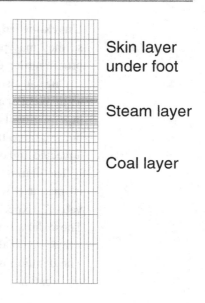

Figure 5.2

Illustration of error from mathematical approximation of a physical problem. Several assumptions have to be made to arrive at the skin-steam-coal layer for computation. For example, the edge effects are ignored, allowing the problem to be 1D. The steam layer is obviously not uniform throughout the surface over the foot but here a constant value has been assumed. Fire-walking image is from http://media.collegepublisher.com/media/paper871/stills/437422f81da14-64-2.jpg. Reproduced with permission. Mesh shown here is from the work of Cargioli *et al.* (2002).

well-defined regions with constant properties. These assumptions lead to physical approximation error.

Computer round-off error Computer round-off errors develop with the representation of floating point numbers on a computer and the accuracy at which numbers are stored. With advanced computer resources, numbers are typically stored with 16, 32 or 64 bits. Round-off errors are not considered significant when compared with other errors. If computer round-off errors are suspected of being significant, they can be tested by running the code at higher precision or on a computer known to store floating point numbers at a higher precision. For today's computers and software, round-off errors are generally not a concern.

Iterative convergence error Iterative convergence errors exist because the iterative methods used in the simulation must have stopping points eventually. Such an error scales to the variation in the solution at the completion of the simulations.

Discretization errors Discretization errors occur from the representation of the continuous field by the governing differential equations by a discrete field of space (such as finite-difference or finite-element) and time. The discrete spatial domain is known as the grid or mesh. The temporal discreteness is manifested through the time step taken. Discretization error is also known as numerical error and *truncation error*. A consistent numerical method will approach the continuum representation of the equations and zero discretization error as the number of grid points increases and the size of the grid spacing tends to zero. As the mesh is refined, the solution should become less sensitive to the grid spacing and approach the continuum solution. This is grid convergence. Such thinking also applies to the time step.

Discretization errors are of most concern to a modeling software user during an application. Discretization errors are of major concern because they depend on the quality of the grid; however, it is often difficult to precisely indicate the relationship between grid quality and an accurate solution prior to beginning a simulation. The level of discretization error depends on grid quality (resolution, aspect ratio etc.) and the physics of the problem. Included in the discretization error are errors due to a solution not properly converging with respect to the iterations, i.e., not reaching iterative convergence.

Computer programming errors Programming errors are "bugs" and mistakes made in programming or writing a code. They are the responsibility of programmers. Such errors are discovered by systematically performing verification studies of subprograms of the code and the entire code, reviewing the lines of code and performing validation studies of the code. Programming errors should be removed from the code prior to release.

Usage errors Usage errors are due to the application of a code in a less-than-accurate or improper manner. Usage errors may actually show up as modeling and discretization errors. The user sets the models, grid, algorithm and inputs used in a simulation, which then establishes the accuracy of the simulation. Usage errors may not be as evident. The potential for usage errors increases as more options are available in the software. Usage errors should be controllable through proper training and analysis.

The user may intentionally introduce modeling and discretization error as an attempt to expedite a simulation at the expense of accuracy. This may be proper in the conceptual stage of a design study where more general information is needed with less accuracy. Even in the later stages, there may not be proper computational resources to simulate at the proper grid density, and this may lead to usage errors. Thus, one has to understand the level of accuracy accompanying the results.

5.2 Estimating error: validation of the model

The errors discussed in the previous section cannot be eliminated entirely but they can be controlled. We need to understand the magnitude of the errors in a particular solution. This process is called *validation of the model* (see, for example, Ilegbusi *et al.*, 1999). The primary reason behind the difficulties in a validation process is that the *true* solution that we want to compare with our simulated solution in order to estimate the error is generally not available (if it were, we would have no reason to go for a numerical solution in the first place). As a side issue, the process of validation can also lead to important insights into the process which can help us improve the model itself.

As we use more commercial software with increasingly user-friendly interfaces, the underlying mathematical details get hidden more and more. This is good news in general, especially for the beginner. However, validation of a model can become more difficult this way, compared with a situation where one has complete access to the computer code and therefore is familiar with all the underlying details of the solution process. We can generally assume the benefits of commercial software far outweigh the risks.

5.2.1 What level of agreement is desirable?

Before we discuss alternative solutions and experimental results against which a model solution can be compared to determine its accuracy, it is important to decide or at least ponder over when such a comparison would be acceptable or "good enough." Trying to achieve perfect agreement is futile and can be very costly. We must always remember that we do not have the true solution available to us. Thus, for example, if we are comparing our results with experimental results, we have to remember that experimental results may also have errors. This is where the specific goals of model building (see Section 1.4) play a very important role. We cannot hope to resolve all the physics of a problem in infinitesimal detail, but *we must resolve the primary physics adequately*, so that the most important parameter defining the goal is resolved at a level demanded by the specific application.

5.2.2 Use your eyes and common sense (qualitative checks)

After verifying that you are in fact using the right geometry, governing equations, boundary and initial conditions, property values etc., the first place to start checking if the results are trustworthy is to display them using the postprocessor and to use your common sense to see if they make any sense. This is also called *checking for reasonableness*. It is particularly true for this step that *we cannot really validate a simulation but only invalidate it*. This is particularly useful in the early stages of the solution. Shaded contour plots are often quite useful for this purpose. To look for problems consider the following points:

(1) Are there unusual variations in the plotted variable, i.e., concentration, where the geometry changes from one region to another? Can we explain any unusual variation?
(2) Do the contours show an unusual increase or decrease in values? Again, can we explain any unusual variation?
(3) Are the values of the plotted variable anywhere in the geometry outside the range between maximum and minimum values, as provided by the boundary and initial conditions? For a transient problem, this should not happen at any time during the entire time period of the simulation.
(4) If the contour shows no variations, i.e., has the same value in the entire region, obviously the computation did not work.
(5) By plotting the flux at the boundaries, make sure that the flux values are indeed zero at the boundaries that you had planned.

5.2.3 Compare with alternative solutions

We can check with an alternative solution method, such as an analytical solution to the problem. An analytical solution, if available, provides perhaps the most rigorous check of the mathematical model (not the physical model) that has been developed. However, if an analytical solution is possible and available, there was probably no need to go through the numerical model in the first place. Thus, hoping for an analytical solution to the same problem may not be very practical. Sometimes an analytical solution is possible for a much simpler (but still related to the current model) governing equation and boundary conditions. Good comparison for such a simpler situation can provide some confidence, but may not guarantee accuracy for an intended situation that is more complex. Like analytical solutions, other numerical solutions, for a situation that is close to the current one, may also be available. Comparisons with such solutions can also provide some estimation of accuracy.

5.2.4 Compare with experimental data

Since practical situations are usually complex, experimental verification is often the gold standard for estimating error. Experimental verification is the most powerful way to compare the accuracy of a model with respect to the actual *physical* situation. Sometimes one is lucky enough that experimental data are available for the exact situation being modeled. More often, the experimental results are available for situations somewhat related, but not the exact situation. Here we have to change our model in response to the experimental conditions so that we can compare directly. For more on this, see Example 5.7.12. After comparing the simplified model with the experimental results, when we change the model back to what we really need we can no longer compare it with the same experimental results. At this point, we hope that the trust we have built in the model for a related scenario actually carries over to this new and intended one (the reasoning that we applied to comparisons with analytical solutions applies here as well).

At the same time, there is no reason to believe that experimental results always provide the absolute and ultimate proof. It is important to remember that they have their own measure of uncertainty. Such uncertainty is not always provided in the published results, but must be considered when comparing simulation results with experiments.

5.2.5 Compare over a large parameter range

The numerical solutions discussed here, obtained for a set of properties and parameter values, are not always expected to work out as we change those values. While generally no solution process is guaranteed to work over all properties and parameter ranges, verification of a model should include a significant range of parameters and property values. The behavior of the model over a large parameter range can also provide additional insight into the model.

5.3 Reducing discretization error: mesh convergence

The numerical method that we use to solve the model provides only an approximate solution and its accuracy increases as more discrete locations are used to compute the solutions, for example, as we make the mesh finer. This process has another obvious name – mesh refinement. Checking for mesh convergence is a critical step in building confidence in the solution obtained. There are no magic formulae that tell us how fine a mesh is appropriate for a given problem. The typical procedure is to repeat the solution for a progressively finer mesh until we see no further changes.

To have a quantitative measure of whether a solution is changing, we need a single measure that we can plot as a function of the number of mesh points

5.3 Reducing discretization error: mesh convergence

or the mesh density. This measure can be, for example, the function value (i.e., concentration) at some critical location, or an average value over the entire domain.

In practice, the sequence for performing error estimation and mesh convergence can depend on the situation. Typically, we iterate between the steps. For example, we could compare the initial computational solution with experimental results to see if they are close, and then do the mesh convergence to see if the computational results change with mesh density. If a finer mesh changes the solution, we continue to refine the mesh until the computational solution no longer changes and we then compare again with the experimental results.

5.3.1 Performing mesh convergence manually

As an example, consider the drug diffusion problem described in Case study III in Chapter 6. The mesh used for obtaining the solution had 375 elements. Three new meshes were created with 15, 120 and 750 elements, as shown in Figure 5.3. The problem, with the same parameter values, was solved again using these new meshes. The contour plots of drug concentration obtained as the solution from the analysis are shown in Figure 5.3. It can be observed that the contours differ slightly and, especially near the interface of patch and skin (top part on the left boundary), the difference is appreciable. The concentration profile appears scattered near the interface, which indicates that geometries with low mesh density (15, 120 elements) are unable to resolve the solution in this case. Geometries with higher mesh density (375 and 750 elements) show comparable concentration profiles leading to the conclusion that an accurate solution is obtainable with a mesh density of 375 elements and further mesh refinement is not required. This can be further justified by looking at the plot of mean concentration versus number of elements (Figure 5.4), which shows that there is no change in the solution as the number of elements increases from 375 to 750. This is the typical procedure in mesh convergence analysis and thus to determine if further refinement of the mesh is required to obtain a more accurate solution.

5.3.2 Performing automated mesh convergence: adaptive meshing

In the last section, we saw that we can change a mesh manually in order to improve the accuracy of a solution. In contrast, adaptive meshing is the automated changing of the mesh to follow the largest changes in the solution, to improve its accuracy. In adaptive mesh refinement, the solver creates a new mesh automatically based on the solution obtained on an initially coarser mesh. For the solver to implement

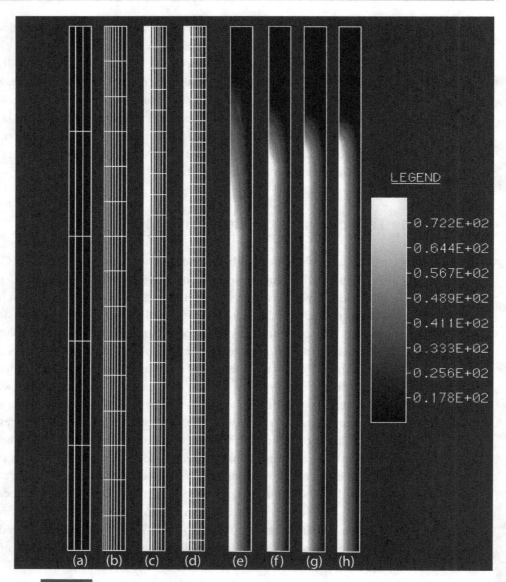

Figure 5.3

Mesh convergence analysis of the drug diffusion problem (Case study III). (a), (b), (c) and (d): mesh with 15, 120, 375 and 750 elements, respectively; (e), (f), (g) and (h): contour plots at the end time ($\approx$1 week) corresponding to the mesh in (a), (b), (c) and (d), respectively.

the adaptive meshing algorithm, it is important not only to obtain the total discretization error, but errors associated with each element in the mesh. The solver then identifies the elements where the errors are large and refines the mesh in those areas. In the absence of the availability of the exact solution to the problem, the

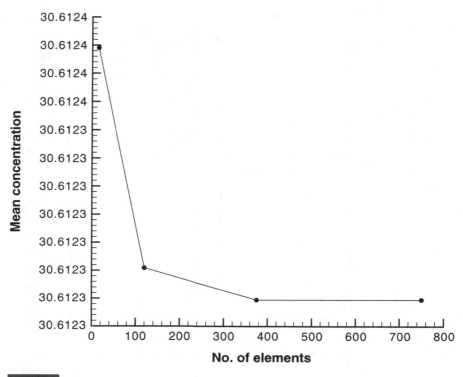

Figure 5.4

Plot of mean drug concentration over the domain at the end time, against the number of elements in the mesh for the drug diffusion problem. The solution is considered to have converged with respect to the mesh when the mean concentration stops changing as more elements are added. For this problem, variation in the mean concentration is small over the range of elements shown, perhaps due to the simple physics of diffusion involved, where even small numbers of elements can resolve the physics.

errors are typically determined by evaluating the residual of the partial differential equation being solved. Please refer to Section 10.11 for additional discussion of how these errors are calculated.

We now look at an example of adaptive mesh refinement. Adaptive mesh refinement was performed for the radiofrequency cardiac ablation Case study (Case study IX, Chapter 6).

Since adaptive mesh refinement in COMSOL is limited to steady-state problems only, we have solved the problem for the steady-state case. For details about implementation in COMSOL, please refer to Case study IX. Figure 5.5 shows the temperature contours obtained for the initial coarse mesh and after successive mesh refinements performed automatically by the solver. For the initial coarse mesh with 65 elements, the temperature contours near the left bottom of the geometry (i.e., near the electrode) are not very smooth. The coarse mesh is unable to resolve the

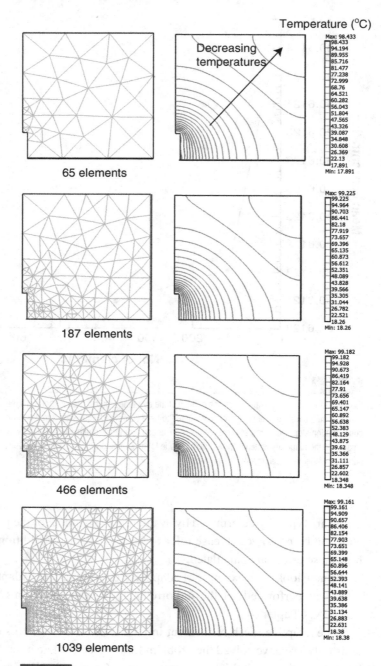

Figure 5.5

Illustration of adaptive mesh refinement, showing four iterations (65, 187, 466 and 1039 elements). The maximum variation in solution is near the electrode (left bottom corner) and hence, each iteration of adaptive mesh refinement adds more elements in that region compared to the rest of the domain, to obtain an improved solution.

high temperature gradients present in this region. As adaptive mesh refinement proceeds, more smaller elements are added near the electrode to resolve the steep temperature gradients. The refined meshes after three successive adaptive meshing procedures contain 187, 466 and 1039 elements, respectively. Due to better resolution of the temperature gradients in these meshes compared with those in the original coarse mesh, the observed temperature contours near the electrode are much smoother (Figure 5.5).

> **Box 5.1 Software implementation**
>
> Implementation in COMSOL of the adaptive meshing discussed here can be seen in Case study IX.

5.4 Estimating uncertainty and relating to design: performing sensitivity analysis

Sensitivity analysis consists of checking the sensitivity of the process model to the material property values, model parameters, boundary conditions and so on. See also the discussion in Section 9.4 on sensitivity with respect to material properties data. There are at least three reasons to consider sensitivity analysis (in the discussion below, the term 'parameter' has a general meaning and refers to property values, model parameters, etc.):

(1) to see how the results might change due to uncertainty of the input parameters. For example, a material property value may not be very reliable and we may not have the time or resources to measure it more accurately. By varying this parameter over a reasonable range, one can obtain the expected variation in the final results for the possible variation in the input parameter.
(2) to see if the behavior of the model makes sense with respect to change of input parameters. This is part of building confidence in the model.
(3) in designing a product or a process, the final goal of modeling is often optimization in relation to various model parameters. Sensitivity analysis can provide the information needed for optimization.

The concept of sensitivity analysis can be used in a more general sense. For example, a solution technique can also be changed to see if it has any effect on the results obtained.

5.4.1 How to perform sensitivity or uncertainty analysis

Sensitivity to (or uncertainty of) the model parameters in a model prediction is not automatically implemented in most commercial software, although some software is beginning to include such analysis. In the absence of in built ways to calculate such sensitivity, there can be two approaches to performing sensitivity analysis using numerical software.

Obtaining range This is the simplest approach but it is not comprehensive. Here the simulation is run for the two extreme values of the parameter, leading to two sets of results which correspond to the two extreme values. It provides no information on which result is more likely.

We can obtain sensitivity to a single parameter by performing simulations that vary this parameter only, while keeping other parameters constant. This can be repeated for multiple parameters, considered individually, i.e., always keeping the remaining parameters constant. From this information, sensitivity to multiple parameters when these parameters vary simultaneously we can deduce. For example, in a heat transfer process, temperature, T, can depend on the thermal conductivity, k, and specific heat, c_p. Let $\partial T/\partial k$ be the sensitivity of T to k, computed from how much the temperature at a chosen location changes when the thermal conductivity, k, is changed while keeping specific heat, C_p, and all other parameters constant. Likewise, let $\partial T/\partial C_p$ be the sensitivity of T to C_p, obtained by changing only C_p. It can be shown that if δk and δC_p are the uncertainties in k and C_p, respectively, for a particular process model, the uncertainty in ΔT when simultaneously considering the uncertainties in k and C_p, is given by

$$\Delta T = \sqrt{\left(\frac{\partial T}{\partial k}\delta k\right)^2 + \left(\frac{\partial T}{\partial C_p}\delta C_p\right)^2} \tag{5.1}$$

An example of uncertainty computation that simultaneously considers uncertainty in multiple parameters can be seen in the cryosurgery work of Rabin (2003). The range of input parameters used in this study, collected from variations in experimental data, is shown in Figure 5.6. Thus, Figure 5.6(a) shows the variation (and therefore the uncertainty), in percentage form, of thermal conductivity data at different temperatures, from the literature. Likewise, Figure 5.6(b), (c), (d), and (e) show the variation in specific heat of ice, the apparent specific heat of tissue during freezing, thermal diffusivity and blood flow, respectively. When considering uncertainties in all these parameters simultaneously, the uncertainty in temperature calculation, computed using an analog of Eqn. (5.1), is shown in Figure 5.7.

5.4 Estimating uncertainty and relating to design: performing sensitivity analysis

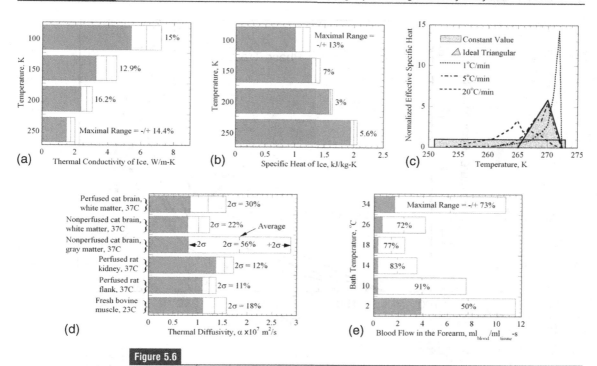

Figure 5.6

Range of input parameters obtained from the literature in the context of modeling a cryosurgical process. (a) Thermal conductivity; (b) specific heat; (c) apparent specific heat; (d) thermal diffusivity; (e) blood flow. Reprinted from Rabin (2003), with permission from Elsevier.

Obtaining distributions Monte Carlo simulation (repeated simulation of a process for a range of parameter values chosen from a probability distribution) may be possible provided that the computation times are not prohibitive. Monte Carlo techniques can be used to observe the effect of variability in the input parameters on the results obtained from a computational model. We use Case study III (Chapter 6) to demonstrate the use of Monte Carlo simulations. The case study looks at the delivery of drug from a birth control patch to tissue and from there to the blood stream. The input parameter for the model is the diffusivity of the drug in the tissue. The diffusivity may vary from person to person and also with the type of drug used. We use the Monte Carlo simulations to determine how the change in diffusivity affects the amount of drug delivered to the blood stream. The diffusivity of the drug is assumed to vary from 80% to 120% of the original diffusivity value (D). This is approximated using a normal distribution with a mean value equal to the original diffusivity, D, and a standard deviation of $0.067D$ (so that most of the values lie within the assumed range of 80–120%). A total of 1000 different diffusivity values are considered from this distribution, as shown in

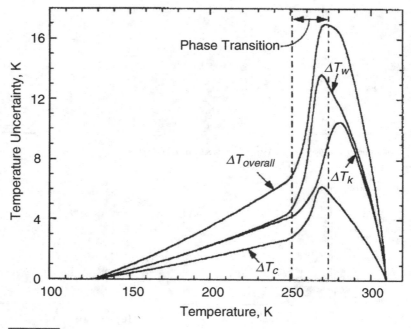

Figure 5.7

Uncertainty in temperatures calculated in a cryosurgical process after 10-minutes of freezing, due to uncertainty in input parameters, illustrated in Figure 5.6. Here ΔT_i represents the absolute difference between the temperatures calculated using average properties and temperatures calculated using a variation in property i only. The overall uncertainty, $\Delta T_{overall}$ is calculated using Eqn. (5.1). Reprinted from Rabin (2003), with permission from Elsevier.

Figure 5.8(a). The computational model was solved using these 1000 different diffusivity values and the amount of drug delivered in each case was obtained. The frequency plot for the amount of drug delivered to the blood stream obtained from the Monte Carlo simulations is shown in Figure 5.8(b). The amount of drug delivered plot also follows a bell-shaped normal distribution, indicating that the drug delivered is directly proportional to the diffusivity of the drug.

Box 5.2 Software implementation

For the implementation of Monte Carlo simulations of the problem in COMSOL, refer to Case study III.

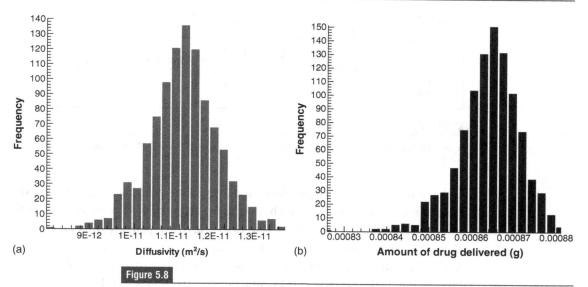

Figure 5.8

Monte Carlo simulation to obtain the sensitivity to drug diffusivity of drug delivery from a patch (Case study III). (a) Frequency plot of 1000 different values of diffusivity of the drug in the tissue, chosen randomly from a normal distribution. (b) Frequency plot of the amount of drug delivered to the blood stream for different diffusivity values chosen in (a).

5.5 Objective functions: simple optimization

When we are varying model parameters in a simulation, as was discussed under sensitivity analysis, we often work under various constraints. For example, in a cryosurgical process, we may be trying to freeze and destroy the tumor tissue but we have to protect the good tissue. From the simulation results showing temperatures, we can qualitatively decide under what conditions we will have maximum destruction of the tumor tissue and minimal damage to the good tissue. Our decision capability would improve if we could quantify the extent of the desired and the undesired. This can be achieved by formulating the problem as an *optimization* (maximization or minimization) problem, where we set up objective functions which we try to maximize or minimize.

In this context, it is helpful to think of problems as being *forward* or *inverse*, as shown in Figure 5.9. For the most part, when we mention modeling, we mean a forward problem where the geometry, governing equation, boundary conditions and properties are given and we are solving for temperature, concentration, fluxes and the like. Commercial CFD (computational fluid dynamics) and related software are typically set up to solve problems this way. A forward problem is also known as a *normal* problem. An inverse problem is when some solution element is desired, and we have to come up with the right problem formulation that will provide

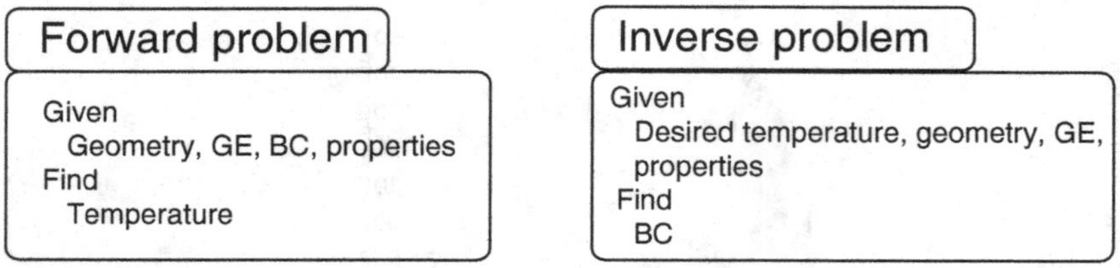

Forward problem

Given
  Geometry, GE, BC, properties
Find
  Temperature

Inverse problem

Given
  Desired temperature, geometry, GE, properties
Find
  BC

Figure 5.9
A forward problem (typical of what we do most of the time in this book) contrasted with an inverse problem.

that solution. For example, in trying to destroy a tumor using heat, we know the desired temperature at a location but we need to figure out what boundary condition, heating duration etc. will lead to that desired temperature. *It is easy to see that most product, process or procedure design problems are really inverse problems.* These inverse problems lead to optimization problems, discussed earlier in this section. Unfortunately, most commercial softwares are not designed to solve inverse or optimization problems in a straightforward manner. The choice, when using commercial software, is to do the forward problem repeatedly, for different parameters (as discussed under sensitivity analysis), and proceed toward optimization in an informal manner by checking when the objective function is maximized (or minimized).

Optimization is a vast discipline and we will only mention the general concept here using a simple example. Consider the Case study II on cryosurgery of a wart, described in Chapter 6. The goal here is to use a cryogen to freeze and thus destroy as much of the wart as possible, while damaging as little healthy skin as possible. For destruction, we assume the temperature of the wart region needs to be between $-40\,°C$ and $-43\,°C$ and, to reduce damage to normal skin, we assume its temperature needs to be above $5\,°C$. We can set up a simple objective function, J, as

$$J = \overbrace{\sum_i F_w(T_i)}^{\text{Nodes to be frozen}} + \overbrace{\sum_j F_n(T_j)}^{\text{Nodes not to be frozen}} \quad (5.2)$$

where the subscripts i, j stand for nodes in the wart (to be frozen) and nodes in the normal tissue (not to be frozen), respectively. The functions F_w and F_n are defined, based on our desired temperature ranges, as:

$$F_w(T) = \begin{cases} T + 40 & T > -40 \\ 0 & -43 \leq T \leq -40 \\ -43 - T & T < -43 \end{cases} \qquad F_n(T) = \begin{cases} 0 & 5 \leq T \\ 5 - T & T < 5 \end{cases}$$

$$(5.3)$$

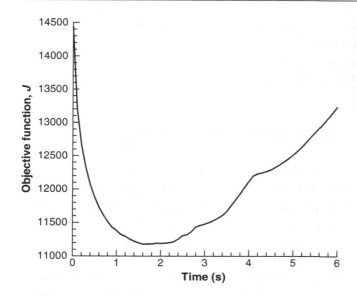

Figure 5.10

Objective function, defined by Eqn. (5.2), computed as a function of freezing time for cryosurgery of a wart (Case study II). The region around the lowest value of objective function, i.e., around a freezing time of 1.8 s, would provide the preferred freezing times.

As defined above, the objective function has a minimum value for the desired situation and therefore it needs to be minimized. If we think of optimization in relation to the duration of the procedure, computation of J with respect to the time involved in the procedure is shown in Figure 5.10. As expected, initially, there is little freezing and therefore the value of the objective function is high. As more of the wart freezes, the value decreases (following Eqn. (5.3)) until it reaches a minimum beyond which more of the normal skin tissue starts to freeze, thus increasing the value of the objective function. More examples of objective functions are shown in Problems 5.7.10 and 5.7.11.

> **Box 5.3 Software implementation**
>
> Implementation in COMSOL of the optimization problem discussed here can be seen in Case study II.

5.6 When things don't work: debugging

It is rare for a model to work on the first run. It is equally hard to pinpoint why the model did not run or to come up with comprehensive and foolproof directions to make the model work. Nevertheless, there are some fairly obvious items we all try as we proceed. Below is a list of such items in a somewhat logical order:

- is the computer solving the problem you want it to solve?
- what problem would be easier to solve?
- add in the complexities, one at a time.
- make sure the model is really working.

5.6.1 Is the computer solving the problem you want it to solve?

You may know this if you have ever done computer programming: the computer never solves what you wish to solve. Rather, it solves what you have asked it to solve (not necessarily what you think you have asked). If you made any mistakes in specifying the problem that you want to solve, the computer would still be trying to correctly solve the wrong problem. Some general checks are:

Check input file The place to start when things don't work is the set of commands that the computer is actually using, as opposed to what we assume it should be using. For COMSOL, do File > Generate Report and look into the report file, going through every item, making sure you know exactly what it means and that it is indeed what you had intended.

Geometry and mesh The numbers that you input as the length dimension in COMSOL are assigned units based on the units you choose when you started using Model Navigator > Settings. Make sure all inputs have values that are consistent with these units. By default, SI units are used.

Governing equations Make sure you are using only the governing equations you need and only the terms you need.

Boundary conditions This is usually where trouble originates. Make sure of the following:
- only one condition is specified for every boundary entity and for every equation you are solving.
- these are indeed the boundary conditions that you want to specify.
- you did not specify boundary conditions for interior regions.

Properties Make sure:
- the units used are correct.
- non-dimensionalization (if used) is correct.

5.6.2 Intermediate steps where things may be going wrong

Checking intermediate results and making sure they make sense is an important step toward debugging.

5.6.3 Can you help the computer?

Here we mean not by changing the problem to an easier problem (the topic of the next section), but by tweaking the method and the solution parameters that the computer uses to solve it. The solution parameters are discussed at length in Chapter 3. Some obvious solution parameters to try out are:

Time step Changing – typically reducing – the time step in a transient problem. Sometimes drastic time-step reduction may be necessary.
Mesh Refining mesh, as discussed in Section 5.3, is another obvious thing to try. One has to balance this against increased computation time.
Solver Changing parameters for the solver, such as what method is being used to solve the set of linear equations, can improve the speed of finding a solution or make a solution possible. In COMSOL, look under Solve > Solver Parameters – see Section 3.5.
Tolerance Tolerances can be reduced to make the computations more accurate – see Section 3.5.

5.6.4 What problem would be the easiest to solve (for you and the computer)?

If you are sure that the input file, i.e., the Report file in COMSOL, is correct (that the computer has been told to solve the problem that you wanted it to), there are probably issues with the numerical complexities. Quite often, the best choice now is to simplify the problem to ease the numerical complexities. So, we would look for a problem that is related, but simpler. One has to be careful in deciding which easier problem to solve. For example, one needs to avoid starting from a problem that is simpler, but will not really lead to the final computation. This is hard to decide, in general.

Two ways for simplifying the problem are:

(1) Simplify the current model to the lowest level possible. In this scenario, keep the geometry and mesh the same, while changing the governing equations, boundary conditions and properties to their simplest possible situations.
(2) Start from the most simplified problem. This may include starting with an even simpler geometry and a different mesh.

5.6.5 Adding the complexities

Once a simple model works, we start to add in complexities. Some general guidelines can be useful, such as:

- adding only one complexity (making just one change) at a time. If this does not work, you need to resolve it, before going further.
- beware that any little change can add to computational complexity.
- check for the entire range of parameters, i.e., check for the entire time duration of simulation in a transient problem after changing diffusivity.

5.6.6 Make sure it is really working before you stop

When you think a simulation works, you really have to be sure about it. The section on validation discusses at length how one can attempt to ensure the validity of the simulation. Remember, when the simulation is working, it is doing so only within the range of parameters that you have tried. It can often encounter difficulty outside this range. However, it is not necessary in general to try the model outside the ranges of parameters of interest.

References

Apiou-Sbirlea, G. and J-P. L'Huiller (1998). Simulating and optimizing of argon laser iridectomy. Influence of irradiation duration on the corneal and lens thermal injury. Part of the EUROPTO Conference on Lasers in Opthalmology, Stockholm, Sweden, Sept. 1998. SPIE Vol. 3564.

Bischoff, J. C., N. Merry and J. Hulbert (1997). Rectal protection during prostate cryosurgery: Design and characterization of an insulating probe. *Cryobiology*, **34**:80–92.

Cargioli, T., Gaites, C., Van Fleet, Geoff, Liem, A. and Lyubchenko, L. (2002). The Mysteries of Firewalking Revealed. On the Web at http://ecommons.library.cornell.edu/handle/1813/204.

Cook, R. D. (1995). *Finite Element Modeling for Stress Analysis*, New York: John Wiley & Sons, Ltd.

Gurzo, M. Ho, J., Lally, S and Selig, M. (2000). Accidental freezing of the tongue to metal poles. On the Web at http://ecommons.library.cornell.edu/handle/1813/277.

Ilegbusi, O. J., M. Iguchi, and W. Wahnsiedler (1999). *Mathematical and Physical Modeling of Material Processing Operations*. Boca Raton, Florida, CRC Press.

Jolly, A., Gupta, S., Auerbach, E., Loaknauth, N. and Drost, L. (2002). Laser hair removal. On the Web at http://ecommons.library.cornell.edu/handle/1813/2000.

Lang, J., B. Erdmann and M. Seebass (1999). Impact of nonlinear heat transfer on temperature control in regional hyperthermia. *IEEE Transactions of Biomedical Engineering*, **46**(9):1129–1138.

Mehta, U. B. (1998). Credible computational fluid dynamics simulations. *AIAA Journal*, **36**, No. 5, May 1998, pp. 665–667.

Ng, E. Y. K. and N. M. Sudharsan (2001). An improved three-dimensional direct numerical modeling and thermal analysis of a female breast with tumour. *Proceedings of the Institution of Mechanical Engineers*, **215**(H):25–37.

Rabin, Y. (2003). A general model for the propagation of uncertainty in measurements into heat transfer simulations and its application to cryosurgery. *Cryobiology*, **46**:109–120.

Rim, J. E., P. M. Pinsky and W. W. van Osdol (2005). Finite element modeling of coupled diffusion with partitioning in transdermal drug delivery. *Annals of Biomedical Engineering*, **33**(10):1422–1438.

Sarntinoranont, M., R. K. Banerjee, R. R. Lonser and P. F. Morrison (2003). A computational model of direct interstitial infusion of macromolecules into the spinal cord. *Annals of Biomedical Engineering*, **31**:448–461.

Slater, J. W. (2003). Uncertainty and error in CFD simulations. Document on the Web at http://www.grc.nasa.gov/WWW/wind/valid/tutorial/errors.html.

Tsuda, N., K. Kuroda and Y. Suzuki (1996). An inverse method to optimize heating conditions in RF-capacitive hyperthermia. *IEEE Transactions in Biomedical Engineering*, **43**(10):1029–1037.

5.7 Problems

5.7.1 Short questions

(1) Name some of the most important steps you will take to verify the results from a simulation before accepting it.

(2) Do the computational steps that you are doing replace experimentation? Explain.

(3) Most of the biomedical modeling projects involve difficulty in finding exact data for the system you are trying to model. How would you compensate (what specifically you will do) for the lack of exact data?

(4) Describe in words how one would use the model itself to determine whether more accurate properties data are needed.

(5) How would you go about choosing the range of parameters for sensitivity analysis? Can you just assume any range? Please explain.

(6) When you find that a solution is particularly sensitive to a certain input parameter, how would you use this information to improve the quality of your solution?

(7) You might already have considerable experience with simulations not working, even when you feel you have a pretty good idea about what you are doing. When a simulation is not working, list four steps that are typically the most important that you will try in order to debug what could be going wrong. Please list them in the order in which you will try them.

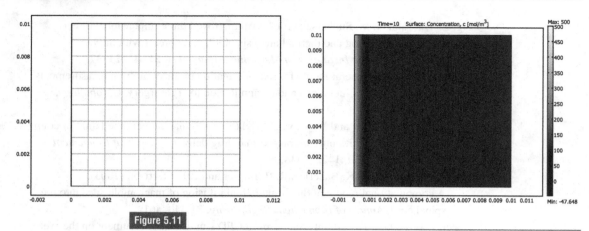

Figure 5.11

Mesh and surface plot for Problem 5.7.2.

(8) When we obtain a computer solution for any physical problem, we usually make a number of errors. Under each type of error (designated (a)–(d) noted earlier) mention (1) what that error is, and (2) some thoughts on how to minimize it: (a) physical approximation error (discuss in relation to the processes in the Problems section in Chapter 1); (b) computer round-off error; (c) iterative convergence error; (d) discretization error.

5.7.2 Qualitative checks

In order to study diffusion of a drug from a patch into the skin, a 2D transient diffusion equation is solved in COMSOL with the following parameters: geometry is a square of size 1 cm × 1 cm; diffusivity of the drug is 1×10^{-9} m^2/s; boundary conditions are given by a constant concentration of 500 g/m^3 on the left boundary while the other boundaries are insulated; initial concentration of the drug is 0 g/m^3; total duration used for simulation is 10 sec; time step size is 0.1 sec and solver setting is direct. The mesh used in the simulation and the concentration contours at the final time are shown below. (1) Discuss what is unphysical in the concentration profile. (2) What parameter(s) would you change to improve the accuracy of the solution?

5.7.3 Qualitative checks

A 2D heat conduction problem was *intended* to be solved for the following parameters: initial temperature is 325 K; boundary conditions are given by the left boundary at 300 K and all other boundaries are insulated. The steady-state

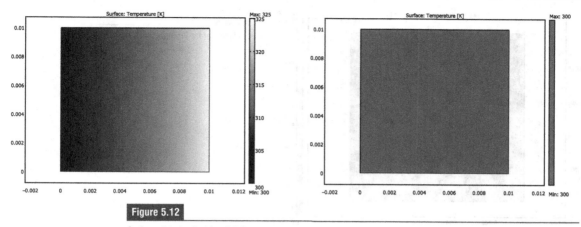

Figure 5.12

Surface plots for Problem 5.7.3.

temperature contours below were obtained by two people. (1) Which *contour plot* is incorrect and why? (2) One parameter is wrongly specified which leads to the incorrect profile. Which parameter?

5.7.4 Qualitative checks

A 2D heat conduction problem was solved with the following conditions: initial temperature is 350 K; boundary conditions are given by the left boundary at 450 K while all other boundaries are insulated; total duration of heating is 120 s; time step size is 1 sec; density is 1000 kg/m^3; specific heat is 2000 J/kg K; thermal conductivity varies from 0.6 at 300 K to 0.1 at 400 K and remains constant outside this temperature range. This was specified in COMSOL as a linear interpolation function as

300K	0.6
350K	0.35
400K	0.1

The temperature contour in Figure 5.13 was obtained. (1) What is wrong with the contour? Be specific. (2) How would you correct it?

5.7.5 Physical approximation error

Discuss physical approximation errors in each of the problem formulation exercises presented at the end of Chapter 1.

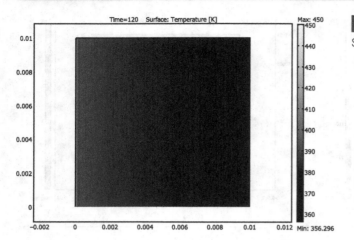

Figure 5.13

Surface plot for Problem 5.7.4.

The following problems have one or more components of: (1) checking for mesh convergence; (2) qualitative check for the results; (3) comparison with experiment; (4) performing sensitivity analysis; and (5) performing optimization. They are taken from published literature and student projects. The detailed governing equations, boundary conditions etc. are omitted here for brevity (they are not required to answer the questions below), for further information the reader is referred to the corresponding publication.

5.7.6 Qualitative checks: laser heating to treat glaucoma

The use of lasers to coagulate tissue (ciliary processes) has found widespread application in glaucoma therapy. Figure 5.14 shows a schematic of the eye section showing the placement of the laser. Heating leads to vaporization and tissue removal, creating a small opening in the iris which equalizes the fluid pressure between the anterior and posterior chambers. Equations representing heat transfer and fluid flow (in aqueous and vitreous humors) are used in formulating the problem, details of which can be seen in Apiou-Sbirlea and L'Huiller (1998).

The computed temperature profiles, at the nodes numbered in the mesh, are shown in Figure 5.15. (1) Explain why the amplitude (maximum minus minimum) of the temperature change is much lower at the cornea than at the iris. (2) Why does the peak temperature occur later than in the iris? (3) From a heat transfer standpoint, for the same average power, why does a high-powered laser (0.88 W, 0.04 s) produce more localized burns and is therefore more effective, as illustrated in the temperature profiles?

5.7 Problems

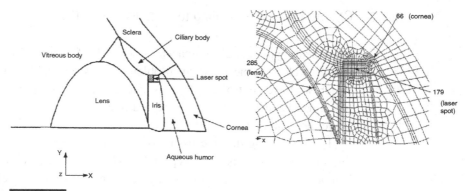

Figure 5.14

Schematic and finite element mesh for studying laser heating to treat glaucoma. Reprinted from Apiou–Sbirlea and L'Huiller (1998), with permission from SPIE.

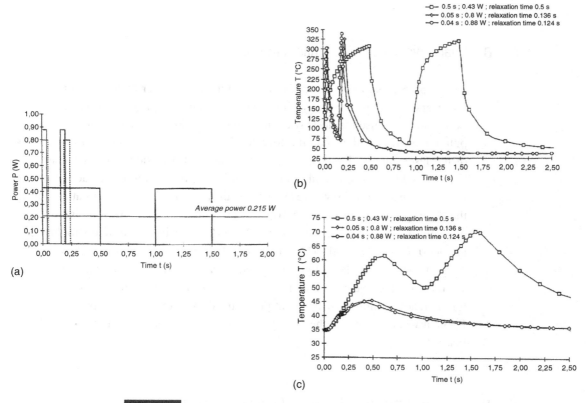

Figure 5.15

Results from the computation related to Figure 5.14. (a) Three different laser shots with the same average power; (b) temperature at the center of laser spot in iris (node 179 in Figure 5.14); (c) temperature at the corneal surface closest to iris (node 66 in Figure 5.14). Reprinted from Apiou-Sbirlea and L'Huiller (1998), with permission from SPIE.

5.7.7 Optimization: laser heating to treat glaucoma (2)

Laser heating to treat glaucoma was described in Section 5.7.6. Figure 5.14 shows the finite element mesh for this problem. The two variables in laser heating over which we have control are the power level and the pulse duration or duty cycle (for a fixed average power). We would like to find the optimum combination of power level and pulse duration.

(1) Considering that the tissue to be destroyed needs to be heated to 300 °C and the good tissue is to be kept below 40 °C, define an objective function that can be used to evaluate various heating scenarios (power levels and pulse durations). (2) State whether the objective function is to be maximized or minimized. (3) Using software such as COMSOL, provide the detailed steps to be followed in computation and explain how you would try to reach the optimum combination of power level and pulse duration.

5.7.8 Qualitative checks: laser hair removal

Figure 5.16(a) shows a cross-section of skin with a laser shining from the top. Laser energy is preferentially absorbed in the hair follicle *(not in the rest of the skin)* and this is used to destroy the unwanted hair. Figure 5.16(b) shows the computational mesh. Other than the follicle, the rest of the hair does not absorb any of the laser energy. The laser is cycled on and off. Consider the laser heat generation to decay exponentially into the follicle starting from its surface (note: no laser heat absorption anywhere else). (1) Is the periodic drop in temperature during the heating process in Figure 5.16(c) an error in computation? Why is this drop different at the two nodes? Explain. (2) Node 2193 in the follicle should reach about 60 °C whereas node 411, which is in the skin, should not reach above 50 °C. As you can see, the laser light is cycled on and off to reach the temperature. If we increase the cycle time and reduce the power level so the desired temperature is reached more slowly. What problems do you expect in the procedure? (3) Explain what happens if the laser power is increased and cycle time is shortened instead.

5.7.9 Mesh convergence: freezing of tongue on metal pole

On a cold day, can the tongue stick to a very cold metal plate (such as a metal pole) outside (Figure 5.17(a))? Sticking would happen if the bottom layer of the tongue

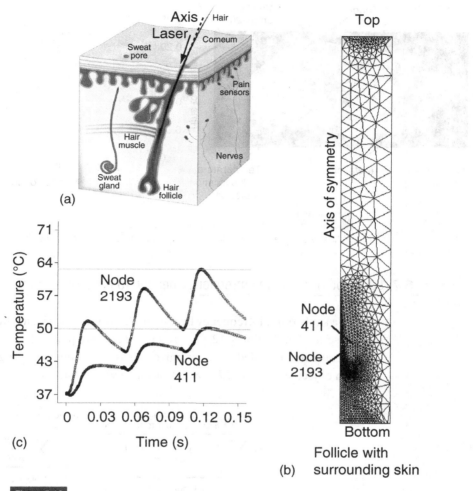

Figure 5.16

(a) Cross-section of skin showing the follicle; (b) finite element mesh of the cross-section of the skin with follicle; (c) temperatures at Node 2193 (upper line) represents follicle temperature, and node 411 (lower line) represents skin temperature (outside of follicle). Mesh and temperature computation from the work of Jolly *et al.* (2002).

froze. As they so often do when solving problems, students in this project wanted to know if they had a mesh that was fine enough to resolve the problem (Gurzo *et al.*, 2000). To do this, the freezing time of the bottom layer of tongue was calculated for different meshes, as shown in Figure 5.17(b). (1) Has convergence been achieved for the smallest mesh size? Discuss why or why not. (2) If convergence has not been achieved yet, when do we stop? (3) What would be another parameter (instead of the freezing time) to observe convergence?

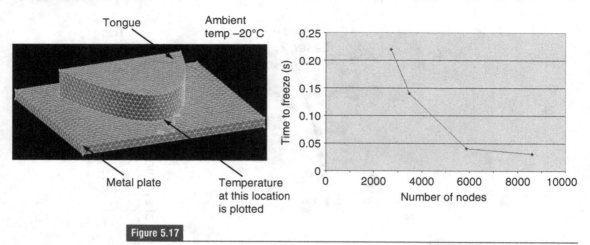

Figure 5.17

(a) Mesh of a tongue on a metal plate. (b) Mesh convergence study. Gurzo et al. (2000).

5.7.10 Optimization: radiofrequency heating to destroy a tumor

In a study of radiofrequency heating to destroy a tumor inside the lower abdomen, described in Section 1.12.12 (Tsuda et al., 1996), placement of electrodes around the abdomen needed to be optimized so as to minimize heating of normal tissue. An objective function, J, for this problem was defined as

$$J = \underbrace{\overbrace{\sum_i F_T(T_i)}^{\text{Nodes to be heated}} + \overbrace{\sum_j F_N(T_j)}^{\text{Nodes not to be heated}}}_{\text{Interior region}} + \underbrace{\overbrace{\lambda \sum \frac{\partial^2 \phi}{\partial r^2}}^{\text{Penalty function for electrode placement}}}_{\text{Boundary region}} \quad (5.4)$$

where the subscripts i, j stand for nodes in the tumor (to be heated) and nodes in the normal tissue (not to be heated), respectively. The functions F_T and F_N are defined as:

$$F_T(T) = \begin{cases} T - 45 & T > 45 \\ 0 & 43 \leq T \leq 45 \\ 43 - T & T < 43 \end{cases} \quad F_N(T) = \begin{cases} T - 40 & T > 40 \\ 0 & 30 \leq T \leq 40 \\ 30 - T & T < 30 \end{cases}$$

$$(5.5)$$

It is not desirable to have drastic variations in voltage, ϕ, from point to point, for realistic placement of electrodes. To bring this constraint in, the quantity $\lambda \partial^2 \phi / \partial r^2$ is added, where r is the location variable. This optimization problem (minimization of J) was achieved using an inverse analysis with custom-written code, unlike what

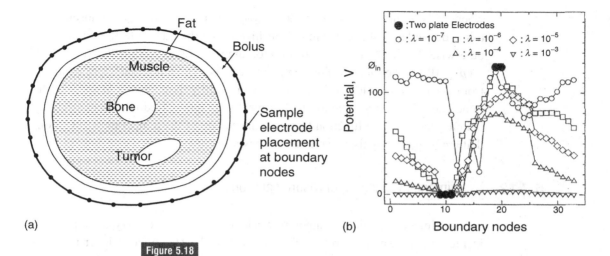

Figure 5.18

(a) Cross-section of the abdomen showing tumor, finite elements and the placement of electrodes; (b) optimized placement of electrodes for various optimization parameters. From Tsuda *et al.* 1996 (©1996 IEEE). Reproduced with permission.

we can achieve using most commercial software. Results from this analysis are shown in Figure 5.18 – we get different optimum solutions depending on the chosen value for λ.

Provide the detailed steps to achieve the optimized location of electrodes using forward analysis such as in COMSOL; i.e., how would you use COMSOL to achieve the optimized location of electrodes? Assume a value for λ. No need to provide the exact governing equations or boundary conditions. Note that it may be difficult to assign boundary conditions to individual nodes in commercial software, in which case the boundary can be divided into regions and a voltage applied to a region.

5.7.11 Optimization: radiofrequency heating to destroy a tumor (2)

Another example of an objective function is given in the study of Lang *et al.* (1999) for modeling hyperthermia of deep-seated tumors. They defined an ideal temperature distribution to be 42–43 °C within the tumor and below 42 °C in healthy tissue, and defined an objective function as

$$J = \underbrace{\int (T_{\text{ther}} - T)^2 \, dV}_{\substack{\text{in the tumor} \\ T < 43°C}} + \underbrace{\int (T - T_{\text{healthy}})^2 \, dV}_{\substack{\text{outside the tumor} \\ T > 42°C}} + p \underbrace{\int (T - T_{\text{lim}})^2 \, dV}_{\substack{\text{outside the tumor} \\ T > T_{\text{lim}}}} \quad (5.6)$$

where V is volume, $T_{therm} = 43\,°C$, $T_{healthy} = 42\,°C$, and T_{lim} is chosen to be tissue dependent – $42\,°C$ for more sensitive tissue (bladder, intestine) and $44\,°C$ otherwise. To ensure high penalization for temperatures exceeding the limits, they set $p = 1000$. The squares of the temperature values guarantee that large deviations from desired temperatures, i.e., *hot spots* in healthy tissue and *cold spots* in the tumor, contribute large amounts to the objective function. Redo Problem 5.7.10 by defining an objective function that uses the square of the temperature deviations that are weighted by tissue volume, as illustrated in Eqn. (5.6).

5.7.12 Experimental validation: cryosurgical procedure

To experimentally validate temperature calculations in cryosurgery, a setup shown in Figure 5.19 was used by Bischoff *et al.* (1997). The silicone layer provides insulation between the prostate and rectum and the insulation capability of the silicone layer was the primary objective of this study. (1) Show how you expect the temperature profile to change for a thicker silicone material. (2) Comment on how good the comparison is between the experimental results and the numerical computation. (3) To improve the comparison, i.e., to bring the experimental and computational results closer, consider improving both computation and experiment. Can you comment on which one (computation or experiment) is more in need of improvement and how you might go about it? If further details are needed, they can be found in the reference cited above.

5.7.13 Inverse problem: transdermal drug delivery with penetration enhancer

Transdermal drug delivery (see schematic in Figure 5.20(a)) can be improved with the use of permeation enhancers that are added to a drug (Rim *et al.*, 2005). The enhancers diffuse with the drug, but also increase the diffusivity of the drug itself and the partition coefficient ($= c_{skin}/c_{patch}$) of the drug between the drug reservoir (patch in figure) and the skin. A fundamental task in the development of transdermal formulations is that of finding the optimal combination of drug and penetration enhancer that achieves the required drug flux for the duration of the patch application. The goal in this study was to determine whether the increased drug flux with increased concentration, observed experimentally (Figures 5.20(b),(c)), is due to increasing diffusivity of the drug or due to increasing value of the partition coefficient at the patch-skin interface. The solid lines in Figure 5.20(b) and Figure 5.20(c), superimposing the experimental data, are the computations assuming either diffusivity increasing with enhancer concentration or partition coefficient increasing with enhancer concentration, respectively.

5.7 Problems

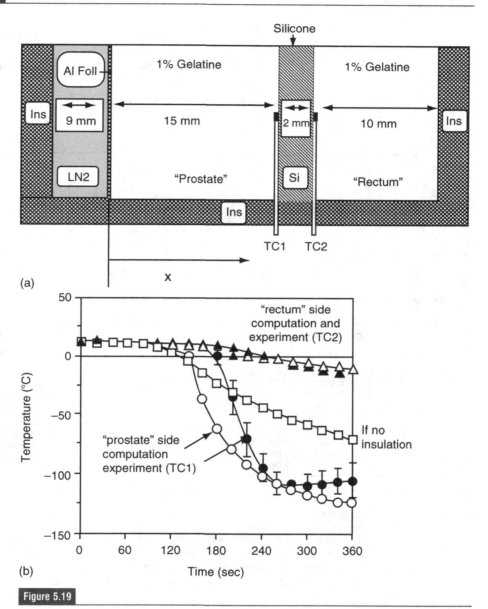

Figure 5.19

(a) Schematic of experimental setup to study cryosurgery showing a styrofoam insulating container with a compartment for liquid nitrogen on the extreme left, followed by a slab of gelatine representing "prostate," a slab of silicone that is to provide the insulation and another gelatine slab representing "rectum"; (b) computed temperature profiles (hollow symbols) compared with experimental temperature profiles (solid symbols) measured using thermocouples TC1 and TC2. Reprinted from Bischoff *et al.*, copyright (1997), with permission from Elsevier.

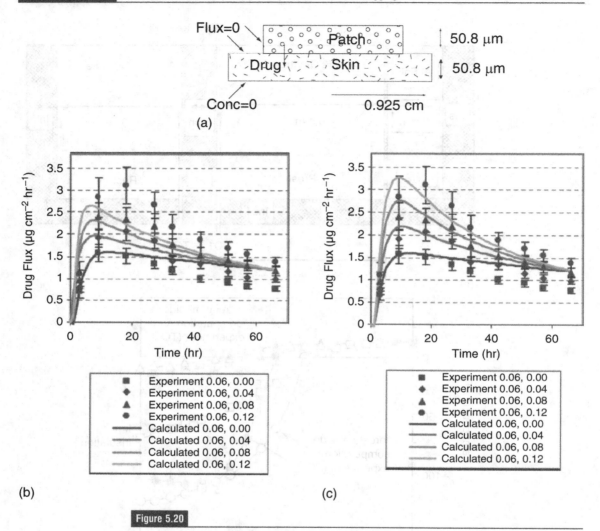

Figure 5.20

(a) Schematic of a patch containing a drug; computed transient flux of drug through the lower boundary of epidermis for drug initial concentration of 0.06 and various enhancer concentrations (0.00–0.12) showing the effects of (b) diffusivity increasing with enhancer concentration; (c) partition coefficient increasing with enhancer concentration. Reprinted from Rim et al. (2005), with kind permission from Springer Science + Business Media.

(1) Why does the flux increase and then decrease? Would the flux behave the same way if we set a constant concentration boundary condition at the patch-skin interface? Explain why or why not. (2) Here the author is trying to estimate the input parameters (diffusivity or partition coefficient) by comparing the predicted flux with the experimental measurement. Which set of parameters matches the data better? (3) Before doing the model, the enhancement mechanism for various enhancer concentrations was thought to work predominantly through the change

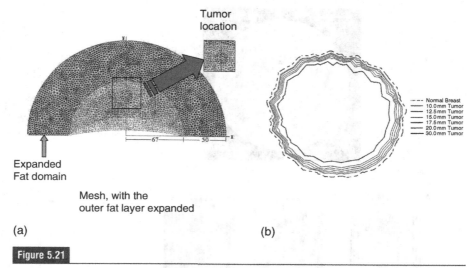

Figure 5.21

Left figure shows computational mesh for breast tissue with tumor location. Right figure shows computed temperature contours. From Ng and Sudharsan (2001). Reproduced with permission.

in drug diffusivity rather than the drug partition coefficient. Is this view supported by the two graphs shown? (4) In this example, is modeling a useful tool with which to analyze the mechanism of enhancement?

5.7.14 Inverse problem: detection of a breast tumor

We are concerned with finding the location and size of a breast tumor based on breast thermography. Figure 5.21(a) shows the computational mesh of a 2D section of a female breast in supine condition with an assumed location of the tumor. Figure 5.21(b) shows the computed temperature contours for the assumed location, using the normal forward analysis (i.e., given tumor size, location etc., temperature contours are computed). Consider the inverse problem of finding the location and size of the breast tumor based on temperature contours, such as in Figure 5.21(b), obtained experimentally using breast thermography. The difficulty here is that a large tumor located deeper inside can give a temperature contour similar to that of a smaller tumor located close to the surface.

(1) Describe the steps that you will take to solve this inverse problem of finding the location of the tumor given the temperature contours from experimentation, when all you have access to is forward analysis capability, as in COMSOL. In other words, provide the sequence of detailed computational steps that you will undertake to determine the tumor size and its location. There is no need to specify

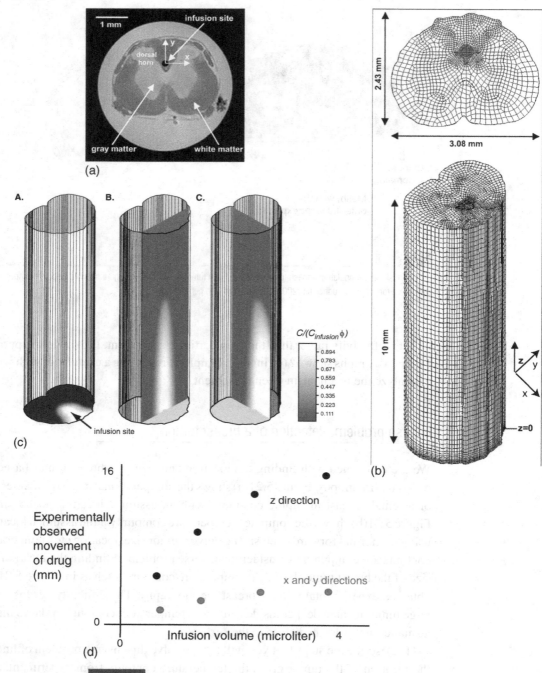

Figure 5.22

(a) 2D NMR image of the rat spinal cord; (b) mesh for the 3D spinal cord; (c) infused albumin concentration (ratio of concentration to maximum concentration) for certain permeability values; (d) experimentally observed movement in z direction (not to scale). From Sarntinoranont et al. (2003), with kind permission from Springer Science + Business Media.

any equation, describe this in words. (2) Discuss the primary difficulty with the procedure you are suggesting and how you might try to resolve it.

5.7.15 Inverse problem: convection-enhanced interstitial diffusion

Convection-enhanced interstitial diffusion can deliver macromolecular drugs to large tissue volumes of the central nervous system, as desired in newly developing approaches for the treatment of cancer, Parkinson's disease and chronic pain therapy. Figure 5.22(a) shows a 2D section of a rat spinal cord and the location where infusion occurs. Figure 5.22(b) shows the full 3D geometry of the same spinal cord with mesh. The governing equation for species concentration includes the diffusion term and the convection term (Darcy flow due to infusion pressure, with velocity given by $u = -k/\mu(\Delta P/\Delta x)$ where k is permeability and μ is viscosity). Note that the diffusivities in all directions can be considered to be the same but the permeabilities are not the same in all directions. Some specific value of infusion pressure led to the velocities in Figure 5.22(c).

(1) How dominant do you think convection is in this problem as compared with diffusion? Provide the reasoning for your answer. (2) There is little data available on the variation of permeability in the two directions (z and x (or y)). The authors had measured data (Figure 5.22(d)) for movement of the drug, as noted by the leading edge of the profile, in both the z and x directions. Using COMSOL as the solver, how would you try to guess the permeabilities in the two directions? (3) When the previous step is completed, your predictions will be close to the experimental values. Explain clearly whether this necessarily means the model is validated using experimental data.

5.7.16 Objective function: radiofrequency ablation

For the radiofrequency ablation problem for heart arrhythmia, described in Section 1.12.11, (1) define an objective function to evaluate the success of the ablation process, making all necessary assumptions; (2) describe qualitatively how this objective function would behave as you increase the voltage (and correspondingly decrease the duration of heating).

II Case studies

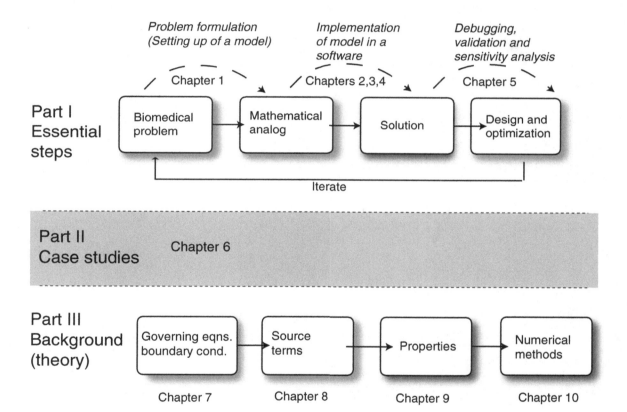

6 Case studies

6.1 Introduction

The case studies are easy-to-follow step-by-step instructions for solving a variety of biomedical transport problems using COMSOL Multiphysics. A number of case studies have been developed which introduce the user to modeling in biomedical engineering using COMSOL. The individual case studies are as follows (they are in no particular order):

> **Box 6.1**
>
> I   Thermal ablation of hepatic tumors
> II Cryosurgery of a wart
> III Drug delivery from a patch
> IV Drug delivery in therapeutic contact lenses
> V   Elimination of nitrogen from the blood stream during deep sea diving
> VI Flow in human carotid artery bifurcation
> VII Radioimmunotherapy of metastatic melanoma
> VIII Burn injury in blood-perfused skin
> IX Radiofrequency cardiac ablation
> X   Laser irradiation of human breast tumor

Table 6.1 lists the modeling features of each case study. The following sequence operates for each case study, as appropriate:

- **Problem description** The biomedical problem to be solved is introduced. A mathematical analog of the physical process is then developed and the geometry, governing equations, such as heat transfer, diffusion etc., to be solved, along with the initial and boundary conditions, are discussed. The parameters, such as density, specific heat capacity, diffusivity, viscosity etc. needed to solve the problem are also listed.

Table 6.1 Case studies.

Case study	Page	Geometry	Mesh	GE, IC, materials properties	BC	Postprocessing
I	186	1D (implemented as 2D with no variation in one direction)	Structured (mapped) mesh	Transient heat transfer; constant properties	Temperature; insulation	Transient temperature at a point; temperature profile at a given time
II	201	2D axisymmetric	Unstructured (free) mesh	Transient heat transfer; thermal conductivity and specific heat are functions of temperature	Convective; insulation; symmetry	Transient temperature at a point; temperature profile at a given time; optimization
III	221	2D axisymmetric	Structured mesh	Flux; symmetry; insulation; concentration; semi-infinite	Transient diffusion; constant properties; Monte Carlo simulation	Transient concentration at a point; concentration profile at a given time
IV	243	2D axisymmetric	Structured mesh	Transient diffusion with sink; constant properties	Symmetry; insulation; semi-infinite	Transient concentration at a point; amount lost/absorbed as a function of time
V	251	2D	Structured mesh	Steady diffusion with convection; fully developed (parabolic) velocity profile	Flux; insulation; concentration; convective flux	Concentration profile; flux at the wall
VI	258	3D	Mixed mesh	Transient Navier–Stokes; time-dependent fully developed inlet velocity profile	Inlet velocity; symmetry; no slip	Velocity plot at different cross-sections

VII	270	Sphere (implemented as 1D using GE in spherical coordinates)	Structured mesh	Three transient diffusion equations with reactions; constant properties	Symmetry; insulation; semi-infinite	Concentration at different times
VIII	280	2D axisymmetric	Structured mesh	Transient bio-heat heat transfer; transient diffusion with source for burn injury; temperature-dependent heat and species source	Convective; temperature; symmetry; semi-infinite; insulation;	Temperature and concentration contour plots at given times
IX	288	2D axisymmetric	Unstructured mesh; adaptive mesh refinement	Transient bio-heat heat transfer; Laplace equation for electric potential; temperature-dependent heat source	Insulation; symmetry; concentration	Transient temperature at a point
X	303	2D axisymmetric	Unstructured and structured meshes	Transient bio-heat heat transfer; moving mesh (radius is a function of time); Transient Navier–Stokes; transient diffusion with convection and source	Convective; insulation; symmetry; mesh velocity; mesh displacement; pressure; no slip; concentration; convective flux	Average temperature as a function of time plot, text export; concentration contour plot at a given time

- **Problem type specification** The user starts the actual modeling in the software. Instructions to set up COMSOL to solve the problem with the required geometry and governing equations are listed.
- **Geometry creation** Step-by-step instructions to create the specified geometry for the problem are included here.
- **Meshing** Describes the method to mesh the geometry.
- **Defining material properties** The material properties for the problem being solved are specified in the software. For example, for a heat transfer problem, density, specific heat capacity and thermal conductivity of the material are specified.
- **Defining initial and boundary conditions** The initial and boundary conditions are specified.
- **Specifying solver parameters and solving** The solver needed for the problem and the appropriate parameters are set up. For example, for a transient problem, the appropriate solver is selected and the time over which the problem needs to be solved is specified. The problem is finally solved in this step.
- **Postprocessing** Postprocessing of the results obtained is performed. Instructions to obtain different types of visualizations, such as line plots, contour plots etc., are discussed in different case studies.

6.2 How to use the case studies

The different case studies have been so selected that they represent a wide variety of problems in biomedical engineering (see Table 6.1). These case studies use a variety of different geometries, governing equations and boundary conditions, and postprocessing approaches. The first three case studies have written instructions along with illustrations in each step required to solve the problem in COMSOL. These three case studies are designed so that by the time users with no prior experience with the software complete them, they will have developed a general understanding of the software and will be comfortable to work with it independently. The subsequent case studies are generally for more complex problems and have detailed written instructions which are easy to follow.

As discussed earlier, each case study has its own unique set of attributes which differ in the preprocessing, processing and/or postprocessing stage from other case studies. Users can use the different implementation details described in the various case studies for implementing different parts of their specific biomedical problem. For example, if someone wants to implement a 3D heat transfer problem and wants to obtain average temperature variation in the domain as a function of time, they can refer to the two case studies (Case studies VI and X) that describe these implementations. For the user to be able to access these resources without going through the entire case study, the case studies are classified in Table 6.1 according to the specific implementations contained within them.

6.3 Additional case studies from the work of students at Cornell University

Students from the "Computer-Aided Engineering: Applications to Biomedical Processes" course at Cornell University have developed a number of case studies in modeling since the course started in 1996. Below is a sample of some of the projects, with the corresponding year noted within parentheses. Please remember that these were class projects, i.e., they are not research projects, and were completed over less than a semester by undergraduates in an engineering curriculum. Thus, gross simplifications have had to be made in order to remain within the limits of time, modeling resource and technical background.

> **Box 6.2**
> Details of all of these modeling projects, including the complete report, are available at http://courses.cit.cornell.edu/bee4530/≫ Past Projects.

Thermal therapy
- Thermal imaging and analysis for breast tumor detection (2007)
- Heating of nanoshells by near-infrared radiation: a rapid and minimally invasive method for destroying tumors (2006)
- Ice therapy on injured muscle (2006, 2004)
- Laser interstitial thermo-therapy in hepatic tissue (2006)
- Heat loss in the carotid artery during selective brain cooling (2006)
- Cryosurgical approach to lung cancer (2004)
- Laser tumor excision (2004)
- Transmyocardial laser revascularization (TMR): an alternate treatment for end stage coronary artery disease (2003)
- Cryopreservation of the kidney (2003)
- Cryopreservation of umbilical cord tissue for stem cell harvesting (2003)
- Radio frequency ablation as treatment for cardiac arrythmia (2003)
- Hyperthermic ablation of hepatic tumors by inductive heating of ferromagnetic alloy implants (2003)
- Radiofrequency ablation to destroy kidney tumors (2003)
- Temperatures in the brain during suspended animation (2003)
- Cryogenic treatment of the common wart (2002)

- The effects of topical heating for therapeutic uses (2002)
- Laser hair removal (2002)
- Radiofrequency ablation for treatment of osteoid osteoma tumors (2002)
- Thermal capsulorraphy for the treatment of acute or chronic shoulder instability (2002)
- Comparison of vaccine freezing methods (2001)
- Ferromagnetic thermal ablation of prostate tumor (2001)
- Heat treatment of an enlarged prostate (2001)
- Cold therapy analysis in structurally damaged tissue (2000)
- Cryogenic freezing of the entire prostate gland (2000)
- Heart cryopreservation (2000)
- Prostate cryosurgery with various numbers of probes (2000)
- Renal tissue preservation: cooling the human kidney for optimal transport conditions (1999)

Thermal comfort
- The effect of the diving/wet suit on the survival time in cold water immersion (2007)
- Tracheal burning from hot air inhalation (2007)
- Testicular thermal damage and infertility from laptop use (2007)
- Magnetic resonance induced heating in a vascular stent (2007)
- The effect of a cooler on heat loss from a horse post-exercise (2006)
- Efficacy of winter sports performance apparel (2006)
- Assessment of flame retardant materials on human skin (2005)
- Heat transfer inside an igloo (2003)
- Heat loss through the head (2002)
- Thermal effects of laser eye correction on the cornea (2002)
- Development of frostbite in the fingers (2002, 2000)
- Heat transfer during firewalking (2002)
- Efficacy of a commercial heating patch (2002)
- Accidental freezing of the tongue to metal poles (2000)
- Thermal burning and its effects on human skin (2000)
- Heating effects of dental drilling (1999)

Drug delivery
- Optimizing release from reservoir microcapsules (2007)
- Oral transmucosal delivery of fentanyl citrate for breakthrough cancer pain relief (2007)
- Patch immunization: transcutaneous vaccination for the cholera toxin and optimization of immunization cycles (2007)
- pH-dependent drug delivery systems (2007)

- Design of autosyringe injection (2006)
- Hormone delivery system: the contraceptive ring (2006)
- An analysis of the Ortho Evra birth control patch (2006)
- Drug-eluting stents: design for prevention of angioplastic restenosis (2006)
- Angina patch: drug delivery for chest pain (2006)
- Chemotherapy: drug diffusion through solid tumor (2005)
- The dermal diffusion of methyl salicylate in over-the-counter pain relief cream (2005)
- Development of a drug delivery system with a constant rate of release (2005)
- Drug delivery mechanism and efficiency of liposomes into skin (2005)
- Effect of drug-eluting stents in coronary arteries (2005)
- Drug delivery through transdermal scopolamine patch (2005)
- Diffusion and binding of radio-labeled antibodies in a tumor (2004)
- Ortho Evra: effect of the patch in women of varying weight (2004)
- Whitening strips: drug delivery in the tooth (2004)
- Role of therapeutic contact lenses in drug delivery (2004)
- Transdermal scopolamine drug delivery systems for motion sickness (2003)
- Optimization of nicotine patch placement (2002)
- Drug delivery in the brain (2001)
- Optimization of chemotherapy for lung cancer (2001)

Transport of blood, oxygen and other materials inside the body
- Laser irradiation of tumors for the treatment of cancer: an analysis of blood flow, temperature and oxygen transport (2007)
- Ex vivo maintenance of heart viability (2006)
- Effects of moisture content on oxygen diffusion through a contact lens (2005)
- Oxygen flow through continuous wear contact lens-cornea system with increasing protein layer build-up (2005)
- Nitrogen elimination in the alveoli (2005)
- Moisture loss during LASIK corrective eye surgery (2004)
- In vitro scaffold construction for a bio-artificial liver (2004)
- Albuterol uptake in bronchioles (2002)
- Delayed effects of hydrofluoric acid burn (2001)
- Airflow during pharyngeal bolus transport (2001)
- Heat and moisture transport in the nasal cavity (2000)
- Atherosclerosis, blood flow, and the carotid artery (1999)
- Angiography (1999)

I Thermal ablation of hepatic tumors

Heated probes can be used to kill tumors in the liver as well as in other parts of the body. Such a procedure is preferable over open surgery because it is minimally invasive. However, thermal ablation must be planned accurately so that the majority of the tumor is heated to 50 °C for destruction, and there is minimum damage to the healthy tissue.

Problem formulation

In this case study, we are interested in finding the time it would take for the probe to heat and kill a tumor that reaches 0.75 cm from the surface of the liver. The first step is to simplify the physical problem. The schematic of the actual system is shown in Figure 6.1(a). We will create a very simplified model for this first case study. We assume that the entire heating probe is in contact with the tumor to be destroyed and the heating is only along the x-direction. The problem can therefore be solved in 1D. However, for better illustration of the process we use a 2D computational domain, as shown in Figure 6.1(b), with no variation in the y-direction.

Governing equations Starting from the general heat transfer equation:

$$\underbrace{\frac{\partial T}{\partial t}}_{\text{transient}} + \underbrace{u\frac{\partial T}{\partial x}}_{\text{convection}} = \underbrace{\frac{k}{\rho C_p}\frac{\partial^2 T}{\partial x^2}}_{\text{conduction}} + \underbrace{\frac{Q}{\rho C_p}}_{\text{source}} \tag{6.1}$$

For our problem, the temperatures are dependent on time; there is no fluid flow and there are no source terms. The only mode of heat transfer is by conduction. So the problem is a transient conduction problem with no convection and heat source. Therefore, the governing equation that needs to be solved is:

$$\frac{\partial T}{\partial t} = \frac{k}{\rho C_p}\frac{\partial^2 T}{\partial x^2} \tag{6.2}$$

Notice here that the governing equation is written in 1D only. There is no variation in the other (y) direction.

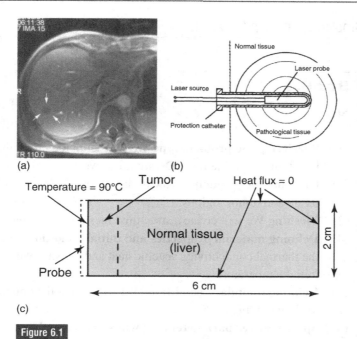

Figure 6.1

(a) A liver tumor (between the arrows); (b) schematic of the process; (c) the computational domain showing the boundary conditions. Note that the probe is not included in the computational domain. Schematic reproduced with kind permission from Springer Science+Business Media: Ma *et al.* (2004).

Boundary conditions The boundary conditions are shown in Figure 6.1(b). To simplify the problem, the probe is assumed to be at a constant temperature of 90 °C placed at the left edge of the tumor. The right edge of the normal tissue is considered to be far away and hence the heat flux is zero. As discussed earlier, there is no temperature variation in the y direction and therefore the boundary condition on the top and bottom surfaces are insulation (or heat flux set to zero).

Input parameters The density, thermal conductivity and specific heat are 1060 kgm^{-3}, 0.512 W (mK)$^{-1}$ and 3600 J (kgK)$^{-1}$, respectively. The properties of the normal tissue and tumor are assumed to be the same.

Reference

Ma, N., X. Gao, and X. X. Zhang. (2004). Two-layer simulation model of laser-induced interstitial thermo-therapy. *Lasers in Medical Science*, **18**(4): 184–189.

Implementation in COMSOL

> **Box 6.3**
>
> **Steps**
>
> (1) **Specifying the problem type** We will setup the problem as a transient heat transfer problem in 2D with no convection and heat source.
> (2) **Creating the geometry** We will draw a rectangle to model the normal tissue and tumor regions.
> (3) **Meshing** We will create a structured mesh in the domain.
> (4) **Defining material properties and initial conditions** We will specify the thermal conductivity, specific heat and density. We will also input the initial temperature.
> (5) **Defining boundary conditions** We will define the boundary conditions as shown in Figure 6.1(c).
> (6) **Specify solver parameters** We will solve the problem for 180 s.
> (7) **Postprocessing** We will plot the temperature history at a point. We will also plot the temperature distribution in the domain after 180s.
> (8) **Save and exit** We will finally save the file and exit.

Step 1: specifying the problem type

> **Box 6.4**
>
> The example in this case is transient heat conduction in a 1D setting. We will first set COMSOL up for solving this type of problem.

I Thermal ablation of hepatic tumors

(1) Start COMSOL by double clicking on the icon on the Desktop.

(2) Select 2D next to Space Dimension (Note: COMSOL can do 1D problems, however to give you a better understanding of COMSOL we will model the problem as 2D).

(3) Click on COMSOL Multiphysics >> Heat Transfer >> Conduction >> Transient Analysis. Transient Analysis under conduction is selected as we intend to solve a time-dependent conduction problem (Equation 6.2).

(4) Click on the Settings tab.

(5) Set the Unit system to SI.

(6) Click OK. COMSOL Window opens up.

(7) Under File, click on Save as ...

(8) Create your folder, specify the file name (e.g., cond.mph) and save it as .mph file. Save often to prevent losing your work.

Step 2: creating the geometry

> **Box 6.5**
>
> The geometry in this case is a 1D slab which is 6 cm wide (along x-axis). Since we are modeling the problem as 2D, we arbitrarily assume the height of the slab to be 2 cm (i.e., along y-axis).

(1) Click on Draw >> Specify Objects >> Rectangle. Rectangle window opens up.

(2) Specify width as 0.06 and height as 0.02. These are the dimensions of the slab in m.

(3) Click on OK.

(4) Click on Zoom Extents to fit the geometry in the window.

The geometry that is created is shown in the figure.

Step 3: meshing

> **Box 6.6**
>
> Meshing is dividing the geometry into small elements. We can mesh the face (2D area) directly. However in this case, we will mesh the edges (sides) first and then the face. This method is used to control the number of elements in certain parts of the geometry, such as the boundaries and interfaces. In many cases we need a finer mesh near the boundary and so we mesh the edges accordingly and obtain a non-uniform mesh. However in this case we will create a uniform mesh.

(1) Under Mesh, click on Mapped Mesh Parameters . . .

(2) Click on the Boundary Tab.

(3) In Boundary Selection, select 1 and 4 by left clicking and holding the Ctrl key. We specify 20 elements each on the left and right edges.

(4) Check the box for Constrained edge element distribution.

(5) Click on Number of edge elements, and type in 20 in the box below.

(6) For boundary 2 and 3, use number of edge elements as 50. We specify 50 elements each on the top and bottom edges.

(7) Press the 'Remesh' Button on the bottom. The mesh that is obtained is shown in the figure.

(8) Click 'OK'.

The mesh obtained is shown (right).

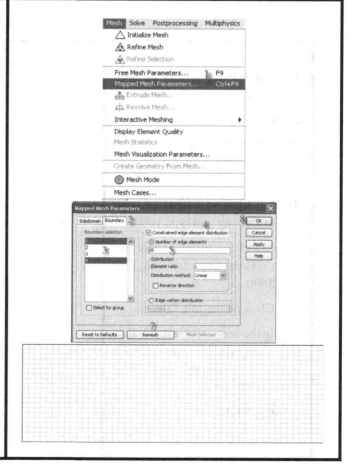

Step 4: defining material properties and initial condition

> **Box 6.7**
>
> We are solving the heat equation and we need to provide the solver with the appropriate material property values required for the analysis. These properties are:
>
> (i) *Density* (ρ): 1060 kgm^{-3}
> (ii) *Thermal conductivity (k)*: 0.512 W (mK)$^{-1}$
> (iii) *Specific heat* (C_p): 3600 J (kgK)$^{-1}$
>
> The tissue and tumor are initially at a temperature of 37 °C. We will also specify this in COMSOL in this step.

(1) Under Physics, click on Subdomain Settings . . .

(2) Click on 1 to select the slab.

(3) Left click on the text field next to Thermal Conductivity and type 0.512.

(4) Left click on the text field next to Density and type 1060.

(5) Left click on the text field next to Heat Capacity and type 3600.

(6) Click on the Init Tab.

(7) In the box under Initial Value, fill in 310. The slab is initially at a temperature of 37 °C (= 310 K).

(8) Click OK.

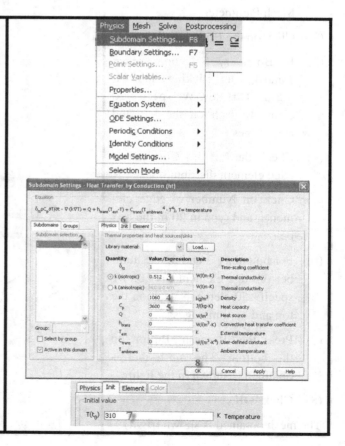

Step 5: defining boundary conditions

> **Box 6.8**
>
> The probe is modeled as a constant temperature at the surface of the tumor (i.e., on the left side of the geometry). Therefore, boundary conditions for the problem are:
> On the left boundary: Temperature = 90 °C
> All other boundaries are insulated. Therefore, heat flux = 0.
> We now specify these boundary conditions to the solver.
> The default boundary condition for the COMSOL solver is thermal insulation so we only need to change the boundary conditions for the left side.

(1) Under Physics, click on Boundary Settings ...

(2) Click on 1 in the Boundary Selection box.

(3) Change the Boundary Condition to Temperature.

(4) Set T_0 to 363. The left boundary of the slab is at 363 K (= 90 °C).

(5) Click on OK.

Step 6: specifying solver parameters

> **Box 6.9**
> We now specify the time interval for which the time-dependent problem will be solved. We choose a thermal ablation time of three minutes (180 seconds) to find out if the tumor is sufficiently heated so as to be destroyed in that time.

(1) Under Solve, click on Solver Parameters.

(2) Under the General tab, select Transient under Analysis if it is not already selected.

(3) Select Time dependent under Solver.

(4) In the Times: box, type in 0:1:180. This tells the solver to start at 0 seconds, then save the solution every second until it reaches 180 seconds.

(5) Click OK.

(6) Click on Get Initial Value under Solve. This step initializes the solver with the value provided when the initial conditions were specified (Step 4).

(7) Under Solve, click on Solver Manager.

(8) Click on the Solve For tab.

(9) Select T for temperature if it is not already selected. By selecting T, we are directing the solver to solve for the temperature.

(10) Click on the Output tab.

(11) Select T for temperature if it is not already selected. By selecting T, we are directing the program to save the temperature values.

(12) Press Solve. Once Solve is pressed, the solver solves the transient heat transfer equation (Equation 6.2). It will take approximately 1–2 min to solve. We are now ready to do postprocessing (analyzing the results).

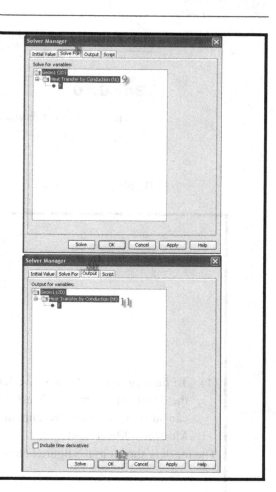

Step 7: postprocessing

> **Box 6.10**
> Postprocessing is the visualization and manipulation of the results obtained from the simulation.

Displaying the mesh

(1) To display the mesh, simply click on the Mesh Mode button. Mesh mode can also be selected by clicking on Mesh >> Mesh Mode.
The mesh is shown in the figure below.

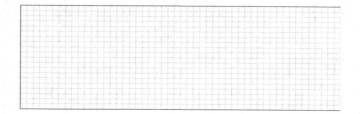

Plotting temperature versus time at a particular coordinate

To determine if the tumor was sufficiently heated within three minutes, we plot the temperature history at point (0.0075, 0.01) at the end of the tumor to see how the temperature varies with time at that location.

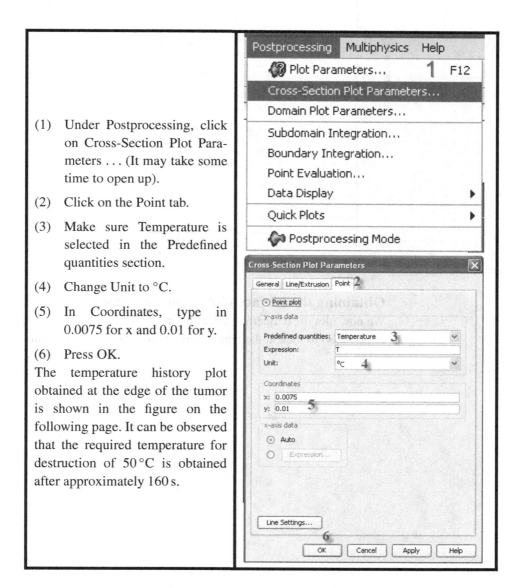

(1) Under Postprocessing, click on Cross-Section Plot Parameters ... (It may take some time to open up).

(2) Click on the Point tab.

(3) Make sure Temperature is selected in the Predefined quantities section.

(4) Change Unit to °C.

(5) In Coordinates, type in 0.0075 for x and 0.01 for y.

(6) Press OK.

The temperature history plot obtained at the edge of the tumor is shown in the figure on the following page. It can be observed that the required temperature for destruction of 50 °C is obtained after approximately 160 s.

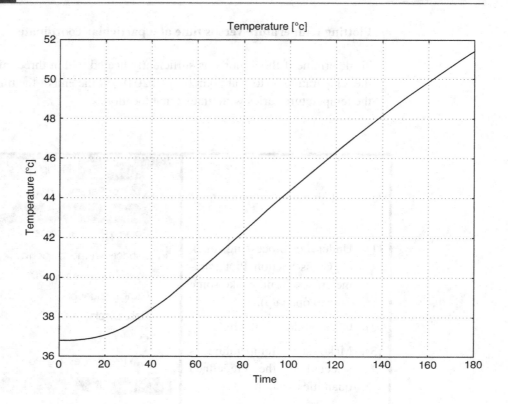

Obtaining the surface plot at the last time step

We now plot the temperature contour in the tumor and tissue at the end time (i.e., after 180 s) to look at the temperature distribution in the domain.

(1) Under Postprocessing click on Plot Parameters.

(2) Click on the General tab.

(3) Check the box for Surface under Plot Type.

(4) Next to Solution at time: select 180.

(5) Click on the Surface tab.

(6) Select Temperature next to Predefined quantities, if it is not already selected.

(7) Change Unit to °C.

(8) Click on OK.

The surface plot is shown on the next page. The vertical line inside the domain shows the extent of the tumor. It can be observed that the temperatures on the right side of the line are below 50 °C and we can therefore say that there is no significant damage to the normal tissue.

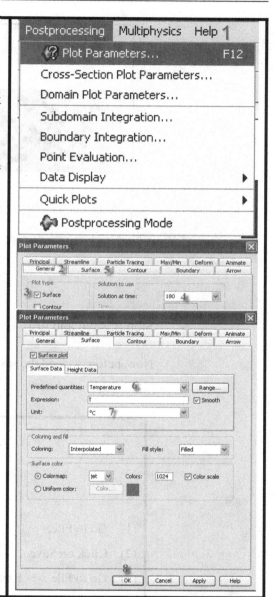

Case studies

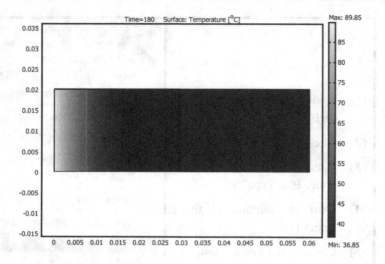

Step 8: save and exit

Now, before we end the session we need to save the files for future use.

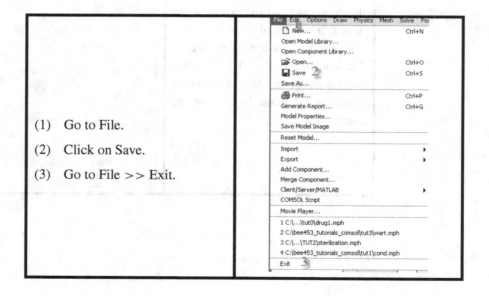

(1) Go to File.
(2) Click on Save.
(3) Go to File >> Exit.

II Cryosurgery of a wart

This case study analyzes the process of cryogenic wart treatment by optimizing the temperature and the duration of liquid jet that is applied to the surface of common warts (Cuneo *et al.*, 2002). The goal is to destroy as much of the wart as possible while damaging as little healthy skin as possible.

Problem formulation

The wart as pictured in Figure 6.2(a) is simplified as a hemispherical protrusion from a flat skin surface. We assume homogenous properties and perfectly symmetrical geometry of the wart and skin. Since we want to destroy only the wart and very little of the surrounding tissue, if any, we can restrict our computational region in the normal tissue to be a symmetric region around the wart. The geometry can then be considered as axisymmetric with the axis as shown in Figure 6.2(b) and therefore, can be modeled in two dimensions so that the wart becomes a quarter-circle attached to a flat skin slab. The geometry can be swept 360 degrees around the axis to obtain the three-dimensional representation (Figure 6.2(a)).

Governing equations The problem involves heat transfer only and the mode of heat transfer is conduction. There is no heat transfer by convection inside the normal tissue and wart system and no heat source term. Therefore the governing equation that needs to be solved is

$$\frac{\partial T}{\partial t} = \frac{1}{\rho C_p}\left[\frac{1}{r}\frac{\partial}{\partial r}\left(kr\frac{\partial T}{\partial r}\right) + \frac{\partial}{\partial z}\left(k\frac{\partial T}{\partial z}\right)\right] \quad (6.3)$$

Notice here that since the problem is axisymmetric, the partial space derivative terms are written with respect to variables r and z instead of x and y.

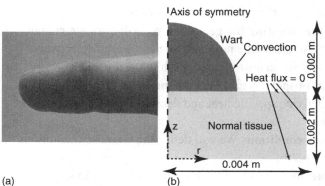

Figure 6.2

(a) Two common warts on the surface of the skin on a finger; (b) the computational domain showing subdomains (wart, normal tissue) and boundary conditions. Wart figure from http://en.wikipedia.org/wiki/Image:Wart_ASA_animated.gif

Boundary conditions The liquid jet is applied to the surface of the wart at $-196\,°C$ with a heat transfer coefficient of $5000\,\text{W m}^{-2}\text{K}^{-1}$. The application of the liquid jet is included as a convective boundary condition, as shown in Figure 6.2(b). The top surface of the normal tissue adjacent to the wart is insulated to the surrounding air. Additionally, the bottom and right edges of the normal tissue are assumed to be at a considerable distance so that there are no variations in temperature in those areas. Hence, a zero heat flux boundary condition is applied on the right and bottom edges (Figure 6.2(b)).

Input parameters Input parameters needed for the model, which include thermal conductivity and specific heat capacity, are shown in Figure 6.3. The rapid change in thermal properties in the temperature range 0 to $-5\,°C$ (as seen in Figure 6.3) is due to freezing of water present in the tissues. The densities of the normal tissue and wart were $1000\,\text{kgm}^{-3}$ and $1500\,\text{kgm}^{-3}$, respectively.

Reference

Cuneo, K., LeBarron, J., Reynolds, J., Tiberio, C. and Yoo, S. (2002). Cryogenic Treatment of the Common Wart On the web at http://ecommons.library.cornell.edu/handle/1813/212

Implementation in COMSOL

> **Box 6.11 Steps**
>
> (1) **Specifying the problem type** We will setup the problem as a heat transfer problem with no convection and heat source.
> (2) **Setting the grid and creating the geometry** We will draw a rectangle and a quarter-circle to model the normal tissue and wart, respectively.
> (3) **Meshing** We will create an unstructured mesh in the domain.
> (4) **Defining material properties and initial conditions** We will specify the thermal conductivity, specific heat and density. We will also input the initial temperature.
> (5) **Defining boundary conditions** We will define the boundary conditions as shown in Figure 6.2(b).
> (6) **Specify solver parameters** We will solve the problem for 15 s.

II Cryosurgery of a wart

> (7) **Postprocessing** We will plot the temperature history at point. We will also plot the temperature distribution in the domain after 15s.
> (8) **Save and exit** We will finally save the file and exit.
>
> **Steps – optimization**
>
> (1) **Open file** We will open the file created in the last step.
> (2) **Define objective function** We will define the objective function for the problem using integration coupling variables.
> (3) **Solve** We will solve the problem again to evaluate the objective function.
> (4) **Plot objective functions versus time** We will plot the objective function as a function of time to determine the time at which it is a minimum.

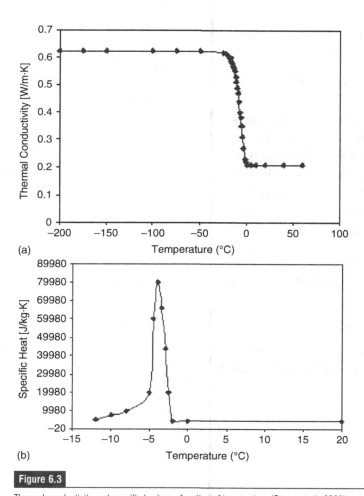

Figure 6.3

Thermal conductivity and specific heat as a function of temperature (Cuneo *et al.*, 2002).

Step 1: specifying the problem type

> **Box 6.12**
>
> The only mode of heat transfer is by conduction. Therefore the problem is a transient conduction problem with no convection and heat source.

(1) Start COMSOL by double clicking on the icon on the Desktop.

(2) Select Axial Symmetry (2D) next to Space Dimension.

(3) Click on COMSOL Multiphysics >> Heat Transfer >> Conduction >> Transient Analysis. Transient Analysis under conduction is selected as we intend to solve a time-dependent conduction problem (Equation 6.3).

(4) Click on the Settings tab.

(5) Set the Unit system to SI.

(6) Click OK. COMSOL Window opens up.

(7) Under File, click on Save as…

(8) Create a folder, specify the file name (e.g., wart.mph) and save it as an .mph file.

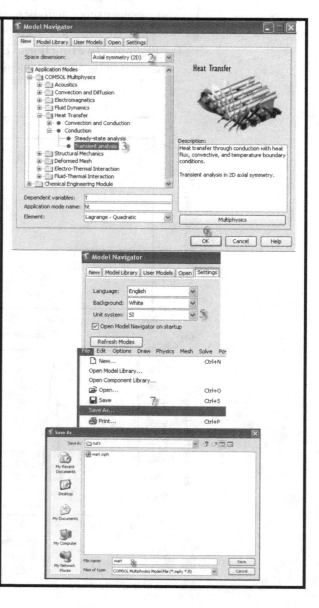

Step 2: setting the grid and creating the geometry

> **Box 6.13**
> The geometry in this case is a rectangle with a quarter-circle and is axisymmetric. The rectangle and the quarter-circle represent the skin and the wart, respectively. We will first draw the rectangle and then the quarter-circle.

Part A: rectangle

(1) Click on Draw >> Specify Objects >> Rectangle. Rectangle window opens up.

(2) Specify width as 0.004 and height as 0.002. These are the dimensions of the skin in m. Only half of the skin and wart is considered as it is symmetric.

(3) Click on OK.

(4) Click on Zoom Extents to fit the geometry in the window.

The rectangle is shown in the figure.

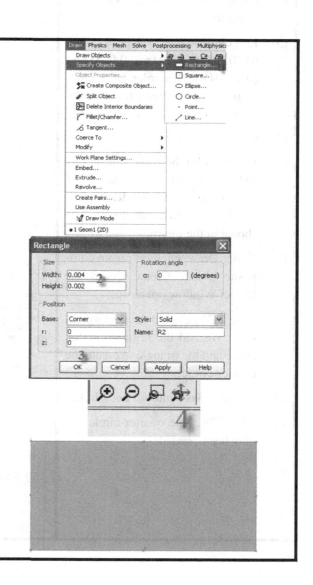

Part B: quarter-circle

(1) Click on Zoom Out to minimize the view.

(2) Click on the second degree Bezier curve tool.

(3) Left click on the coordinate points (0.002, 0.002), (0.002, 0.004) and (0, 0.004), in that order. Right click on (0, 0.004) to create the segment of the circle shown in the figure.

(4) Click on the line tool.

(5) Left click on the coordinate points (0.002, 0.002), (0, 0.004) and (0, 0.002), in that order. Right click on (0, 0.002) to create the triangle shown in the figure.

(6) Under Draw, click on create composite object. Create Composite Object Window opens up.

(7) Type CO1+CO2 in the set formula field. We add the segment of the circle and the triangle to create the quarter-circle.

(8) Uncheck Keep interior boundaries.

(9) Click OK. The quarter-circle has now been created.

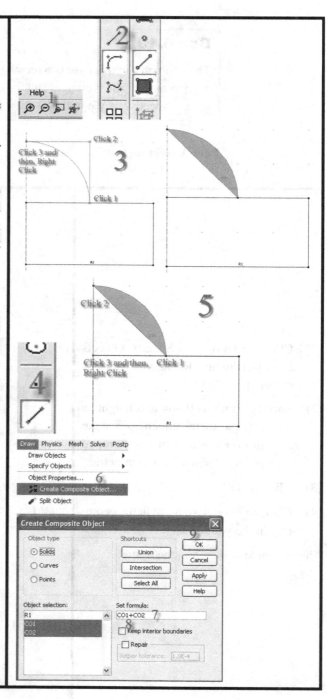

Step 3: meshing

> **Box 6.14**
> Meshing is dividing the geometry into small elements. We will create meshes with the same mesh density for both the skin and the wart faces.

(1) Under Mesh (on the top toolbar), click on Free Mesh Parameters.

(2) Select Subdomain 1 representing the skin and Subdomain 2 representing the wart, by left clicking and holding the Ctrl key.

(3) Next to Maximum element size, type in 0.0001.

(4) Click on Remesh.

(5) Then click on OK.

Your mesh should now look like the figure below.

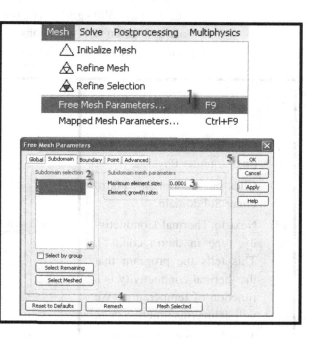

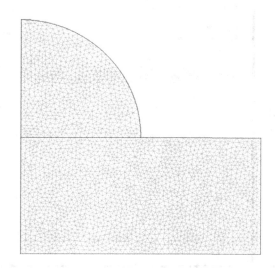

Step 4: defining material properties and initial conditions

> **Box 6.15**
>
> We are solving for the heat transfer equation. The material properties required for the analysis are, therefore, thermal conductivity, specific heat and density. The density of the normal tissue layer is assumed to be 1000 kgm^{-3} and that of the wart as 1500 kgm^{-3}. The thermal conductivity and specific heat constant are both taken to be functions of temperature, as specified in Figure 6.3. The temperature inside the skin and the wart is 37 °C (= 310 K) initially.

(1) Click on Subdomain Settings ...under Physics.

(2) Click on Subdomain 1 (skin) under Subdomain.

(3) Next to Thermal Conductivity, type in therm_cond(T). This tells the program that the thermal conductivity is a function of temperature. We will define this function later.

(4) Type 1000 in the density field.

(5) Next to Heat Capacity, type in heat_cap(T).

(6) Click on the Init tab and under Initial Value, fill in 310.

(7) Select 2 (for wart) under Subdomain in the Physics tab and repeat steps 3–6 using 1500 in the density field and the same values in the other fields.

(8) Click OK.

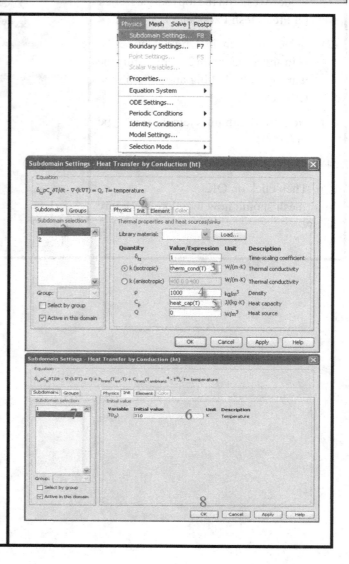

We will now define the functions, therm_cond(T) and heat_cap(T).

(1) Under Options, click on Functions...

(2) Click on the New...button.

(3) Next to Function name: type in therm_cond.

(4) Check Interpolation. By checking this, we are directing the solver to interpolate values of thermal conductivity between different temperatures.

(5) Select Table next to the Use data from box.

(6) Click OK.

(7) Fill in the values as shown. These approximate values are obtained from the graph shown in Figure 6.3(a). The first column (under x) shows the values of temperature in K and f(x) represents the corresponding thermal conductivity values.

(8) Click on the New...button. This is done to define the function, heat_cap.

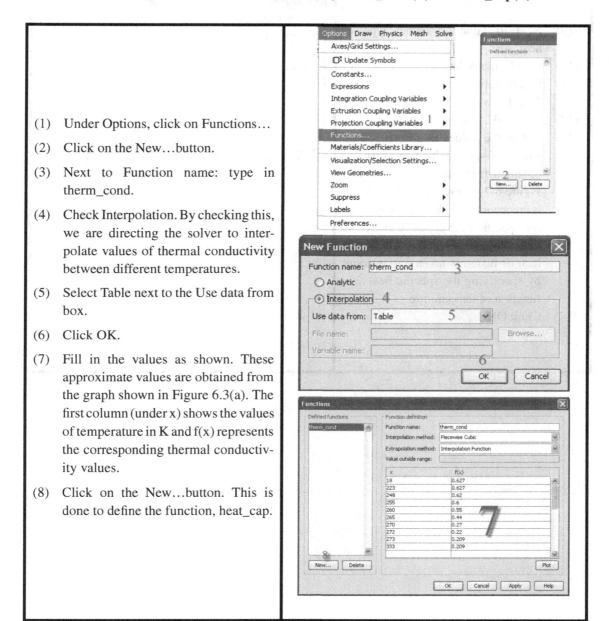

(9) Next to Function name: type in heat_cap.
(10) Check Interpolation.
(11) Select Table next to the Use data from box.
(12) Click OK.
(13) Fill in the values as shown. These values of specific heat capacity are obtained from the graph shown in Figure 6.3(b). Freezing of the tissue is incorporated in the model by using the specific heat capacity values as given by Figure 6.3(b). The latent heat of fusion is taken into account by specifying the specific heat as a function of temperature.
(14) Click OK.

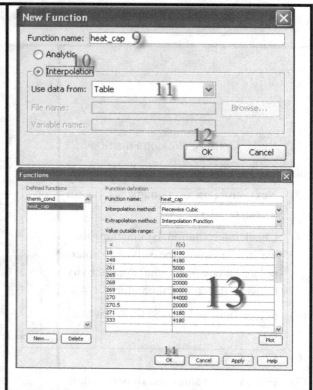

Step 5: defining boundary conditions

> **Box 6.16**
> The boundary conditions for the problem are shown in Figure 6.2(b). The curved surface of the wart has a convective boundary condition and all other boundaries have zero heat flux condition (insulated boundaries). The default boundary condition in COMSOL is insulation, so we need to specify boundary conditions for only the curved surface.

(1) Under Physics, select Boundary Settings...

(2) Select boundary 7. This is the wart's curved surface.

(3) Select Heat flux under boundary condition.

(4) Next to Heat transfer coefficient, input 5000.

(5) Next to T_{inf} type 77.

(6) Click OK.

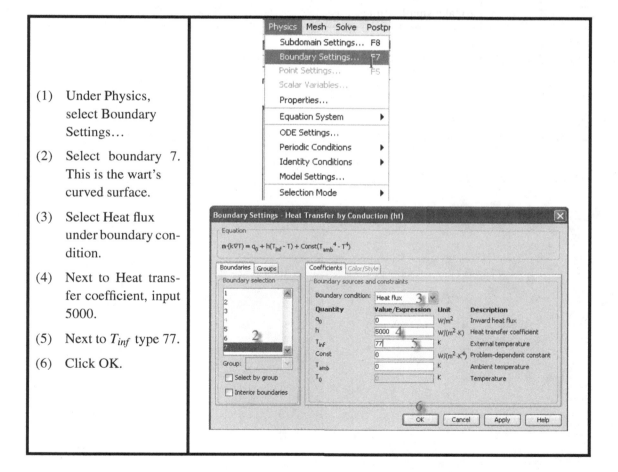

Step 6: specify solver parameters

> **Box 6.17**
>
> We need to specify the time interval of the process as well as the times at which the solution will be stored by the solver. We will solve the problem for 15 s. As you can see from Figure 6.3, the thermal properties change very rapidly near 0 °C. Therefore to obtain an accurate solution, the time steps taken by the solver should be small so that the temperature dependence of thermal conductivity and specific heat is resolved precisely. By default, the solver calculates the time step size internally. In this example, we will force it to take smaller time steps.

(1) Under Solve, click on Solver Parameters.

(2) Under the General tab, select Transient under Analysis if it is not already selected.

(3) Select Time dependent under Solver.

(4) In the Times: box, type in 0:0.1:15. This instructs the solver to start at 0 seconds and save the solution every 0.1 seconds until it reaches 15 seconds.

(5) Click on the Time Stepping tab. We will now set the solver to use a maximum time step size of 0.001.

(6) Check Manual tuning of step size.

(7) Enter Initial time step and Maximum time step both as 0.001.

(8) Click OK.

(9) Press Solve Problem under the solve menu. Once Solve is pressed, the solver solves the transient heat transfer equation (Equation 6.3). It will take approximately 30–35 min to solve. After that we are ready to postprocess the results.

Step 7: postprocessing

Plotting temperature versus time at wart interior

We will now plot the temperature history at point (0.0015, 0.003) in the interior of the wart, to see how the temperature varies with time at that location.

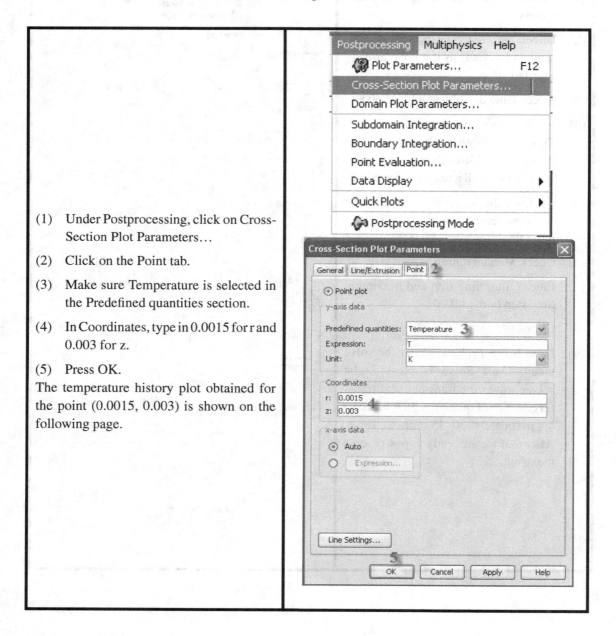

(1) Under Postprocessing, click on Cross-Section Plot Parameters...

(2) Click on the Point tab.

(3) Make sure Temperature is selected in the Predefined quantities section.

(4) In Coordinates, type in 0.0015 for r and 0.003 for z.

(5) Press OK.

The temperature history plot obtained for the point (0.0015, 0.003) is shown on the following page.

II Cryosurgery of a wart

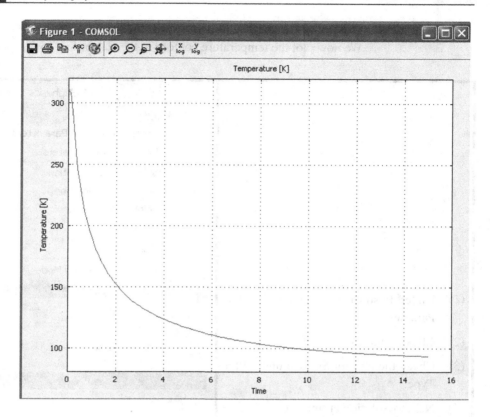

Obtaining the surface plot at a specific time

We now plot the temperature contour in the slab at $t = 15$ s.

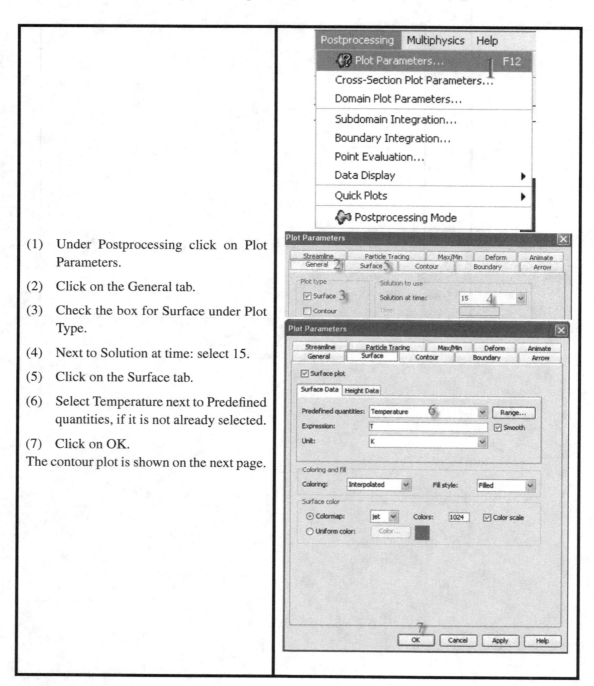

(1) Under Postprocessing click on Plot Parameters.

(2) Click on the General tab.

(3) Check the box for Surface under Plot Type.

(4) Next to Solution at time: select 15.

(5) Click on the Surface tab.

(6) Select Temperature next to Predefined quantities, if it is not already selected.

(7) Click on OK.

The contour plot is shown on the next page.

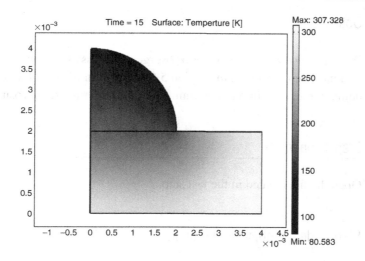

Remember, the aim was to obtain very low temperatures inside the wart to destroy it, and at the same time we did not want to damage the healthy cells inside the skin due to low temperatures (below −40 °C). It can be confirmed from the figure that after 15 s of application, temperatures in the wart are very low thereby destroying it. However, some of the normal tissue is also destroyed as temperatures go below −40 °C.

Step 8: save and exit

Now, before we end the session we need to save the files for future use.

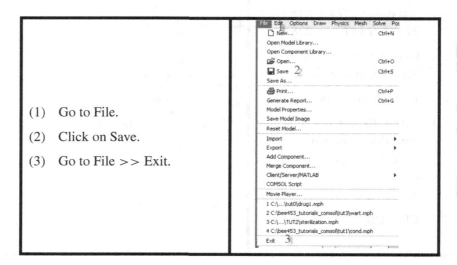

(1) Go to File.

(2) Click on Save.

(3) Go to File >> Exit.

Optimization

We will now try to optimize the cryosurgery process using computations. Optimization has been discussed in Section 5.5. We will implement the objective function defined in that section to determine the optimum time for cryosurgery.

Step 1: open file

Open the file created in the last step.

Step 2: define objective function

(1) Click on Options >> Integration Coupling Variables >> Subdomain Variables.

(2) Click on subdomain 1 under Subdomain selection.

(3) Under Name, type J1.

(4) Under Expression, type the objective function for the normal tissue, $(278 - T) * (T < 278) + 0 * (T >= 278)$.

(5) Click on subdomain 2 under Subdomain selection.

(6) Under Name, type J2.

(7) Under Expression, type the objective function for the wart, $(T - 233) * (T > 233) + 0 * (T >= 230) * (T <= 233) + (230 - T) * (T < 230)$.

(8) Click OK.

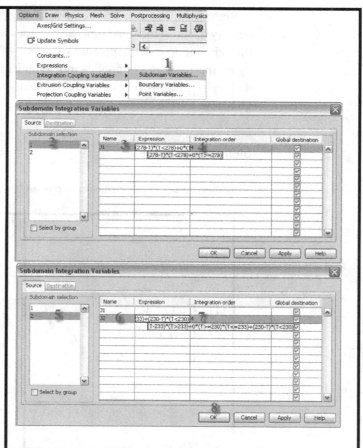

Step 3: solve

Repeat Step 6 from the last section to solve again.

Step 4: plot objective function versus time

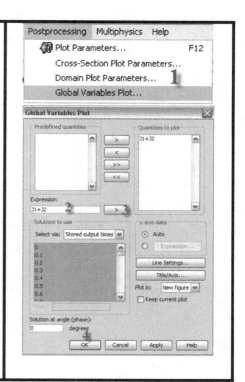

(1) Under Postprocessing, click on Global Variables Plot...

(2) Under Expression, type J1+J2.

(3) Click on > next to Expression so that J1+J2 displays under Quantities to plot.

(4) Click OK.

The objective function obtained as a function of time is shown on the following page. It can be observed that the objective function is minimum close to 2s.

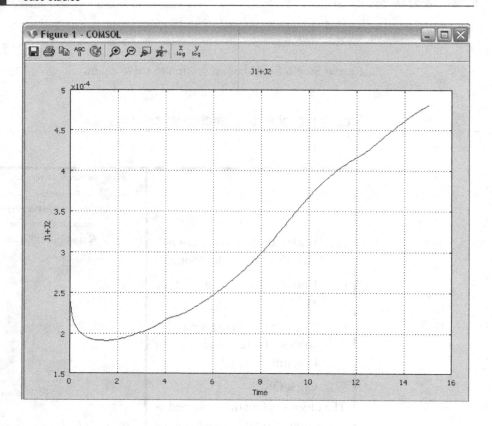

III Drug delivery from a patch

Here is another case study to demonstrate how to setup up a problem in COMSOL and solve it. The example deals with the analysis of a birth control patch (Fields *et al.*, 2006). The contraceptive patch is an effective method for birth control. The patch can be placed on multiple parts of the skin and the drugs are then transferred through the tissue by diffusion and finally to the blood stream. The blood stream in turn transports these drugs to various parts of the body. Norelgestromin and ethinyl estradiol are the two drugs used in the patch. In this example, we model the movement of norelgestromin through the tissue.

Problem formulation

Although the actual shape of the patch is not circular (Fig. 6.4(a)), we model the patch as circular to simplify the problem (Fig. 6.4(b)). Once the patch is considered circular, the problem can be solved as an axisymmetric one and does not need to be solved in 3D (Fig. 6.4(c)). Also in this case study we are primarily interested in looking at how the drug diffuses in the tissue layer over time, and not the characteristics of the patch material. Hence to further simplify the problem, the amount of drug supplied to the skin surface by the patch is assumed to be given by a constant flux (8.849×10^{-7} g/m²s). The patch radius is 24 mm. We model 6 mm of the skin around it to see how the drug spreads out from the edge of the patch. The thickness of the tissue is taken as 1.2 mm and the drug diffuses out into the blood stream beyond the tissue layer at the bottom. It is assumed that the entire drug reaching the edge of the tissue is taken away by the blood stream.

Governing equations Drug delivery in the tissue is only due to diffusion and so the governing equation for the problem is:

$$\frac{\partial c}{\partial t} = D \left[\frac{1}{r} \frac{\partial}{\partial r} \left(r \frac{\partial c}{\partial r} \right) + \frac{\partial^2 c}{\partial z^2} \right] \tag{6.4}$$

where c is the concentration of the drug and D is its diffusivity.

Boundary conditions The patch provides a constant drug flux of 8.849×10^{-7} g/m²s on the top left boundary representing the patch. On the bottom boundary, as discussed above, it is assumed that the blood carries the drug completely and so the concentration of the drug is zero. The top right boundary representing the skin layer does not allow loss of drug (is impermeable).

222 Case studies

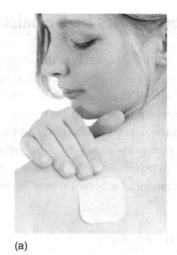

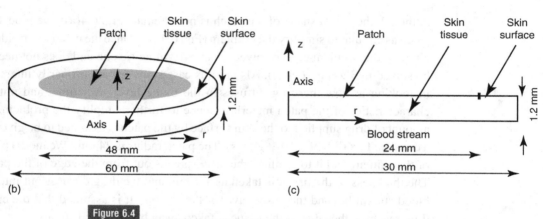

Figure 6.4

(a) The actual situation, (b) three-dimensional view of the skin-patch system, (c) axisymmetric geometry used for the computations.

Input parameters The diffusivity of Norelgestromin in the tissue is 1.11×10^{-11} m^2/s.

Reference

Fields, R., Fisher, E., Kramer, S., Kwan, E. and Wong, A. (2006). An Analysis of the Ortho Evra Birth Control Patch. On the web at http://ecommons.library.cornell.edu/handle/1813/3058

Implementation in COMSOL

We will first solve the problem in COMSOL. We will then perform Monte Carlo simulations (discussed in Section 5.4.1) to study the effect of change in the diffusivity of the drug on the amount of drug delivered to the blood stream.

> **Box 6.18 Steps**
>
> (1) **Specifying the problem type** We will setup the problem as a transient diffusion problem.
> (2) **Setting the grid and creating the geometry** We will draw a rectangle to represent the tissue and a line to model the patch.
> (3) **Meshing** We will create a structured mesh in the domain.
> (4) **Defining material properties and initial conditions** We will specify the diffusivity of the drug in the tissue. We will also enter the initial concentration of drug in the tissue as zero.
> (5) **Defining boundary conditions** We will supply the flux boundary condition for the patch and zero concentration boundary condition for the bottom edge of the tissue.
> (6) **Specify solver parameters** We will solve the problem for 1 week.
> (7) **Postprocessing** We will plot the drug concentration history as a function of time at point. We will also plot the distribution of the drug in the domain after 3 days.
> (8) **Save and exit** We will finally save the file and exit.
>
> **Steps – Monte Carlo simulation**
>
> (1) **Open file** We will open the file created in the last step.
> (2) **Save as .m file** We will then save the COMSOL file, which is in .mph format originally, as a .m file.
> (3) **Start editor and open .m file** We will open a text editor to edit the .m file.
> (4) **Modify the .m file in COMSOL script editor** We will add commands in the .m file.
> (5) **Run the .m file to perform Monte Carlo simulations** We will finally run the file.

Case studies

Step 1: specifying the problem type

> **Box 6.19**
>
> The problem in this case is transient diffusion in a 2D axisymmetric setting. We will first set COMSOL up for this type of problem.

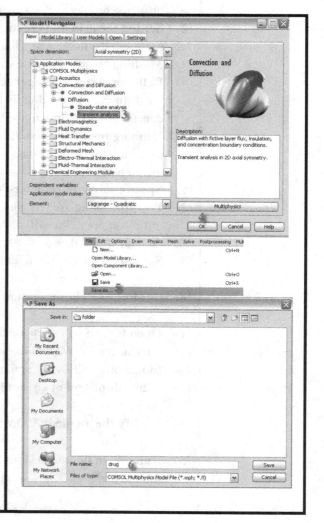

(1) Start COMSOL by double clicking on the icon on the Desktop.

(2) Select Axial Symmetry (2D) next to Space Dimension.

(3) Click on COMSOL Multiphysics >> Convection and Diffusion >> Diffusion >> Transient Analysis. Transient Analysis under diffusion is selected as we intend to solve a time-dependent diffusion problem (Equation 1).

(4) Click OK. COMSOL Window opens up.

(5) In the COMSOL window under File, click on Save as...

(6) Create an appropriate folder and save your work there. Specify the file name (e.g., drug.mph) and save it as a .mph file.

Step 2: creating the geometry

(1) Click on Draw >> Specify Objects >> Rectangle. Rectangle window opens up.

(2) Specify width as 0.03 and height as 0.0012. These are the dimensions of the tissue (in m) over which the patch is placed.

(3) Click on OK.

(4) Click on Zoom Extents to fit the geometry in the window.

(5) Click on Draw >> Specify Objects >> Line. Line window opens up.

(6) Specify r as *0 0.024* and z as *0.0012 0.0012*. This line with endpoints (0, 0.0012) and (0.024, 0.0012) represents the patch.

(7) Click on OK.

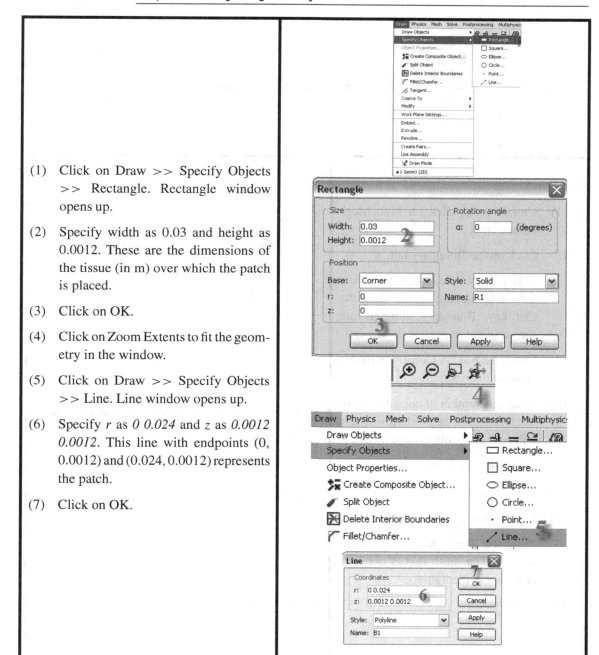

Step 3: meshing

> **Box 6.20**
>
> We will mesh the sides first and create a uniform mapped mesh.

(1) Under Mesh, click on Map Mesh Parameters...

(2) Click on the Boundary tab.

(3) In Boundary Selection, select 1 and 5 by left clicking and holding the Ctrl key. Boundary numbers are shown in the figure.

(4) Check the box for Constrained edge element distribution.

(5) Click on Number of edge elements, and type in 5 in the box below. There will be 5 elements in edges 1 and 5.

(6) For Boundary 2 use 125 elements.

(7) For boundary 3 use 100 elements.

(8) For Boundary 4 use 25 elements.

(9) Press the 'Remesh' Button on the bottom. The mesh that is obtained is shown in the figure.

(10) Click 'OK'.

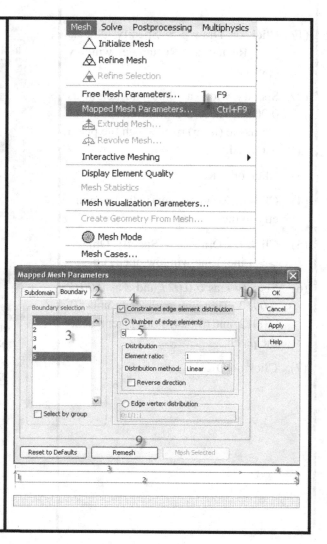

Step 4: defining material properties and initial conditions

> **Box 6.21**
> We are solving a transient diffusion equation for drug delivery and so we need to provide the diffusivity of the drug. The diffusivity of the drug in the tissue is 1.11×10^{-11} m²/s. Concentration of the drug in the tissue is zero initially.

(1) Under Physics, click on Subdomain Settings...

(2) Click on 1 to select the domain. The domain represents the tissue patch system.

(3) Enter D isotropic as $1.11\mathrm{E}{-}11$.

(4) Click on the Init tab.

(5) Retain the default value of 0 under Initial value. Concentration of the drug in the tissue is zero initially.

(6) Click on OK.

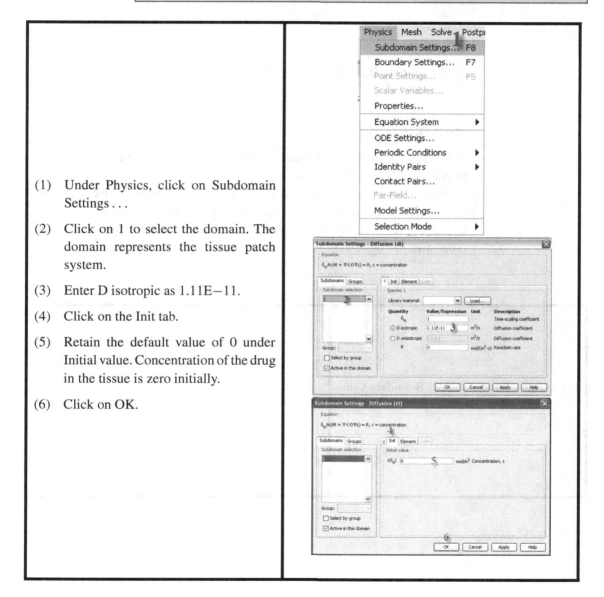

Case studies

```
                         Patch                Top skin
                Flux = 8.849 × 10⁻⁷ g/m²s     Flux = 0
     Axis     ┌─────────────────────────────┐  Right skin
     Flux = 0 │                             │  Flux = 0
              └─────────────────────────────┘
                       Bottom skin
                    Concentration = 0
```

Figure 6.5

Boundary conditions for the problem.

Step 5: defining boundary conditions

> **Box 6.22**
>
> In this step, we define the boundary conditions for the problem. The patch provides a constant flux of 8.849×10^{-7} g/m²s on the top left boundary. On the bottom boundary, it is assumed that the blood carries the drug completely and so the concentration of the drug is zero. The other boundaries are insulated and so the flux is zero. All the boundary conditions for the problem are shown in Figure 6.5.

(1) Under Physics, select Boundary Conditions...	Physics menu showing: Subdomain Settings... F8, **Boundary Settings... F7**, Point Settings..., Scalar Variables..., Properties..., Equation System ▶, ODE Settings..., Periodic Conditions ▶, Identity Pairs ▶, Model Settings..., Selection Mode ▶

(2) Click on Boundary 3 to select the boundary representing the patch as shown in the figure.

(3) Set the Boundary Condition to Flux.

(4) Next to Inward Flux, type in 8.849E−7. Note that the unit of flux for the problem is in g/m^{2}s. COMSOL displays the units in the SI system. However, we will keep track of the units throughout the problem.

(5) Click on Boundary 2. Boundary 2 is the bottom line in the geometry.

(6) Set the Boundary Condition to Concentration.

(7) Next to Concentration, type in 0.

(8) Click OK.

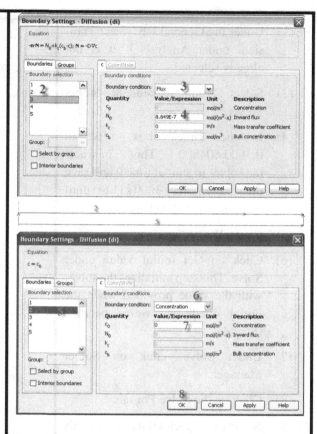

Step 6: specify solver parameters

(1) Under Solve, click on Solver Parameters.

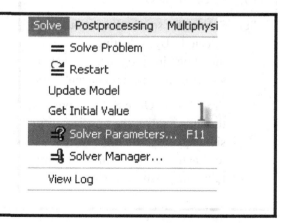

Case studies

(2) Under the General tab, select Transient under Analysis if it is not already selected.

(3) Select Time dependent under Solver.

(4) In the Times: box, type in 0:3600:604800. This instructs the solver to start at time 0 and save the solution every 3600 s (1 hr) until it reaches 604 800 (1 week).

(5) Click OK.

(6) Click on Get Initial Value under Solve. This step initializes the solver with the value provided when the initial conditions were specified (Step 4).

(7) Under Solve, click on Solver Manager.

(8) Click on the Solve For tab.

(9) Select c for species if it is not already selected. By selecting c, we are directing the solver to solve for the species concentration.

(10) Click on the Output tab.

(11) Select c for species if it is not already selected. By selecting c, we are directing the program to save the values of the species concentration.

(12) Press Solve. Once Solve is pressed, the solver solves the transient diffusion equation. We are now ready to look at the results, or in other words, to do the postprocessing.

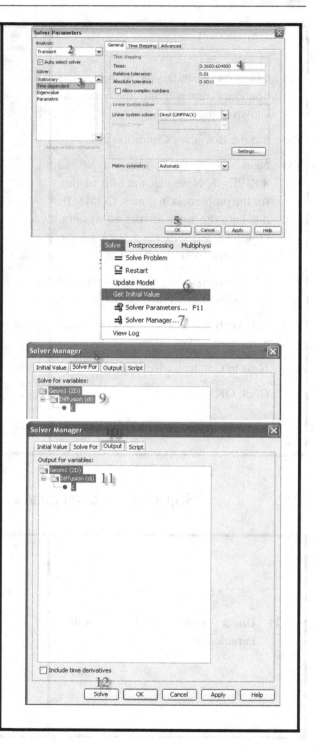

Step 7: postprocessing

Plotting drug concentration versus time at a particular coordinate

We now plot the drug concentration history at a coordinate point (0.0144, 0.0006) in the interior of the tissue.

(1) Under Postprocessing, click on Cross-Section Plot Parameters...
(2) Click on the Point tab.
(3) Make sure Concentration is selected in the Predefined quantities section.
(4) Under Coordinates, enter r as 14.4E–3 and z as 0.6E–3.
(5) Press OK.

The plot is shown on the next page.

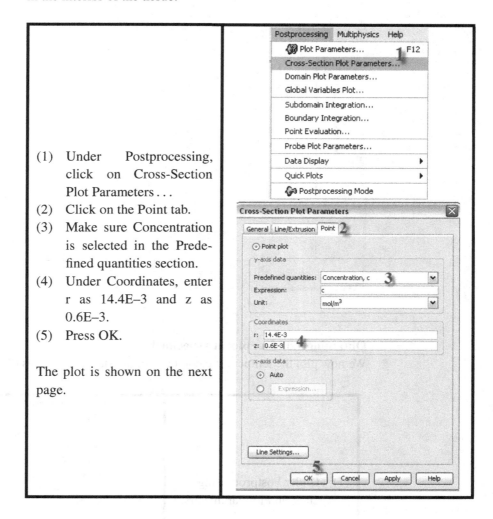

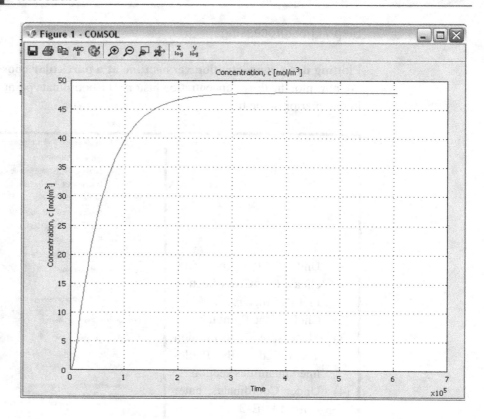

Obtaining a surface plot at a specified time

We now plot the temperature contours in the slab after 3 days (259 200 s).

(1) Under Postprocessing click on Plot Parameters.

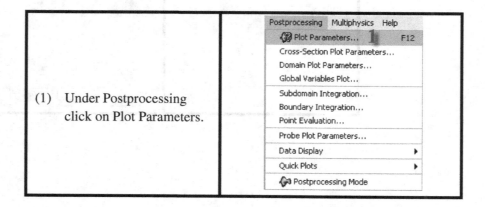

III Drug delivery from a patch

(2) Click on the General tab.

(3) Check the box for Surface under Plot Type.

(4) Next to Solution at time: select 2.592e5.

(5) Click on the Surface tab.

(6) Select Concentration next to Predefined quantities, if it is not already selected.

(7) Click on OK.

The surface plot that is obtained is shown below.

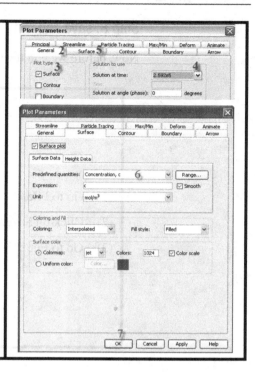

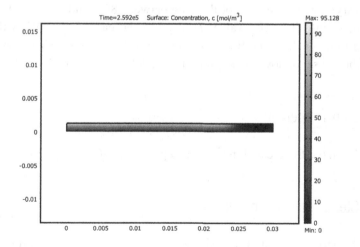

Step 8: save and exit

Now, before we end the session we need to save the files for future use.

(1) Go to File.

(2) Click on Save.

(3) Go to File >> Exit.

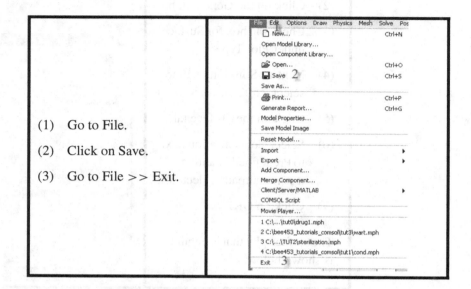

Monte Carlo Simulation

We will now perform Monte Carlo simulation (discussed in Section 5.4.1) by varying the diffusivity of the drug between 80% and 120% of 1.11×10^{-11} m^2/s (original diffusivity value, D, used in the case study). We will generate 1000 different values of diffusivity in the 80–120%D range in a normal distribution and then observe how the amount of drug delivered to the blood stream changes.

Step 1: open file

Open the file created in the last step in COMSOL.

Step 2: save as .m File

To perform Monte Carlo simulations, we need to use scripting in COMSOL. In other words, we will use COMSOL Script where we will write a small code to do the analysis. The first step is to export the COMSOL file in a text format, known as .m format, so that we can add our own set of commands and modify the file in COMSOL Script.

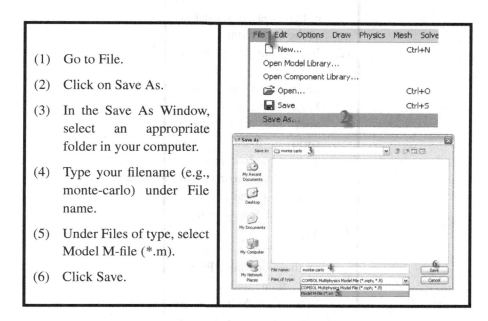

(1) Go to File.

(2) Click on Save As.

(3) In the Save As Window, select an appropriate folder in your computer.

(4) Type your filename (e.g., monte-carlo) under File name.

(5) Under Files of type, select Model M-file (*.m).

(6) Click Save.

Step 3: start editor and open .m File

We now open the .m file created in the last step in a text editor. In this example, we use the COMSOL Script editor to open the file. However, you can also use any other editor such as Notepad, WordPad or MATLAB editor to open the .m file and add the set of commands that will be shown later.

Step 4: modify the .m file in COMSOL Script editor

Once the .m file is opened in an editor (COMSOL Script editor in this case), we can make the required changes in the file to perform Monte Carlo simulations. The modified code is shown below. Note that the lines that start with % are comment lines and are generated by COMSOL to describe the different components of the .m file. Some comments (shown in bold) have been inserted in the code to provide the reader with explanations for the different parts of the file and the modifications made in the original .m file. The comment lines are not required for solving the problem.

236 Case studies

(1) Start COMSOL Script either from the Desktop or the Start menu.

(2) In the Command Prompt, specify the path of the .m file saved in the last step. For example, if you saved your file in C:\monte-carlo folder, type `path('C:\monte-carlo', path)` next to C>> in the command prompt and press Enter.

(3) Type `edit monte-carlo.m` in the Command Prompt and then press Enter. The COMSOL Script editor opens up with the monte-carlo.m file saved in the last step. In the next step, we will modify this file.

% **We first remove all the variables from the memory.**

`clear all`

% **The original .m file starts here.**

`% COMSOL Multiphysics Model M-file`

`% Generated by COMSOL 3.4 (COMSOL 3.4.0.248, $Date: 2007/10/10 16:07:51 $)`

`flclear fem`

```
% COMSOL version
clear vrsn
vrsn.name = 'COMSOL 3.4';
vrsn.ext = '';
vrsn.major = 0;
vrsn.build = 248;
vrsn.rcs = '$Name: $';
vrsn.date = '$Date: 2007/10/10 16:07:51 $';
fem.version = vrsn;

% Geometry
g1=rect2('0.03','0.0012','base','corner','pos',{'0','0'},'rot','0');
g2=curve2([0,0.024],[0.0012,0.0012]);

% Analyzed geometry
clear c s
c.objs={g2};
c.name={'B1'};
c.tags={'g2'};
s.objs={g1};
s.name={'R1'};
s.tags={'g1'};
fem.draw=struct('c',c,'s',s);
fem.geom=geomcsg(fem);

% Create mapped quad mesh
fem.mesh=meshmap(fem, ...
'edgelem',{1,[0 0.2 0.4 0.6 0.8 1],2,[0 0.0080 0.016
0.024 0.032 0.04 0.048 0.056 0.064 0.072 0.08 0.088
0.096 0.104 0.112 0.12 0.128 0.136 0.144 0.152 0.16
0.168 0.176 0.184 0.192 0.2 0.208 0.216 0.224 0.232
0.24 0.248 0.256 0.264 0.272 0.28 0.288 0.296 0.304
0.312 0.32 0.328 0.336 0.344 0.352 0.36 0.368 0.376
0.384 0.392 0.4 0.408 0.416 0.424 0.432 0.44 0.448
0.456 0.464 0.472 0.48 0.488 0.496 0.504 0.512 0.52
0.528 0.536 0.544 0.552 0.56 0.568 0.576 0.584 0.592
0.6 0.608 0.616 0.624 0.632 0.64 0.648 0.656 0.664
0.672 0.68 0.688 0.696 0.704 0.712 0.72 0.728 0.736
0.744 0.752 0.76 0.768 0.776 0.784 0.792 0.8 0.808
0.816 0.824 0.832 0.84 0.848 0.856 0.864 0.872 0.88
```

```
0.888 0.896 0.904 0.912 0.92 0.928 0.936 0.944 0.952
0.96 0.968 0.976 0.984 0.992 1],3,[0 0.01 0.02 0.03
0.04 0.05 0.06 0.07 0.08 0.09 0.1 0.11 0.12 0.13 0.14
0.15 0.16 0.17 0.18 0.19 0.2 0.21 0.22 0.23 0.24 0.25
0.26 0.27 0.28 0.29 0.3 0.31 0.32 0.33 0.34 0.35 0.36
0.37 0.38 0.39 0.4 0.41 0.42 0.43 0.44 0.45 0.46 0.47
0.48 0.49 0.5 0.51 0.52 0.53 0.54 0.55 0.56 0.57 0.58
0.59 0.6 0.61 0.62 0.63 0.64 0.65 0.66 0.67 0.68 0.69
0.7 0.71 0.72 0.73 0.74 0.75 0.76 0.77 0.78 0.79 0.8
0.81 0.82 0.83 0.84 0.85 0.86 0.87 0.88 0.89 0.9 0.91
0.92 0.93 0.94 0.95 0.96 0.97 0.98 0.99 1],4,[0 0.04
0.08 0.12 0.16 0.2 0.24 0.28 0.32 0.36 0.4 0.44 0.48
0.52 0.56 0.6 0.64 0.68 0.72 0.76 0.8 0.84 0.88 0.92
0.96 1], 5,[0 0.2 0.4 0.6 0.8 1]}, ...
'hauto',5);
```

% We now generate 1000 different numbers in a normal distribution using the *randn* command, scale the numbers and correlate them to diffusivity so that −1 corresponds to 0.8*D*, 0 corresponds to *D*, and +1 corresponds to 1.2*D*.

```
randn_scaled=(1/3)*randn(1000,1);
D_all = (1+randn_scaled*0.2)*1.11E-11;
```

% Since we started with a normal distribution, the range of values obtained from the *randn* command can potentially be from $-\infty$ to $+\infty$. Therefore the values of diffusivity obtained above can also be negative or zero which is not possible in a real situation. We, therefore, remove all the negative values of diffusivity and count the number of positive values of diffusivity.

```
j=0;
for i=1:1000
if (D_all(i,1)>0)
j=j+1;
D_pos(j,1)=D_all(i,1);
end
end
counter=j;
```

% We now start a loop to solve the drug diffusion problem for the different values of diffusivity generated above.

```
for i = 1:counter
```

```
D=D_pos(i,1);

% (Default values are not included)
```

% **This part of the .m file contains information about the governing equations to solve, boundary conditions and input parameters.**
```
% Application mode 1
clear appl
appl.mode.class = 'FlDiffusion';
appl.mode.type = 'axi';
appl.assignsuffix = '_di';
clear bnd
bnd.N = {0,8.849E-7,0};
bnd.type = {'N0','N','C'};
bnd.ind = [1,3,2,1,1];
appl.bnd = bnd;
clear equ
```
% **We set the diffusivity to be used by the solver for the different values generated above, one at a time.**
```
equ.D = D;
equ.ind = [1];
appl.equ = equ;
fem.appl1 = appl;
fem.sdim = {'r','z'};
 fem.frame = {'ref'};
fem.border = 1;
fem.outform = 'general';
clear units;
units.basesystem = 'SI';
fem.units = units;

% ODE Settings
clear ode
clear units;
units.basesystem = 'SI';
ode.units = units;
fem.ode=ode;

% Multiphysics
```

```
fem=multiphysics(fem);

% Extend mesh

fem.xmesh=meshextend(fem);

% Solve problem
fem.sol=femtime(fem, ...
'solcomp',{'c'}, ...
'outcomp',{'c'}, ...
'tlist',[0:3600:604800], ...
'tout','tlist');

% Save current fem structure for restart purposes
fem0=fem;

% Plot solution
postplot(fem, ...
'tridata',{'c','cont','internal','unit',
'mol/m^3'}, ...
'trimap','jet(1024)', ...
'solnum','end', ...
'title','Time=6.048e5 Surface: Concentration,
c [mol/m^3]', ...
'axis',[-0.0014999999664723875, 0.031499999295920136,
-0.011743630874307097, 0.012943630931304038]);

% Plot in cross-section or along domain
 postcrossplot(fem,0,[14.4E-3;0.6E-3], ...
'pointdata',{'c','unit','mol/m^3'}, ...
'title','Concentration, c [mol/m^3]', ...
'axislabel',{'Time', 'Concentration',' c [mol/m^3]'});

% Plot solution
 postplot(fem, ... 'tridata',{'c','cont','internal',
'unit',
'mol/m^3'}, ...
'trimap','jet(1024)', ...
'solnum',73, ...
'title','Time=2.592e5 Surface: Concentration,
```

```
c [mol/m^3]', ...
'axis',[-0.0014999999664723875, 0.031499999295920136,
-0.012434323450653496, 0.013634323507650438]);
```

% **We now use the subdomain integration tool in COMSOL to calculate the amount of drug in the tissue, Am_tissue, after 604 800 s (application time).**
```
% Integrate
Am_tissue=postint(fem,'2*pi*r*(c)', ...
'unit','mol', ...
'dl',[1], ...
'solnum','end');
```

% **The total amount of drug diffusing out from the patch, Am_total, is calculated knowing the flux (8.849E-7 g/m^{2}s), area of the patch ($\pi 0.024^2$ m^2) and the time of application (604 800 s).**
```
Am_total = 8.849E-7*pi*24E-3^2*604800;
```

% **Knowing the total amount of drug supplied by the patch and the amount of drug in the tissue, the amount of drug delivered to the blood stream, Am_deli, is calculated.**
```
Am_deli=Am_total-Am_tissue;
```

% **The different values of diffusivity used and the corresponding amount of drug delivered is written in two columns of the array *Res*.**
```
Res(i,1) = D;
Res(i,2) = Am_deli;
```

% **The loop for solving the problem with different values of diffusivity ends here.**
```
end
```

% **Finally the array, *Res*, is written in an Excel file and exported by specifying the location and file name.**
```
xlswrite('C:\monte-carlo\monte-carlo-res.xls',Res)
```
Save the .m file after the modifications shown above are made.

Step 5: run the .m file to perform Monte Carlo simulations

(1) Go to the Command Prompt window, type `run monte-carlo.m` (use the filename that you used to save the modified .m file in the last step) and then press Enter. The solver will take about 1–2 h to solve the problem.

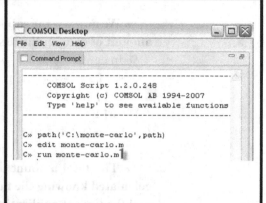

The Excel file with the different values of diffusivity in the range 80–120%D will be saved in the folder location that you specified in the code (C:\monte-carlo in this case). The data can be used for generating a frequency plot of amount of drug delivered to the blood stream for different diffusivity values. The plot is shown and discussed in Section 5.4.1.

IV Drug delivery in therapeutic contact lenses

Glaucoma is an optical condition caused by pressure build-up in the eye and is the leading cause of blindness. Current methods to treat glaucoma include medicated eye drops and oral medication, which are both inefficient methods of administration. Most of the medication in eye drops does not reach the target tissue. In addition, when taken orally, much of the drug circulates in the bloodstream instead of reaching the eye. This is a potential problem since drugs used to treat glaucoma, such as timolol maleate, are also prescribed to elevate hypertension. To avoid possible side effects, researchers have developed a novel method of drug delivery which involves enclosing the drug in the contact lens to be worn directly over the eye. The drug-encapsulated contact lens can deliver the drug to the target tissue more effectively (Gulsen and Chauhan, 2004).

Problem formulation

In this case study, we will develop a model to study the diffusion of the drug from the contact lens to the aqueous humor in the eye (Fung *et al.*, 2004). The function of the drug, timolol maleate, is to lower the fluid pressure in the eye by reducing the production of aqueous humor. The schematic of the eye is shown in Figure 6.6(a). To simplify the problem, the geometry of the eye is assumed to be cylindrical. The curvature of the eye is neglected. Five layers are considered, as shown in Figure 6.6(b): contact lens, tear film, cornea which in turn is composed of the epithelium (10%) and stroma (90%) and the aqueous humor. The dimensions have been taken from Johnson and Johnson contact lens specification sheet, Davidson and Kuonen (2004) and Edwards and Prausnitz (1998). The aqueous humor is assumed to be wide enough so that it can be treated as semi-infinite and therefore, the flux of drug is zero at the end of the aqueous humor.

Governing equation The governing equation for the problem is the mass transfer equation in cylindrical coordinates:

$$\frac{\partial c}{\partial t} = D_{d,layer} \left[\frac{1}{r} \frac{\partial}{\partial r} \left(r \frac{\partial c}{\partial r} \right) + \frac{\partial^2 c}{\partial z^2} \right] - r_{d,layer} \tag{6.5}$$

Here c is the concentration of drug in a particular layer, $D_{d,layer}$ represents the diffusivity of the drug in that layer and $r_{d,layer}$ represents the loss of drug in the same layer. The loss of drug happens in the tear film and aqueous humor regions, and it is zero in the other layers. We are ignoring any possible effects coming from partition coefficients between any two layers.

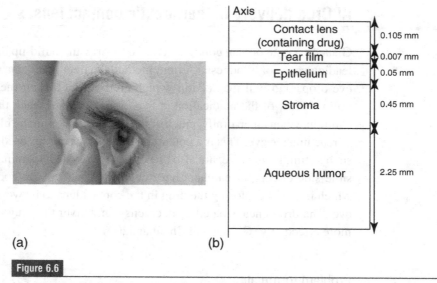

Figure 6.6

(a) A contact lens being applied onto the eye, (b) computational domain for the contact lens drug delivery system. Zero-flux boundary condition is used on all surfaces.

Table 6.2 Input parameters.

Diffusivity	Value (cm^2/s)	Source
Contact lens	9.9 E−9	(Hiratani and Alvarez-Lorenzo, 2004)
Tear film	5 E−5	
Cornea–epithelial	6.022 E−7	(Higashiyama et al., 2004)
Cornea–stroma	8.72 E−9	(Higashiyama et al., 2004)
Aqueous humor	5 E−5	
Drug removal rate	**Value (s^{-1})**	**Source**
Tear film	1 E−4	
Aqueous humor	0.003	

Boundary conditions The left boundary is the axis of symmetry, the bottom boundary is considered to be semi-infinite and other boundaries are assumed to be insulated, so flux of the drug is zero on all boundaries.

Input parameters The input parameters for the problem are shown in Table 6.2. The initial concentration of the drug in the contact lens is 46.5 mg/ml.

References

Davidson, H. J. and Kuonen, V. J. (2004). The tear film and ocular mucins. *Veterinary Ophthalmology*, **7**(2):71–77.

Edwards, A. and Prausnitz, M. R. (1998). Fiber matrix model of sclera and corneal stroma for drug delivery to the eye. *AIChE Journal*, **44**(1):214–225.

Fung, E. Y., Lee, J., Tong, A., Tran, B. and Yau, Y. Y. (2004). A Study of the Role of Therapeutic Contact Lenses in Drug Delivery. On the web at http://ecommons.library.cornell.edu/handle/1813/142

Gulsen, D. and Chauhan, A. (2004). Ophthalmic drug delivery through contact lenses. *Investigative Ophthalmology & Visual Science*, **45**(7):2342–2347.

Higashiyama, M., Inada, K., Ohtori, A. and Tojo, K. (2004). Improvement of the ocular bioavailability of timolol by sorbic acid. *International Journal of Pharmaceutics*, **272**(1–2):91–98.

Hiratani, H. and Alvarez-Lorenzo, C. (2004). The nature of backbone monomers determines the performance of imprinted soft contact lenses as timolol drug delivery systems. *Biomaterials*, **25**(6):1105–1113.

Johnson & Johnson Vision Care Inc. SUREVUE lenses. http://www.acuvue.com.my/my/products/pi.pdf (2007).

Implementation in COMSOL

> **Box 6.23 Steps**
>
> (1) **Problem type specification** We will setup the problem as a transient diffusion problem with a sink term in axisymmetric setting.
>
> (2) **Geometry creation** We will draw a rectangle with five subdomains to represent the different layers of the eye and the contact lens.
>
> (3) **Meshing** We will create a structured mesh in the domain.
>
> (4) **Defining governing equations, source terms, I.C., B.C.** We will specify the diffusivity of the drug and the sink term in the different layers. The initial concentration of drug in the contact lens will be entered. We will also supply zero-flux boundary condition for all the edges.
>
> (5) **Define postprocessing variable** We will define a variable to calculate the amount of drug in the contact lens and aqueous humor as functions of time.
>
> (6) **Solver setting and solution** We will solve the problem for 20 h.
>
> (7) **Postprocessing** We will plot the drug concentration at a point on the surface of the lens as a function of time. We will obtain plots of the drug lost from the lens and drug absorbed in the aqueous humor as functions of time. We will also save an image obtained during postprocessing.

Problem type specification

1. Open COMSOL Multiphysics.
2. In the **Model Navigator** select Axial Symmetry (2D) in the **Space dimensions** list, then click the **Multiphysics** button.
3. In the list of application modes select: **COMSOL Multiphysics > Convection and diffusion > Diffusion > Transient Analysis** and add it to the model with the **Add** button.
4. Click **OK**.

Geometry creation

(1) Select **Draw > Specify Objects > Rectangle**.
(2) Set **Width** to 0.675, **Height** to 0.225, **Base** to **Corner** and **r** and **z** to 0. Click **OK**. Note that the dimensions are in cm. COMSOL assumes the units in the SI system and displays accordingly, which is wrong in this case. Therefore, we will keep track of the units manually throughout the problem.
(3) Similarly, create four additional rectangles with these specifications:

Width	Height	r	z
0.675	0.045	0	0.225
0.675	0.005	0	0.27
0.675	7E−4	0	0.275
0.675	0.0105	0	0.2757

Meshing

(1) Open the **Mapped Mesh Parameters** dialog box from the **Mesh** menu.
(2) Click the **Boundary** tab.
(3) Select 1 and 12 under **Boundary Selection** holding the Ctrl button, check **Constrained edge element distribution** and set **Number of edge elements** to 40.
(4) Similarly, select other boundaries under **Boundary Selection,** check **Constrained edge element distribution** and set **Number of edge elements** as shown on the following page.
(5) Click on **Remesh**. Click **OK**.

IV Drug delivery in therapeutic contact lenses

Boundary	2,4,6,8,10,11	3,13	5,14	7,15	9,16
Number of edge elements	10	20	30	10	20

Governing equations, source terms, I.C., B.C. – diffusion

Subdomain settings

(1) Open the **Subdomain Settings** dialog box from the **Physics** menu.
(2) Select the subdomains under **Subdomain selection** and enter the values of **D isotropic** and **R** as shown.

Subdomain	1	2	3	4	5
D isotropic	5E−5	8.72E−9	6.022E−7	5E−5	9.9E−9
R	−1E−4*c	0	0	−0.003*c	0

(3) *Initial condition*: Select subdomain 5, click on the **Init** tab and enter $c(t_0)$ as 46.5. Click **OK**.

Boundary settings

All boundaries have species flux as zero, either due to symmetry (left boundaries) or insulation (right boundaries), or are considered to be semi-infinite (bottom boundary). Since, the default boundary condition of the solver is zero flux, we retain the default settings.

Define postprocessing variables

We will calculate the amount of drug (in mg) in the contact lens and aqueous humor as a function of time. Using this data, we can calculate the total amount of drug lost by the contact lens, and the amount absorbed by the aqueous humor when the contact lens is worn for a certain time.

(1) On the **Options** menu, point to **Integration Coupling Variables** and then click **Subdomain Variables**.

(2) Select Subdomain 1 under **Subdomain selection**.
(3) Type ctot_hum under **Name** and 2*pi()*r*c under **Expression**. The quantity 2*pi()*r is multiplied by c (concentration) as the problem is axisymmetric and therefore, we want to calculate the volume integral.
(4) Similarly, select Subdomain 5 under **Subdomain selection**.
(5) Type ctot_lens in **Name** field in the second row and 2*pi()*r*c in the corresponding **Expression** field. Click **OK**. *Make sure that you do not overwrite ctot_hum.*

Solver settings

(1) Click on the **Solver Parameters** button under the **Solve** menu.
(2) Select **Time dependent** in the **Solver** list.
(3) In the **Timestepping** area, type 0:100:72000 in the **Times** edit field.
(4) Click **OK**.

Solution

(1) Compute the final solution by clicking **Solve Problem** under **Solve** on the Main toolbar.

Postprocessing

To generate a plot of concentration of the drug on the surface of the lens as a function of time:

(1) Open the **Domain Plot Parameters** dialog box from the **Postprocessing** menu.
(2) On the **Point** page, select point 6 under **Point selection**. Click **OK**.

This generates the figure on the next page.

To generate the plot of drug lost from the lens and absorbed in the aqueous humor as functions of time:

(1) Open the **Global Variables Plot** dialog box from the **Postprocessing** menu.
(2) Type ctot_lens under **Expression** and click on the arrow (>) next to it to move it to the **Quantities to plot** list on the right.

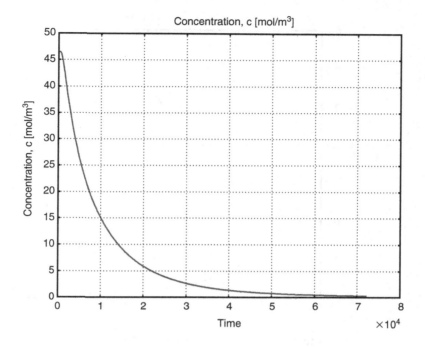

(3) Select ctot_lens in the **Quantities to plot** list. Click **OK**.
(4) Similarly, plot ctot_hum as a function of time (after removing ctot_lens from the **Quantities to plot** list by clicking on the < arrow).

This generates the two figures on the following page.

It can be observed that the amount of drug in the lens falls from 0.7 mg initially to about zero after 20 h (72 000 s). There is an increase in the amount of drug in the aqueous humor as more drug diffuses into the eye from the lens. The peak value of about 9 μg is reached at approximately 12.86 h (46 300 s) and then the amount of drug starts falling due to removal of the drug from the aqueous humor.

Saving an image

To save the image currently being displayed:

(1) Click **File > Export > Image...** to open the **Export Image** dialog box.
(2) Select the appropriate **Output format** that you want (Bitmap graphics for jpg file type and Vector graphics for eps file type), image size and rendering options (these options help you select the different components that you want your image to contain).

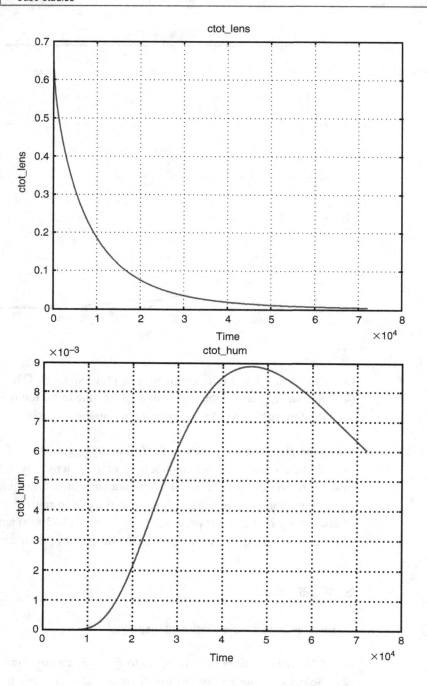

(3) Click **Preview** to view the image with your selected options.
(4) Click on **Export**, specify the file name and select the file type. Click **Export** to save the image.

V Elimination of nitrogen from the blood stream during deep sea diving

During deep sea diving, human beings are exposed to hazards that are unique to the hyperbaric underwater environment, and the physical behavior of gases at higher ambient pressure (Tetzlaff and Thorsen, 2005). Bends and decompression sickness result from physical dissolution of gases (especially nitrogen) in the blood stream at high pressures, followed by rapid exsolution during decompression when the diver returns to the surface (Wlodarczyk, McMillan and Greenfield, 2006). The rate of nitrogen removal from the blood stream is, therefore, critical during deep sea diving. In this case study, we will model the transfer of nitrogen from the blood stream to the alveolus following deep sea diving in order to estimate the nitrogen removal rate (Tomlinson et al., 2005).

Problem formulation

We develop a simple model to study the process of diffusion of nitrogen from the blood stream to the alveolus (Fig. 6.7(a)) through a thin layer of epithelial cells during deep sea diving. We will simplify the problem by assuming that the entire network of blood vessels can be modeled as a layer of blood stream flowing over a thin layer of epithelial cells in 2D. We assume here that the network of blood vessels fully covers the alveoli surface. The simplified geometry used for the model is shown in Fig. 6.7(b). The nitrogen is removed from the blood stream through the layer of cells into the air in the alveolus.

Governing equation   The process was considered to be steady state and hence, the governing equations in the different regions are:

in the blood stream:

$$u\frac{\partial c}{\partial x} = D_{N_2,\text{ blood}} \frac{\partial^2 c}{\partial x^2} \qquad (6.6)$$

in the epithelial cells:

$$D_{N_2,\text{ tissue}}\left(\frac{\partial^2 c}{\partial x^2} + \frac{\partial^2 c}{\partial y^2}\right) = 0 \qquad (6.7)$$

The velocity profile of blood inside the vessel is of the form:

$$u = u_{\max}\left[1 - \left(\frac{y}{Y}\right)^2\right]$$

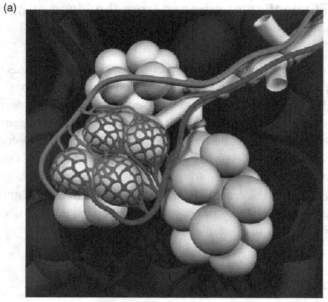

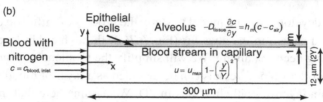

Figure 6.7

(a) Alveoli with blood vessels; (b) a highly simplified schematic used for computation, showing the region of blood flow in a vessel and the surrounding epithelial cells.

Boundary conditions It was assumed that nitrogen gas is exchanged only in the alveoli, so the alveolar air is at ambient nitrogen concentration. The worst case scenario of diving was considered, assuming that the depth of dive was 40 m corresponding to a pressure of 5 atm. The blood stream nitrogen concentration at this pressure was used for the analysis and is given in Table 6.3. The blood stream concentration of nitrogen was applied as a boundary condition on the left wall of the blood stream, as shown in Figure 6.7.

We apply the model to the 16th generation alveolus where the air velocity can be taken as 1.4% that of the bulk intake air velocity. The flux of nitrogen removal from the epithelial cell layer to the air in the alveolus was determined using a mass transfer coefficient:

$$-D_{N_2,\text{tissue}} \frac{\partial c}{\partial y} = h_m (c - c_{\text{air}})$$

Table 6.3 Input parameters.

Parameter	Value
Nitrogen diffusivity in blood, $D_{N_2, \text{blood}}$	3000×10^{-12} m^2/s
Nitrogen diffusivity in epithelial cells, $D_{N_2, \text{tissue}}$	5×10^{-12} m^2/s
Maximum velocity of blood, u_{max}	470×10^{-6} m/s
Length of alveolus, l	300×10^{-6} m
Tissue depth, d	1×10^{-6} m
Blood stream thickness, $2Y$	12×10^{-6} m
Inlet blood nitrogen concentration, $c_{\text{blood,inlet}}$	8.125×10^{-2} kg/m^3
Nitrogen concentration in air, c_{air}	1.3×10^{-2} kg/m^3
Mass transfer coefficient, h_m	3.97×10^{-8} m/s

Input parameters The value of the mass transfer coefficient was calculated based on the alveolar air velocity and is given in Table 6.3. Other input parameters are shown in Table 6.3.

References

Tetzlaff, K., Thorsen, E. (2005). Breathing at depth: Physiologic and clinical aspects of diving while breathing compressed gas. *Clinics in Chest Medicine.* **26**(3): 355–380.

Tomlinson, J., Moore, F., Khan, S., Keegan, J., and Jawahar, R. (2005). Nitrogen Elimination in the Alveoli. On the web at http://ecommons.library.cornell.edu/handle/1813/2618

Wlodarczyk, A., McMillan P. F. and Greenfield S. (2006). High pressure effects in anaesthesia and narcosis. *Chemical Society Reviews.* **35**(10):890–898.

Implementation in COMSOL

Box 6.24 Steps

(1) **Problem type specification** We will setup the problem as a steady convection-diffusion problem in 2D.
(2) **Geometry creation** We will draw two rectangles to represent the epithelial cells and the blood stream.

> (3) **Meshing** We will create a structured mesh in the domain.
> (4) **Defining governing equations, source terms, I.C., B.C.** We will specify the diffusivity of the drug in the two layers and the velocity of blood. We will also supply the boundary conditions shown in Figure 6.7.
> (5) **Solver setting and solution** We will solve the steady-state problem.
> (6) **Postprocessing** We will plot the nitrogen concentration in the domain at steady state. We will also plot the nitrogen flux at the alveolar wall as a function of distance.

Problem type specification

(1) Open COMSOL Multiphysics.
(2) In the **Model Navigator** select 2D in the **Space dimensions** list, then click the **Multiphysics** button.
(3) In the list of application modes select
COMSOL Multiphysics > Convection and diffusion > Convection and diffusion > Steady-State Analysis.
(4) Click **OK**.

Geometry creation

(1) Select **Draw > Specify Objects > Rectangle**.
(2) Set the **Width** to $300\,E-6$, **Height** to $12\,E-6$, **Base** to **Corner** and **y** to $-6\,E-6$. Click **OK**.
(3) Select **Draw > Specify Objects > Rectangle**.
(4) Set the **Width** to $300\,E-6$, **Height** to $1\,E-6$, **Base** to **Corner** and **y** to $6\,E-6$. Click **OK**.

Meshing

(1) Open the **Mapped Mesh Parameters** dialog box from the **Mesh** menu.
(2) Click the **Boundary** tab.
(3) Select 1, 3, 6 and 7 under **Boundary Selection** holding the Ctrl button, check **Constrained edge element distribution** and set **Number of edge elements** to 10.
(4) Select 2, 4 and 5 under **Boundary Selection**, check **Constrained edge element distribution** and set **Number of edge elements** to 50.
(5) Click on **Remesh**. Click **OK**.

Governing equations, source terms, I.C., B.C. – convection and diffusion

Subdomain settings

(1) Open the **Subdomain Settings** dialog box from the **Physics** menu.
(2) Select subdomain 1 and enter these coefficients; when done, click **OK**:

Name	Expression
D isotropic	3000 E−12
R	0
u	470 * (1−y^2/6 E−6^2)
v	0

(3) Select subdomain 2 and enter **D isotropic** as 5 E−12. Click **OK**.

Boundary settings

(1) Select **Boundary Settings** from the **Physics** menu and enter these boundary conditions; when done, click **OK**.

Boundary	1	5	6	All other
Boundary condition to select	Concentration (c_0)	Flux (N_0)	Convective flux	Insulation/ symmetry
Value	8.125 E−2	−3.97 E−8*(c−1.3 E−2)	-	-

Solver settings

We will retain the default solver settings for the problem.

Solution

(1) Compute the final solution by clicking **Solve Problem** under **Solve** on the Main toolbar.

Postprocessing

To generate a surface plot of nitrogen concentration in the domain:

(1) Open the **Plot Parameters** dialog box from the **Postprocessing** menu.
(2) On the **General** page select the **Surface** check box for a surface plot.
(3) On the **Surface** page, select **Concentration, c** in the **Predefined quantities** area. Click **OK**.

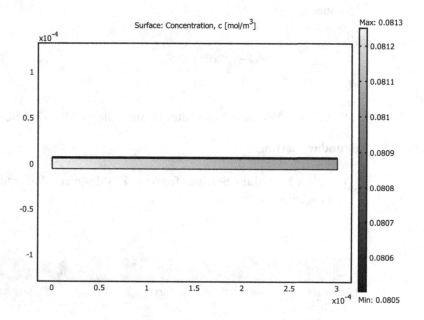

To generate the plot of nitrogen flux at the alveolar wall as a function of distance:

(1) Open the **Domain Plot Parameters** dialog box from the **Postprocessing** menu.
(2) On the **Line/Extrusion** page, select Boundary 5.
(3) Select **Normal total flux, c** in the **Predefined quantities** area. Click **OK**.

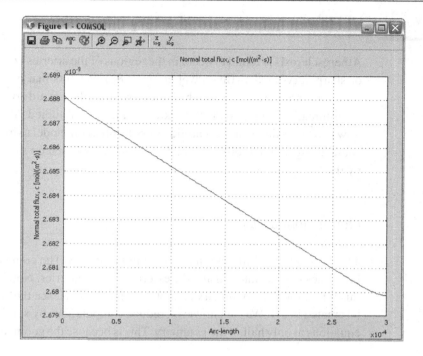

VI Flow in human carotid artery bifurcation

Atherosclerosis is found to occur in the regions of the arteries where flow divisions or sharp curves exist (Bharadvaj *et al.*, 1982). Patients that have atherosclerosis often have plaque build-up on the interior walls of the carotid artery directly before the bifurcation (Figure 6.8(c)). Perktold *et al.* (1991) studied the detailed pulsatile flow in a three-dimensional carotid artery bifurcation model to investigate the flow characteristics in the bifurcation. This example is based on the model developed by them.

Problem formulation

The carotid artery bifurcation is shown in Figure 6.8. The common carotid artery bifurcates into the internal and the external carotid arteries. Notice here that due to this bifurcation, axisymmetry cannot be considered. Hence the geometry needs to be created in 3D. We can, however, reduce computations by solving the governing equations in only half of the geometry. This is because the geometry is symmetrical

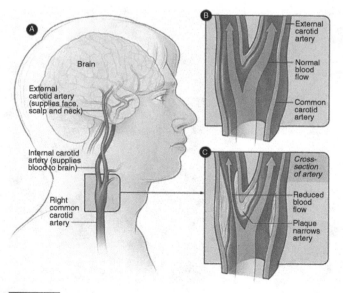

Figure 6.8

Human carotid artery bifurcation: (a) location of the artery in the body; (b) cross-sectional view of the normal system; (c) carotid artery after atherosclerosis. Reproduced from NHLBI (part of the National Institute of Health and the U.S. Department of Health and Human Services) Diseases and Conditions Index topic, Carotid Artery Disease (http://www.nhlbi.nih.gov/).

VI Flow in human carotid artery bifurcation

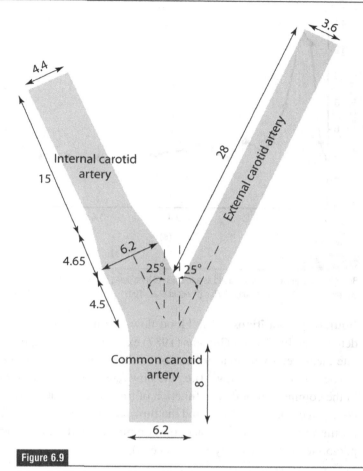

Figure 6.9

Schematic of the human carotid artery bifurcation model (measurements in mm). The geometry is based on the work of Perktold *et al.* (1991). The geometry has been cut in half to reduce computational resources (discussed in the text).

if cut into two parts along the axis. The computational domain used for the model is shown in Figure 6.9. At the inlet of the common carotid, the blood flow is fully developed and time dependent representing a pulse cycle based on the study of Ku and Giddens, 1987. The flow characteristics as a function of time are studied using the model. The blood flow is assumed to be Newtonian for simplicity.

Governing equation Blood flow in the arteries is governed by the continuity and Navier–Stokes equations:

$$\nabla \cdot u = 0 \tag{6.8}$$

$$\rho \left(\frac{\partial u}{\partial t} + u \cdot \nabla u \right) = -\nabla p + \mu \nabla^2 u \tag{6.9}$$

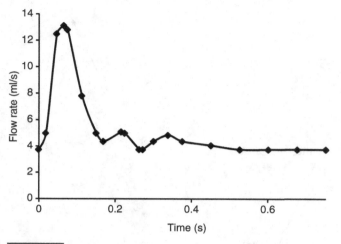

Figure 6.10

Blood flow rate in the common carotid as a function of time using the data from Ku and Giddens (1987). The time for one pulse cycle is assumed to be 0.75 s (Perktold et al., 1991).

Boundary conditions The blood flow rate in the common carotid artery was determined by Ku and Giddens (1987) experimentally. Figure 6.9 shows the flow rate measured by them as a function of time, based on a pulse cycle time of 0.75 s. The average velocity, v_{av} (i.e., the flow rate divided by the cross-sectional area) in the common carotid, as a function of time, is calculated using the radius of the common carotid as 3.1 mm and the flow rates shown in Figure 6.10. The velocity profile across the inlet section of the common carotid is fully developed, and hence parabolic, and is given by Equation 6.10,

$$v(r) = v_0 \left(1 - \frac{r^2}{R^2}\right) \tag{6.10}$$

where $v(r)$ is the velocity at a distance r from the center, v_0 is the velocity at the center of the common carotid inlet and R is the radius of the common carotid.

The velocity at the center of the common carotid and the average velocity are related by:

$$v_0 = 2v_{av} \tag{6.11}$$

The boundary condition at the inlet of the common carotid is thus calculated using the formulation above and applied in COMSOL Multiphysics. At the outlet of the internal and external carotid arteries, the surface traction is zero. At the walls of the arteries, no-slip boundary condition (zero tangential velocity) is applied. Half the geometry is used for computations in COMSOL to reduce computational resources (as discussed earlier). Therefore the symmetry boundary condition is used at the symmetry line.

Table 6.4 Input parameters (all values are taken from Perktold *et al.*, 1991.)

Parameter	Value
Common carotid inlet velocity	Eqn. 6.10 and Fig. 6.10
Common carotid radius, R	3.1 mm
Time for one pulse cycle, t_p	0.75 s
Blood Density, ρ	1000 kgm^{-3}
Blood viscosity, μ	0.0035 Pa s

Input parameters   The input parameters of the model are shown in Table 6.4.

References

Bharadvaj, B. K., Mabon, R. F., and Giddens, D. P. (1982). Steady Flow in a Model of the Human Carotid Bifurcation .1. Flow Visualization, *Journal of Biomechanics*, **15**, pp. 349–362.

Perktold, K., Resch, M., and Florian, H. (1991). Pulsatile Non-Newtonian Flow Characteristics in a 3-Dimensional Human Carotid Bifurcation Model. *Journal of Biomechanical Engineering*-Transactions of the ASME, **113**, pp. 464–475.

Ku, D. N., and Giddens, D. P. (1987). Laser Doppler Anemometer Measurements of Pulsatile Flow in a Model Carotid Bifurcation. *Journal of Biomechanics*, **20**, pp. 407–421.

Implementation in COMSOL

The transient Navier–Stokes equation will be solved in COMSOL. A text (.txt) file needs to be created that contains the velocity at the inlet of the common carotid (calculated as described earlier) as a function of time, so that it can be used as the boundary condition in COMSOL. The format of the file is:

%time
Time values separated by spaces
%velocity
Velocity values separated by spaces

The file can be created in *Notepad*. It should be made sure that there are no line breaks between the data. Remember here that the velocities in the text file refer to the velocities at the center of the common carotid, v_0. The text file should look like this:

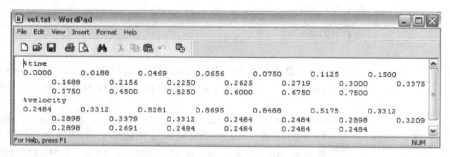

Box 6.25 Steps

(1) **Problem type specification** We will setup the problem for solving the Navier–Stokes equations in 3D.
(2) **Geometry creation** We will create the 3D geometry using tools such as extrusion and revolution. To save computational resources, we will use symmetry to reduce the domain size by half.
(3) **Meshing** We will create a structured mesh in the majority of the domain. The remaining part will be meshed using an unstructured mesh.
(4) **Defining governing equations, source terms, I.C., B.C.** We will specify the density and viscosity of blood. The boundary conditions that include inlet velocity, no slip and symmetry will be supplied. We will also define a function that interpolates the inlet velocity as a function of time.
(5) **Solver setting and solution** We will solve the problem for 0.75 s.
(6) **Postprocessing** We will plot the velocity distribution at 15 different slices obtained after 0.75 s.

Problem type specification

(1) Open COMSOL Multiphysics.
(2) In the **Model Navigator** select 3D in the **Space dimensions** list.
(3) In the list of application modes select
COMSOL Multiphysics > Fluid Dynamics > Incompressible Navier-Stokes > Transient Analysis.
(4) Click **OK**.

Geometry creation

The geometry for the problem is a little complicated and therefore, we create the different parts in steps. We will first create a cylinder representing the common carotid before the bifurcation. Then we create the bent portion of the common carotid that opens into the internal carotid. This is done by creating a circle in 2D and then using the feature in COMSOL Multiphysics that revolves the 2D geometry about an axis to create a 3D geometry. We then create a cylinder which represents the first part of the interior carotid artery. We do this using the extrusion feature in COMSOL, where first a circle is created in 2D and then extended to create a cylinder. After this we create a cone which represents the second part of the interior carotid. Finally, we create another cylinder using extrusion to create the full internal carotid geometry. For the external carotid, we first create a small cylinder at the specified angle from the common carotid. We then extrude a circle to create the full external carotid geometry. We create the external carotid geometry in two parts as mentioned previously, as it will help us create a mapped mesh for a majority of the domain. Additionally, the domain is very large hence, considerable computational resources are required to solve it. The geometry is symmetric about the y-axis (as can be seen in COMSOL). To reduce the problem size, this symmetry can be considered and the equations need to be solved in only half of the domain. Therefore, the final part of the geometry creation step is to remove one half of the geometry that is symmetric. Use **Edit > Undo** to undo the step if you make any mistake. The following steps demonstrate the process:

(1) Select **Draw > Cylinder.** Set the **Radius** to 3.1, **Height** to 8, under **Axis direction vector** select **Cartesian coordinates** and set **x, y, z** as 0, 0 and −1, respectively. Click **OK**.
(2) Open the **Work Plane Settings** dialog box from the **Draw** Menu.
  (a) Click on **Quick** tab.

(b) Select the **Plane** as **x-y** with **z** as 0 (default). Click **OK**. Click on **Zoom Extents** button on the top menu (in the second row). The outline in blue shows the projection of the geometry already created in the selected plane.
(c) Select **Draw** > **Specify Objects** > **Circle**.
(d) Set the **Radius** to 3.1 and retain the default for other settings. Click **OK**.
(e) Select **Revolve** from the **Draw** Menu.
(f) Under **Angle of revolution**, set **a1** as 0 and **a2** as −25.
(g) Under **Revolution axis** set **x** as −12.1 and **y** as 0 for **Point on axis**, and **x** as −12.1 and **y** as 1 for **Axis direction through: Second point**.
(h) Click **OK**. The part of the common carotid with the bend has now been created and is shown in the 3D view.

(3) Open the **Work Plane Settings** dialog box from the **Draw** Menu.
(a) Click on **Face Parallel** tab.
(b) From the **Face selection** list, select face 3 under REV1. Click **OK**. Click on **Zoom Extents** button.
(c) Select **Draw** > **Specify Objects** > **Circle**.
(d) Set the **Radius** to 3.1 and set **x** as 3.1 and **y** as 3.1 under **Position** with **Base** set as **Center**. Click **OK**.
(e) Select **Extrude** from the **Draw** menu.
(f) Select **C2** as the **Object to extrude**.
(g) Under **Extrusion parameters**, set **Distance** as 4.5 and retain the default values for other parameters.
(h) Click **OK**. The first part of the internal carotid has been created and is shown in the 3D view.

(4) Select **Draw** > **Cone**. Set the **Radius** to 3.1, **Height** to 4.65 and **Semi-angle** as 11, under **Axis base point**, set **x** as 12.1*cos(25*pi/180)−4.5*cos(65*pi/180) −12.1, **y** as 0 and **z** as 12.1*sin(25*pi/180)+4.5*sin(65*pi/180), under **Axis direction vector** select **Spherical coordinates** and set θ **(theta)** as −25. Click **OK**. The values of the angles are multiplied by pi/180 since COMSOL takes the argument of the Trigonometric functions in radians and not degrees.

(5) Open the **Work Plane Settings** dialog box from the **Draw** Menu.
(a) Click on **Face Parallel** tab.
(b) From the **Face selection** list, select face 4 under CON1. Click **OK**. Click on **Zoom Extents** button.
(c) Select **Draw** > **Specify Objects** > **Circle**.
(d) Set the **Radius**, **x** and **y** (under **Position**) all to 3.1−4.65*tan(11*pi/180) with **Base** set as **Center**. Click **OK**.
(e) Select **Extrude** from the **Draw** Menu.
(f) Select **C3** as the **Object to extrude**.

(g) Under **Extrusion parameters**, set **Distance** as 15.
(h) Click **OK**. The last part of the internal carotid geometry has now been created and can be seen in the 3D view.

(6) Select **Draw** > **Cylinder**. Set the **Radius** to 1.8, **Height** to 6, under **Axis base point** set **x, y, z** as 1.25, 0 and 0.9, respectively and under **Axis direction vector** select **Spherical coordinates** and set θ **(theta)** as 25. Click **OK**.

(7) Open the **Work Plane Settings** dialog box from the **Draw** Menu.
   (a) Click on **Face Parallel** tab.
   (b) From the **Face selection** list, select face 4 under CYL2. Click **OK**. Click on **Zoom Extents** button.
   (c) Select **Draw** > **Specify Objects** > **Circle**.
   (d) Set the **Radius** to 1.8 and set **x** as 1.8 and **y** as 1.8 under **Position** with **Base** set as **Center**. Click **OK**.
   (e) Select **Extrude** from the **Draw** Menu.
   (f) Select **C4** as the **Object to extrude**.
   (g) Under **Extrusion parameters**, set **Distance** as 28.
   (h) Click **OK**. The final part of the external carotid geometry has now been created and can be seen in the 3D view.

(8) Select **Draw** > **Create Composite Object**. **Set formula** as REV1+CYL2 and uncheck **Keep interior boundaries**. Click **OK**.

(9) Open the **Work Plane Settings** dialog box from the **Draw** Menu.
   (a) Click on **Quick** tab.
   (b) Select the **Plane** as **z-x** with **y** as 0 (default). Click **OK**. Click on **Zoom Extents** button.
   (c) Select **Draw** > **Specify Objects** > **Rectangle.**
   (d) Set the **Width** to 50, **Height** to 40, **Base** to **Corner** and **x** to -10 and **y** to -20. Click **OK**.
   (e) Select **Extrude** from the **Draw** Menu.
   (f) Select **R1** as the **Object to extrude**.
   (g) Under **Extrusion parameters**, set **Distance** as -4.
   (h) Click **OK**. A rectangular block is created to delete one half of the carotid geometry.

(10) Select **Draw** > **Create Composite Object**. **Set formula** as CO1+CON1+CYL1+EXT2+EXT1+EXT3−EXT4 and check **Keep interior boundaries**. Click **OK**.

(11) Select **Edit** > **Select All**.

(12) Select **Draw** > **Modify** > **Scale**. Set **x, y** and **z** all as $1E-3$ under **Scale factor**. Click **OK**. We do this since we created the geometry in m but the actual dimensions are in mm. Click on **Zoom Extents** button.

(13) Under **Model Tree** (at the left side of the screen), right click on **Geom2** and select **Remove**. We remove the 2D geometries that we used to create our 3D model.

Meshing

As mentioned earlier, we try to create a mapped mesh for the domain. We will first create a free mesh on the cross-sectional surfaces and then map it to obtain the mapped volume mesh.

(1) Open the **Free Mesh Parameters** dialog box from the **Mesh** menu.
(2) Select **Extremely fine** under **Predefined mesh sizes** in the **Global** tab.
(3) Click the **Boundary** tab.
(4) Select Boundaries 1, 18 and 28 under **Boundary Selection** holding the Ctrl button.
(5) Click **Mesh Selected**. Click **OK**. The selected surfaces are meshed. We will now create the volume mesh.
(6) Open the **Swept Mesh Parameters** dialog box from the **Mesh** menu.
(7) Select Subdomain 1 under **Subdomain Selection**, check **Manual specification of element layers** and set **Number of element layers** to 40. Click **Mesh Selected**.
(8) Select Subdomains 2 and 3 under **Subdomain Selection**, check **Manual specification of element layers** and set **Number of element layers** to 20. Click **Mesh Selected**.
(9) Similarly, select Subdomain 5 under **Subdomain Selection**, check **Manual specification of element layers** and set **Number of element layers** to 30. Click **Mesh Selected**.
(10) Finally, select Subdomain 6 under **Subdomain Selection**, check **Manual specification of element layers** and set **Number of element layers** to 70. Click **Mesh Selected**.
(11) Open the **Free Mesh Parameters** dialog box from the **Mesh** menu.
(12) Select **Extremely fine** under **Predefined mesh sizes** in the **Global** tab.
(13) Click the **Subdomain** tab.
(14) Select Subdomain 4 under **Subdomain Selection.** Click **Mesh Selected**. We cannot create a mapped mesh in subdomain 4 and hence we create a free mesh.
(15) Click **OK**.

Governing equations, source terms, I.C., B.C. – incompressible Navier–Stokes

Subdomain settings

(1) Select **Incompressible Navier-Stokes (ns)** under **Multiphysics** menu.
(2) Open the **Subdomain Settings** dialog box from the **Physics** menu.
(3) Select all subdomains (hold Ctrl and select each of these) under **Subdomain Selection**, set ρ as 1000 and η as 0.0035.
(4) Click **OK**.

Boundary settings

(1) Make sure that **Incompressible Navier-Stokes (ns)** is selected under the **Multiphysics** menu.
(2) Open the **Boundary Settings** dialog box from the **Physics** menu.
(3) Select Boundary 17 under **Boundary Selection**, change the **Boundary Condition** to Inflow/Outflow velocity and set u_0 as 0, v_0 as 0 and w_0 as vel(t)*(1 − (x^2 + y^2)/0.0031^2).
(4) Select Boundaries 1 and 28 under **Boundary Selection** and change the **Boundary Condition** to Neutral.
(5) Select Boundaries 3, 6, 9, 13, 16 and 25 under **Boundary Selection** and change the **Boundary Condition** to Slip/Symmetry. Click **OK**.

Define function

The function, vel(t), used earlier, needs to be defined before the problem is solved.

(1) Open the **Functions** window from the **Options** menu.
(2) Click on **New**, enter Function name as vel, select **Interpolation**, and select **File** in the drop down menu below **Use data from**.
(3) Click on **Browse** and select the file, vel.txt (or your filename), created earlier. Click **OK**.
(4) Select **Interpolation Function** in the drop down menu next to **Extrapolation method** under Function Definition. Click **OK**.

Solver settings and solution

(1) Open **Solver Parameters** dialog box from the **Solve** menu.
(2) Select **Transient** under **Analysis** in the **General** tab. The **Solver** should automatically be selected as **Time dependent**.
(3) In the **Timestepping** area in the **General** tab, type 0:0.001:0.75 in the **Times** edit field.
(4) Select **Direct (UMFPACK)** under **Linear system solver**.
(5) Under the **Time Stepping** Tab, select **Time steps taken by solver** to Strict. Click **OK**.
(6) Click **Solve**.

Note that the problem is very large and needs considerable memory (more than 4 Gb) to solve it. Additionally, it will take around 4–5 hrs to solve the problem.

Postprocessing

We will plot the velocity distribution in the domain at 0.75 s (end time) at 15 different slices along the z-axis. To generate this plot:

(1) Open the **Plot Parameters** dialog box from the **Postprocessing** menu.
(2) On the **General** page select the **Slice** check box for surface plots at different slices.

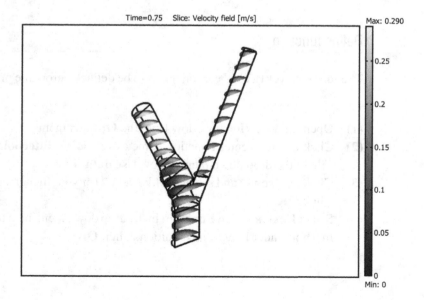

(3) On the **Slice Page**, select **Velocity field, U_ns** in the **Predefined quantities** area.
(4) Set **Number of levels** as 15 under **z levels** in **Slice Positioning** area. This will plot the velocity distribution at 15 slices in the z-direction. You can also specify the **Vector with coordinates** to plot the surface plot at any specified slice.
(5) Click **OK**.

VII Radioimmunotherapy of metastatic melanoma

Metastatic melanoma is the most dangerous form of skin cancer and causes 77% of all skin cancer-related deaths. New methods for its treatment are, therefore, urgently needed. The feasibility of radioimmunotherapy (RIT) in mice, which involves the binding of the antibody to a tumor-associated antigen to deliver a lethal dose of radiation to tumor cells, was established in a study by Dadachova *et al.* (2004). Another study (Schweitzer *et al.*, 2007) looked at the process of radioimmunotherapy in detail with the use of a computer model. In this particular example, we will follow this paper to develop a mathematical model of the process.

Problem formulation

Figure 6.11 shows the process of radioimmunotherapy. The process involves intravenous administration of a radio-labeled antibody in the patient's body which circulates in the plasma and is transported across capillary walls into the normal tissue. The shape of the tumor is assumed to be a sphere. The antibody then diffuses through normal tissue to the tumor to reach the antigen present inside the tumor and binds with it to form an antibody-antigen complex. Also, the antibody-antigen complex that is formed dissociates into free antibody and antigen. Some amount of the antibody in the normal tissue is removed by the lymphatic vessels and there is no significant binding of the antibody in the normal tissue.

Governing equations

The governing equations that describe the set of processes are given by:

(I) Antibody uptake from blood:

$$c_{\text{Abp}}(t) = c_{\text{Ab0}} e^{-\lambda t} \tag{6.12}$$

where c_{Ab0} is the initial plasma antibody concentration after intravenous administration, λ is the plasma kinetics constant and c_{Abp} is the antibody concentration in the plasma as a function of time.

(II) Transport, uptake and clearance of antibody in normal tissue:

$$\frac{\partial c_{\text{Abt}}}{\partial t} = D_{\text{tis}} \frac{1}{r^2} \frac{\partial}{\partial r}\left(r^2 \frac{\partial c_{\text{Abt}}}{\partial r}\right) + k^{\text{bl}} c_{\text{Abp}} - k^{\text{ly}} c_{\text{Abt}} \tag{6.13}$$

where c_{Abt} is the antibody concentration in the normal tissue, D_{tis} is the diffusivity of antibody in the tissue, r is the radial distance from the tumor center, and k^{bl} and k^{ly} are the rate constants for uptake into and removal from tissue, respectively.

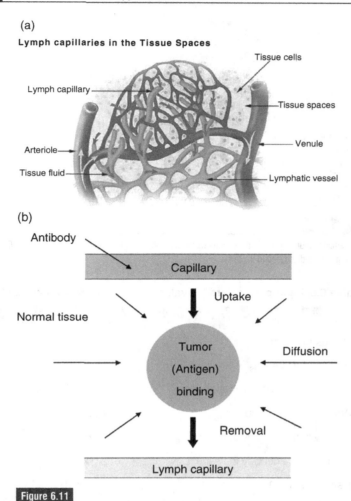

Figure 6.11

(a) The blood and lymph capillary system in the tissue. Figure from http://upload.wikimedia.org/wikipedia/commons/1/19/Illu_lymph_capillary.png; (b) physical description of radioimmunotherapy.

(III) Transport, complex formation and dissociation in tumor:

$$\frac{\partial c_{Ab}}{\partial t} = D_{tum} \frac{1}{r^2} \frac{\partial}{\partial r}\left(r^2 \frac{\partial c_{Ab}}{\partial r}\right) - k_{+1} c_{Ab} c_{Ag} + k_{-1} c_{Ab-Ag} \qquad (6.14)$$

where c_{Ab} is the concentration of free antibody in the tumor, D_{tum} is the diffusivity of antibody in the tumor and k_{+1} and k_{-1} are the rates of the forward and backward reactions, respectively.

(IV) Antigen concentration due to complex formation and dissociation:

$$\frac{\partial c_{Ag}}{\partial t} = n\left(-k_{+1} c_{Ab} c_{Ag} + k_{-1} c_{Ab-Ag}\right) \qquad (6.15)$$

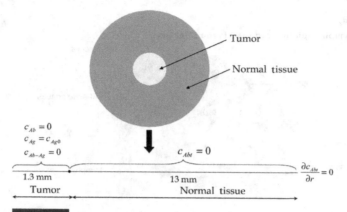

Figure 6.12

Schematic, boundary and initial conditions for the problem. The computations are done in 1D considering spherical symmetry (as discussed in the text).

Table 6.5 Input parameters (all values taken from Schweitzer et al., 2007).

Input parameter	Value
Radius of tumor, r_{tum}	1.3 mm
Radius of tissue, r_{tis}	14.3 mm
Initial antibody concentration, c_{Ab0}	4.94 nM
Antigen concentration, c_{Ag0}	76 000 nM
Diffusivity of antibody in tumor, D_{tum}	4.16×10^{-7} cm^2 s^{-1}
Diffusivity of antibody in tissue, D_{tis}	2.0×10^{-7} cm^2 s^{-1}
Rate constant, λ	2.96×10^{-5} s^{-1}
Rate constant for transcapillary transport, k^{bl}	4.6×10^{-5} mL/(s*mL ECF)
Rate constant for lymphatic clearance, k^{ly}	1.78×10^{-5} mL/(s*mL ECF)
Forward binding rate constant, k_{+1}	5.0×10^4 M^{-1}s^{-1}
Dissociation rate constant, k_{-1}	1.0×10^{-5} s^{-1}
Valence of antibody/antigen binding, n	5

where c_{Ag} is the antigen concentration and n is the valence of 6D2/melanin binding.

(V) Complex concentration due to formation and dissociation:

$$\frac{\partial c_{Ab-Ag}}{\partial t} = k_{+1} c_{Ab} c_{Ag} - k_{-1} c_{Ab-Ag} \qquad (6.16)$$

where c_{Ab-Ag} is the complex concentration.

Boundary conditions The boundary conditions are shown in Figure 6.12. On the left tumor boundary, species flux for all species is zero due to symmetry. On the right normal tissue boundary, species flux is zero as the boundary is considered to be at a large distance. Initially, there is no antibody and complex in the domain. The tumor has an initial antigen concentration of 76 000 nM (Table 6.5).

Input parameters Input parameters are shown in Table 6.5.

References

Dadachova, E., Nosanchuk, J. D., Shi, L., *et al.* (2004). Dead cells in melanoma tumors provide abundant antigen for targeted delivery of ionizing radiation by a monoclonal antibody to melanin. *Proc Natl Acad Sci USA* 2004;**101**:14865–14870.

Schweitzer A. D., Rakesh V., Revskaya E., Datta A., Casadevall A., Dadachova E. (2007). Computational model predicts effective delivery of 188-Re-labeled melanin-binding antibody to the metastatic melanoma tumors with wide range of melanin concentrations. *Melanoma Research*, **17**(5):291–303.

Implementation in COMSOL

As discussed earlier, the shape of the tumor is a sphere. All the computations can, therefore, be done in the spherical coordinate system. Since the concentrations are dependent only on the radial distances, as can be seen in Equations 6.13 and 6.14, a simplified geometry consisting of a line can be used for the model (Figure 6.12). This reduces the amount of computational resources that would be required if we modeled the complete sphere. However, the governing equations need to be solved in the spherical coordinate system. COMSOL solves all the governing equations in the Cartesian system and hence, we need to modify the equations in COMSOL to implement the spherical coordinate system. We take the species transfer equation in the Cartesian (1D) and spherical (with radial dependence only) systems:

$$\frac{\partial c}{\partial t} = \frac{\partial}{\partial x}\left(D_{\text{cart}}\frac{\partial c}{\partial x}\right) + R_{\text{cart}} \tag{6.17}$$

$$\frac{\partial c}{\partial t} = \frac{1}{r^2}\frac{\partial}{\partial r}\left(r^2 D\frac{\partial c}{\partial r}\right) + R \tag{6.18}$$

Equation 6.18 can be multiplied by r^2 on both sides giving:

$$r^2\frac{\partial c}{\partial t} = \frac{\partial}{\partial r}\left(r^2 D\frac{\partial c}{\partial r}\right) + r^2 R \tag{6.19}$$

Equation 6.19 can be implemented in COMSOL as Equation 6.17 using the coefficient of $\partial c/\partial t$ as r^2, $D_{cart} = r^2 D$ and $R_{cart} = r^2 R$. The problem has been solved using this method for conversion to the spherical coordinate system in this example.

> **Box 6.26 Steps**
>
> (1) **Problem type specification** We will setup the problem by selecting three transient convection-diffusion equations to solve for the concentrations of antibody, antigen and complex.
> (2) **Geometry creation** We will draw a line with one division to represent the normal tissue and tumor regions.
> (3) **Meshing** We will create a structured mesh.
> (4) **Defining governing equations, source terms, I.C., B.C.** We will specify the diffusivity of the antibody in the normal tissue and tumor regions. Zero diffusivity will be entered for the antigen and complex. The reactions between the antibody and antigen will be specified. The initial concentration of the antigen inside the tumor will be entered. We will also specify the boundary conditions shown in Figure 6.12.
> (5) **Solver setting and solution** We will solve the problem for 3 days.
> (6) **Postprocessing** We will plot the concentration of the antibody in the normal tissue and tumor at six different times in the same plot.

Problem type specification

(1) Open COMSOL Multiphysics.
(2) In **Model Navigator** select 1D in the **Space dimensions** list, then click the **Multiphysics** button.
(3) Click on **Add Geometry**, replace **x y z** by **r th phi**. We change the nomenclature of the original coordinate system to indicate that we will work in spherical coordinates. The quantities r, th and phi are the user-defined variables for the problem.
(4) In the list of application modes select
COMSOL Multiphysics > Convection and diffusion > Diffusion > Transient Analysis and add it to the model with the **Add** button.
(5) Click on **Add** twice to add two more **Diffusion** modes. We do this as we have three species in our problem.
(6) Click **OK**.

Geometry creation

(1) Select **Draw** > **Specify Objects** > **Line**.
(2) Set **r** as **0 0.13**. Click **OK**. This will create a line of length, 0.13 units. Note that the dimensions are in cm. COMSOL will display the units in the SI system. However, we will keep track of the units throughout the problem.
(3) Similarly, create another line with **r** as **0.13 1.43**. Click **OK**. This will create a line of length 1.3 units.
(4) Click on **Zoom Extents** button on the top menu (in the second row) to fit the geometry in the screen.

Meshing

(1) Open the **Free Mesh Parameters** dialog box from the **Mesh** menu.
(2) Set **Maximum element size** to 0.0005.
(3) Click on **Remesh**. Click **OK**.

Governing equations, source terms, I.C., B.C. – diffusion

Define constants and expressions

The following constants and expressions are defined to be used in Subdomain Settings later.

Constants

(1) Open the **Constants** window from the **Options** menu.
(2) Define the following constants (see next page): *Note here that the descriptions are optional and need not be typed in.*

Expressions

(1) Open the **Expressions** > **Global Expressions** window from the **Options** menu.
(2) Define the following expressions (see next page):

Constants

Name	Expression	Description
kbl	4.60 E − 05	blood uptake
kly	1.78 E − 05	lymph removal
kp1	5.00 E − 05	forward reaction
km1	1.00 E − 05	backward reaction
lamb	2.96 E − 05	rate constant
n	5	valence
cab0	4.94	antibody injected
cag0	76000	initial antigen
Dtiss	2.00 E − 07	diffusivity of antibody in tissue
Dtum	4.16 E − 07	diffusivity of antibody in tumor
tumor_rad	0.13	tumor radius

Expressions

Name	Expression	Description
A_r	r*r	coefficient of the storage term in governing equation
Dtum_r	Dtum*r*r	coefficient to implement equations in spherical coordinates
Dtiss_r	Dtiss*r*r	diffusivity to implement equations in spherical coordinates
Rtiss_r	(kbl*cab0*exp(−lamb*t)−kly*c)*r*r	antibody source – tissue
Rtum_r	(−kp1*c*c2+km1*c3)*r*r	antibody source – tumor
Rag_r	n*(−kp1*c*c2+km1*c3)*r*r	antigen source
Rcm_r	(kp1*c*c2 − km1*c3)*r*r	complex source

Subdomain settings

For species 1 – antibody

(1) Select **Diffusion (di)** under **Geom1** in the **Model tree** menu on the left of the screen OR **1 Diffusion (di)** under the **Multiphysics** menu on the top.

(2) Open the **Subdomain Settings** dialog box from the **Physics** menu.
(3) Select the subdomains under **Subdomain selection** and enter the values of **D isotropic** and **R** as shown, then click **OK**:

Subdomain	1	2
δ_{ts}	A_r	A_r
D	Dtum_r	Dtiss_r
R	Rtum_r	Rtiss_r

For species 2 – antigen

(1) Select **Diffusion (di2)** under **Geom1** in the **Model tree** menu on the left of the screen *OR* **2 Diffusion (di2)** under the **Multiphysics** menu on the top.
(2) Open the **Subdomain Settings** dialog box from the **Physics** menu.
(3) Select Subdomain **2** under **Subdomain selection** and uncheck **Active in this domain**. In this step, we tell the solver that species 2 (antigen) is not present in subdomain 2 (normal tissue).
(4) Select Subdomain **1** (tumor) under **Subdomain selection** and fill in the following values:

Parameter	Value
δ_{ts}	A_r
D	0
R	Rag_r

(5) *Initial condition* Select subdomain **1**, click on the **Init** Tab and enter $c2(t_0)$ as cag0. Click **OK**.

For species 3 – antibody-antigen complex

(1) Select **Diffusion (di3)** under **Geom1** in the **Model tree** menu on the left of the screen *OR* **3 Diffusion (di3)** under the **Multiphysics** menu on the top.
(2) Open the **Subdomain Settings** dialog box from the **Physics** menu.

(3) Select Subdomain **2** under **Subdomain selection** and uncheck **Active in this domain**. Again, we tell the solver that species 3 (complex) is not present in subdomain 2 (normal tissue).
(4) Select Subdomain **1** (tumor) under **Subdomain selection** and fill in the following values; then click **OK**:

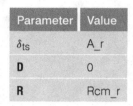

Parameter	Value
δ_{ts}	A_r
D	0
R	Rcm_r

Boundary settings

All boundaries for the different species have species flux as zero due to symmetry, insulation or are considered to be semi-infinite. Since the default boundary condition of the solver is zero flux, we retain the default settings.

Solver settings

(1) Click on the **Solver Parameters** button under the **Solve** menu.
(2) Select **Time dependent** in the **Solver** list.
(3) In the **Timestepping** area, type 0:3600:259200 in the **Times** edit field.
(4) Click **OK**.

Solution

(1) Click **Get Initial Value** under **Solve** menu.
(2) Compute the final solution by clicking **Solve Problem** under **Solve**.

Postprocessing

To generate a plot of concentration of the antibody in the normal tissue and tumor at 6 different times (0, 8, 16, 24, 48 and 72 hrs):

(1) Open the **Domain Plot Parameters** dialog box from the **Postprocessing** menu.

(2) Select **Concentration, c** under **Predefined quantities**. *Note*: to plot antigen concentration, select **Concentration, c2** and for complex concentration, select **Concentration, c3**.
(3) On the **Line/Extrusion** page, select subdomains 1 and 2 under **Subdomain selection**.
(4) On the **General page**, hold the *Ctrl* button, to select the 6 desired times: 0, 28800, 57600, 86400, 172800 and 259200 in the time selection area. Click **OK**.
(5) To view the legends on the figure, click on **Edit plot** button in the figure window, click on **line(0)** under **Axes** and check **Show Legend**.

It can be observed from the figure that antibody concentration increases in the normal tissue as it is taken up from the blood until about 16 h (57 600 s). However, as the antibody diffuses inside the tumor at later times and is removed by the lymph, the concentration of the antibody in the normal tissue keeps going down, as can be seen from the curves for 48 and 74 h. It should also be noted here that the normal tissue was assumed to be semi-infinite. This can be verified from the above figure which shows that there is no change in concentration near the right boundary. Similar plots of antigen and complex concentration (step 2 above) can be drawn to look at their variations with time and space.

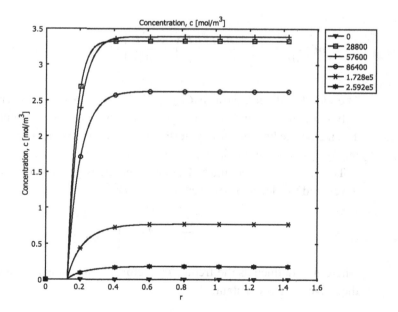

VIII Burn injury in blood-perfused skin

Thermal burns in skin tissue are caused due to exposure to high temperatures for a period of time. The most common skin burns result from surface application of a heat source. The temperature magnitude and duration of exposure determine the extent of burn injury in the tissue (Diller and Hayes, 1983). We will implement the model developed by Diller and Hayes (1983) to quantify the burn injury, in this example.

Problem formulation

Figure 6.13(a) shows the heating of the tissue by applying a cylindrical heated disc. The disc heats the skin surface for a specified time and temperature. The tissue is heated by conduction. Blood perfusion is significant only in the dermis. The aim of the model is to estimate the temperature profile and burn injury due to the heating process in the tissue layers.

Governing equations

The heat transfer in the tissue layers is governed by the bioheat equation:

$$\rho C_p \frac{\partial T}{\partial t} = k \left[\frac{1}{r} \frac{\partial}{\partial r} \left(r \frac{\partial T}{\partial r} \right) + \frac{\partial^2 T}{\partial z^2} \right] + \rho_b C_{p,b} \dot{V}_b^v (T_a - T) + Q \qquad (6.20)$$

The blood perfusion term, $\rho_b C_{p,b} \dot{V}_b^v (T_a - T)$, as discussed earlier, is significant only in the dermis and is set to zero in the epidermal and subcutaneous fat layer. The metabolic heat generation term, Q, is zero in the epidermal and subcutaneous fat layer and negligible in the dermis.

The degree of tissue injury, Ω, due to heating is given by the following equation developed by Henriques and Moritz (1947):

$$\frac{d\Omega}{dt} = A \exp\left[-\frac{\Delta E}{RT}\right] \qquad (6.21)$$

where A is defined as the frequency factor, ΔE as the activation energy and R is the universal gas constant.

Boundary conditions The boundary conditions and the schematic used for the model in COMSOL are shown in Figure 6.13(b). The geometry is modeled as

VIII Burn injury in blood-perfused skin

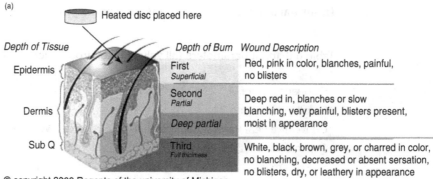

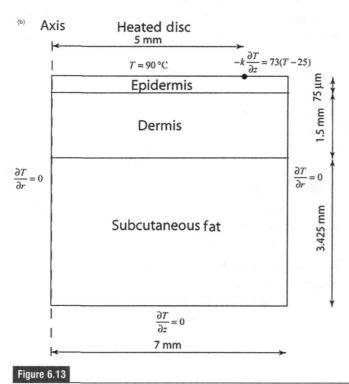

Figure 6.13

(a) Schematic showing the different layers of the skin and the different degrees of burn injury. Copyright University of Michigan. Used with permission. (b) Schematic, boundary and initial conditions for the problem.

axisymmetric, as the disc is cylindrical. The heated disc is at a constant temperature of 90 °C. The top of the skin surrounding the heated disc is cooled by the air through natural conduction. The heat flux at all other surfaces is zero as the left surface is axis and the other surfaces are assumed to be at a very far distance.

Input parameters Input parameters for the problem are listed in Table 6.6.

Table 6.6 Input parameters (all values taken from Diller and Hayes, 1983).

Parameter	Value
Frequency factor, A	3×10^{98} 1/s
Duration of heating, t	15 s
Convective heat transfer coefficient, h	73 Wm^{-2}K^{-1}
Thermal conductivity, k	
Epidermis	0.21 Wm^{-1}K^{-1}
Dermis	0.37 Wm^{-1}K^{-1}
Subcutaneous fat	0.16 Wm^{-1}K^{-1}
Density, ρ	
Blood, ρ_b	1100 kgm^{-3}
Epidermis	1000 kgm^{-3}
Dermis	1000 kgm^{-3}
Subcutaneous fat	1000 kgm^{-3}
Specific heat, C_p	
Blood, $C_{p,b}$	3300 Jkg^{-1}K^{-1}
Epidermis	3181.82 Jkg^{-1}K^{-1}
Dermis	2846.15 Jkg^{-1}K^{-1}
Subcutaneous fat	1975.31 Jkg^{-1}K^{-1}
Arterial blood temperature, T_a	37 °C
Initial tissue temperature, T_i	34 °C
Heating disc temperature, T_d	90 °C
Ambient air temperature, T_∞	25 °C
Dermal blood perfusion rate, $\dot{V}_b^w$	0.024 ml/min/ml tissue

References

Diller K. R. and Hayes L. J. (1983). A finite element model of burn injury in blood-perfused skin. *Journal of Biomechanical Engineering*, **105**, 300–307.

Henriques, F. C., and Moritz, A. R. (1947). Studies of Thermal Injury. 1. The conduction of heat to and through skin and the temperatures attained therein - a theoretical and an experimental investigation, *American Journal of Pathology*, **23**, pp. 531–549.

Implementation in COMSOL

The transient heat conduction equation in COMSOL is solved to obtain the temperatures inside the domain. In order to model the burn injury, the transient diffusion equation in COMSOL is selected. This is because ordinary differential equations (Equation 6.21) cannot be solved directly in COMSOL. The diffusivity of the equation is set to zero and the source term of the diffusion is set as $A \exp\left[-\frac{\Delta E}{RT}\right]$. The boundary condition of the diffusion equation is set as insulation (flux = 0) at all the boundaries.

> **Box 6.27 Steps**
>
> (1) **Problem type specification** We will setup the problem as a transient conduction problem to solve for temperatures in an axisymmetric setting. To solve for burn injury, we will select a diffusion equation.
> (2) **Geometry creation** We will draw a rectangle with three subdomains to represent the different layers of the skin and a line to represent the heated disc.
> (3) **Meshing** We will create a structured mesh in the domain.
> (4) **Defining governing equations, source terms, I.C., B.C.** We will specify the thermal conductivity, specific heat, density and heat source terms in the different layers. We will also specify the source term for the diffusion equation. The initial body temperature will be specified. We will also specify the boundary conditions as shown in Figure 6.13(b).
> (5) **Solver setting and solution** We will solve the problem for 15 s.
> (6) **Postprocessing** We will plot the temperature contours at nine different levels and burn injury contours at five different levels at the end time.

Problem type specification

(1) Open COMSOL Multiphysics.
(2) In the **Model Navigator** select Axial Symmetry (2D) in the **Space dimensions** list, then click the **Multiphysics** button.
(3) In the list of application modes select **COMSOL Multiphysics > Heat Transfer > Conduction> Transient Analysis** and add it to the model with the **Add** button.
(4) Then select

COMSOL Multiphysics > Convection and diffusion > Diffusion > Transient Analysis and add it to the model with the **Add** button. The transient diffusion equation will be used to quantify the burn injury (Ω).

(5) Click **OK**.

Geometry creation

(1) Select **Draw > Specify Objects > Rectangle**.
(2) Set the **Width** to 7 E−3, **Height** to 3.425 E−3, **Base** to **Corner** and **r** and **z** to 0. Click **OK**. Note that the dimensions are in m.
(3) Similarly, create 2 additional rectangles with these specifications:

Width	Height	r	z
7 E−3	1.5 E−3	0	3.425 E−3
7 E−3	75 E−6	0	4.925 E−3

(4) Select **Draw > Specify Objects > Line**. The line will represent the heating disc.
(5) Set **r** as 0 5 E−3 and **z** as 5 E−3 5 E−3. Click **OK**.
(6) Click on **Zoom Extents** button on the top menu (in the second row) to fit the geometry in the screen.

Meshing

(1) Open the **Mapped Mesh Parameters** dialog box from the **Mesh** menu.
(2) Click the **Boundary** tab.
(3) Select 1 and 9 under **Boundary Selection** holding the Ctrl button, check **Constrained edge element distribution**, set **Number of edge elements** to 10, **Element ratio** under **Distribution** to 2 and check **Reverse Direction**.
(4) Similarly, select other boundaries under **Boundary Selection,** check **Constrained edge element distribution**, set **Number of edge elements**, **Element ratio** and **Reverse direction** as shown below.

Boundary	2,4,6	3,10	5,11	7	8
Number of edge elements	210	20	10	150	60
Element ratio	1	20	1	1	1
Reverse direction		check			

(5) Click on **Remesh**. Click **OK**.

Governing equations, source terms, I.C., B.C. – conduction

Subdomain settings

(1) Select **Conduction** under the **Multiphysics** menu.
(2) Open the **Subdomain Settings** dialog box from the **Physics** menu.
(3) Select the subdomains under **Subdomain selection** and enter the values of **k (isotropic)**, ρ, c_p and **Q** as shown:

Subdomain	1	2	3
k	0.16	0.37	0.21
ρ	1000	1000	1000
c_p	1975.31	2846.15	3181.82
Q	0	1.1e3*3.3e3*2.4e − 2/60*(310.15 − T)	0

(4) *Initial condition* Select subdomains 1, 2 and 3 (holding Ctrl and selecting them under **Subdomain Selection**), click on the **Init** tab and enter $T(t_0)$ as 307.15. Click **OK**.

Boundary settings

(1) Make sure that **Conduction** is selected under the **Multiphysics** menu.
(2) Open the **Boundary Settings** dialog box from the **Physics** menu.
(3) Select Boundaries 1, 3 and 5 under **Boundary Selection**, change the **Boundary Condition** to Axial Symmetry.
(4) Select **Boundary** 7 under **Boundary Selection**, change the **Boundary Condition** to Temperature and set T_0 as 363.15.
(5) Select **Boundary** 8 under **Boundary Selection**, change the **Boundary Condition** to Heat Flux and set q_0 as $-73*(T - 298.15)$. Click **OK**.

Governing equations, source terms, I.C., B.C. – diffusion

Subdomain settings

(1) Select **Diffusion** under the **Multiphysics** menu.
(2) Open the **Subdomain Settings** dialog box from the **Physics** menu.
(3) Select subdomains 1, 2 and 3, under **Subdomain Selection**.
(4) Set **D** as 0 and **R** as $3e98*\exp(-6.3e8/(8.314e3*T))$. Click **OK**.

Boundary settings

All boundaries for the species (burn) have species flux as zero due to impermeability. Since the default boundary condition of the solver is zero flux, we retain the default settings.

Solver settings

(1) Click on the **Solver Parameters** button under the **Solve** menu.
(2) Select **Time dependent** in the **Solver** list.
(3) In the **Timestepping** area, type 0:1:15 in the **Times** edit field.
(4) Click **OK**.

Solution

(1) Compute the final solution by clicking **Solve Problem** under **Solve** on the Main toolbar.

Postprocessing

We will plot the temperature contours levels of 85, 80, 75, 70, 65, 60, 55, 50 and 45 °C in the domain at the end time, as shown in Diller and Hayes (1983). To generate this plot:

(1) Open the **Plot Parameters** dialog box from the **Postprocessing** menu.
(2) On the **General** page, check **Contour**.
(3) On the **Contour** page, select Temperature under **Predefined quantities**, select °C as **Unit**.
(4) Under **Contour Levels**, select **Vector with isolevels**, and type 85 80 75 70 65 60 55 50 45. Click **OK**. If you want to print the contour level values on the graph, check **Labels**.

We will now plot the burn injury contour levels of 1E7, 1E5, 1000, 10 and 0.1 in the domain at the end time, as shown in Diller and Hayes (1983). To generate this plot:

(1) Open the **Plot Parameters** dialog box from the **Postprocessing** menu.
(2) On the **General** page, check **Contour**.
(3) On the **Contour** page, select Concentration under **Predefined quantities**.

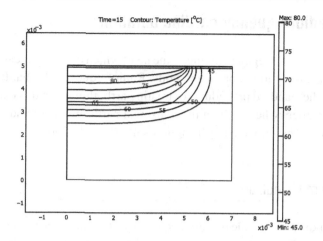

(4) Under **Contour Levels**, select **Vector with isolevels**, and type 1e7 1e5 1e3 1e1 1e−1. Click **OK**. If you want to print the contour level values on the graph, check **Labels**.

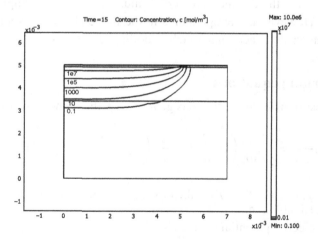

Values of burn injury (Ω) of 0.53, 1 and 10^4 correspond to injury threshold, second and third degree burn, respectively (Diller and Hayes, 1983). The figure shows the level of burns inside the tissue after the skin is heated by the disc for 15 s.

IX Radiofrequency cardiac ablation

Arrhythmia is irregular beating of the heart which causes decreased blood flow and oxygen supply to the brain and body. Radiofrequency (RF) ablation (see Section 8.5) is the standard procedure for treating arrhythmias. In this example, we look at radiofrequency heating in a tissue using a partly modified and simplified version of this process modeled by Tungjitkusolmun *et al.* (2000).

Problem formulation

A cylindrical electrode is introduced into the middle of the tissue where RF ablation is needed. The problem is therefore axisymmetric and the schematic is shown in Figure 6.14. The properties of the tissue are homogenous. The tissue is heated by resistive heating due to Joule heat generation as there is a potential difference between the electrode tip and the outer edge of the tissue. The optimal goal of RF ablation is to increase the temperature of the tissue from 37 °C to more than 50 °C, when the desired myocardial injury takes place (Tungjitkusolmun *et al.*, 2000). However, the temperatures in the tissue should be kept below 100 °C to avoid unwanted phenomena such as boiling, charring etc.

Governing equations

The heat transfer in the tissue is governed by the bioheat equation:

$$\rho C_p \frac{\partial T}{\partial t} = \left[\frac{1}{r}\frac{\partial}{\partial r}\left(kr\frac{\partial T}{\partial r}\right) + \frac{\partial}{\partial z}\left(k\frac{\partial T}{\partial z}\right) \right] + \rho_b C_{p,b} \dot{V}_b^v (T_a - T) + Q \quad (6.22)$$

where $\rho_b C_{p,b} \dot{V}_b^v (T_a - T)$ is the blood perfusion term and Q is the heat generated during RF ablation.

The electric potential is given by:

$$\nabla \cdot (\sigma \nabla V) = 0 \quad (6.23)$$

The heat generated, Q, due to Joule heating is given by:

$$Q = \sigma |\nabla V|^2 \quad (6.24)$$

where V is the electric potential and σ is the electrical conductivity.

IX Radiofrequency cardiac ablation

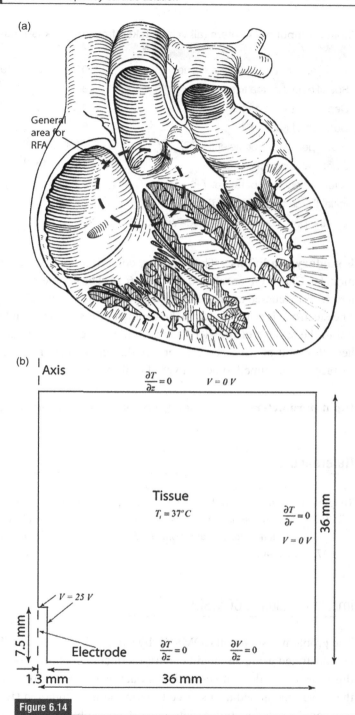

Figure 6.14

(a) Placement of the electrode for RF ablation in the heart; (b) the computational domain showing the boundary and initial conditions for the problem.

Table 6.7 Input parameters (all values taken from Tungjitkusolmun et al., 2000).

Parameter	Value
Thermal conductivity of the tissue, k	$0.4925 + 0.001\,195T$ Wm^{-1}K^{-1}
Specific heat of the tissue, C_p	3200 Jkg^{-1}K^{-1}
Density of the tissue, ρ	1200 kgm^{-3}
Duration of heating, t	60 s
Blood perfusion coefficient, $\rho_b C_{p,b} \dot{V}_b^w$	2000 Wm^{-3}K^{-1}
Electrical conductivity, σ	0.222 S m^{-1}
Arterial blood temperature, T_a	37 °C
Initial tissue temperature, T_i	37 °C
Electric potential at the electrode surface, V	25 V

Boundary conditions The boundary conditions and the schematic used for the model in COMSOL are shown in Figure 6.14. The geometry is modeled as axisymmetric as discussed earlier. The tissue is at a constant initial temperature of 37 °C. The electrode is not included in the geometry for simplicity and the electric potential to the surface of the electrode is implemented as a boundary condition. The heat fluxes at all surfaces are zero as the left surface is the axis and the other surfaces are assumed to be at a very far distance.

Input parameters Input parameters for the problem are listed in Table 6.7.

Reference

Tungjitkusolmun, S., Woo, E. J., Cao, H., Tsai, J. Z., Vorperian, V. R. and Webster, J. G. (2000). Finite element analyses of uniform current density electrodes for radio-frequency cardiac ablation. *IEEE Transactions on Biomedical Engineering*, **47**, pp. 32–40.

Implementation in COMSOL

The problem is solved in COMSOL by two different methods. The first method is demonstrated using an inbuilt electro-thermal interaction module in the solver. In this case, the Joule heat generation (Equation 6.24) is automatically computed by the solver and added as a source term in the heat equation (Equation 6.22). The transient heat conduction equation is solved to obtain the temperatures inside the domain.

In the second method, the electric potential equation (Equation 6.23) is modeled using a steady-state diffusion equation (Laplace equation) as it is basically a Laplace equation. The Joule heating term is then calculated in the solver by specifying the relation between heat generated and potential difference, as given by Equation 6.24. Again, the transient heat conduction equation is solved to obtain the temperatures.

Adaptive mesh refinement will also be demonstrated for the problem at the end. Note that adaptive mesh refinement in COMSOL can be implemented for steady-state problems only and therefore we will implement the steady-state version of this problem.

Box 6.28

Steps – using electro-thermal interaction module

(1) **Problem type specification** We will setup the problem as an electro-thermal interaction problem in an axisymmetric setting.
(2) **Geometry creation** We will draw the computational domain representing tissue.
(3) **Meshing** We will create an unstructured mesh in the domain.
(4) **Defining governing equations, source terms, I.C., B.C.** We will specify the thermal conductivity, specific heat, density and heat source term. We will also specify the electrical conductivity of the tissue and the boundary conditions, shown in Figure 6.14.
(5) **Solver setting and solution** We will solve the problem for 60 s.
(6) **Postprocessing** We will plot the temperature history at the top edge of the probe.

Steps – using conduction and diffusion modules

(1) **Problem type specification** We will solve for temperatures using the conduction equation and electric potential using the diffusion equation.
(2) **Geometry creation** We will draw the same computational domain as above.
(3) **Meshing** We will create the same unstructured mesh.
(4) **Defining governing equations, source terms, I.C., B.C.** We will specify the thermal properties. The electrical conductivity of the tissue will be specified as the diffusivity.
(5) **Solver setting and solution** We will solve the problem for 60 s.

> **Steps – adaptive mesh refinement**
>
> (1) **Modify Mesh** We will open and modify the mesh of the original problem to create a coarse mesh.
> (2) **Change solver settings and solve** Since adaptive mesh refinement can be done only for steady-state problems we will change the solver and solve the problem.
> (3) **Postprocessing using the coarse mesh solution** We will plot the temperature contours obtained from the coarse mesh.
> (4) **Change solver settings for adaptive mesh refinement and solve** We will setup the problem for adaptive mesh refinement and solve it again.
> (5) **Postprocessing using the refined mesh solution** We will plot the temperature contours again to compare with those obtained from the coarse mesh above (in step 3).

Problem type specification

(1) Open COMSOL Multiphysics.
(2) In **Model Navigator** select Axial Symmetry (2D) in the **Space dimensions** list.
(3) In the list of application modes select
   COMSOL Multiphysics > Electro-Thermal Interaction > Joule Heating > Transient Analysis.
(4) Click **OK**.

Geometry creation

(1) Select **Draw > Specify Objects > Rectangle**.
(2) Set the **Width** to $1.3\,E-3$, **Height** to $7.5\,E-3$, **Base** to **Corner** (which is the default setting) and **r** and **z** to 0. Click **OK**.
(3) Select **Draw > Specify Objects > Square**.
(4) Set the **Width** to $36\,E-3$, **Base** to **Corner** and **r** and **z** to 0. Click **OK**.
(5) Select **Draw > Create composite object**.
(6) Enter SQ1-R1 under **Set formula** and click **OK**.
(7) Click on **Zoom Extents** button on the top menu to fit the geometry in the screen.

Meshing

(1) Open the **Free Mesh Parameters** dialog box from the **Mesh** menu.
(2) Click the **Boundary** tab.
(3) Select 2 and 4 under **Boundary Selection** holding the Ctrl button, and enter $5\,E-5$ as the **Maximum element size**. We need more elements near the electrode in the tissue since there is maximum temperature change in this region.
(4) Click on **Remesh**. Click **OK**.

Governing equations, source terms, I.C., B.C. – conduction

Subdomain settings

(1) Select **Heat Transfer by Conduction** under the **Multiphysics** menu.
(2) Open the **Subdomain Settings** dialog box from the **Physics** menu.
(3) Select the subdomain 1 under **Subdomain selection** and enter the values of **k (isotropic)**, ρ, c_p and **Q** as shown:

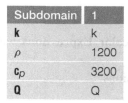

Subdomain	1
k	k
ρ	1200
c_p	3200
Q	Q

(4) *Initial condition* Select subdomain 1, click on the **Init** tab and enter $T(t_0)$ as 310.15. Click **OK**.

Boundary settings

All boundaries for heat transfer have a zero flux condition due to insulation or symmetry. Since a zero flux boundary condition is the default in COMSOL, we will not change any settings.

Governing equations, source terms, I.C., B.C. – Conductive media DC

Subdomain settings

(1) Select **Conductive Media DC** under **Multiphysics** menu.
(2) Open the **Subdomain Settings** dialog box from the **Physics** menu.

(3) Select subdomain 1, under **Subdomain Selection**.
(4) Check σ **(isotropic)** and set σ **(isotropic)** as 0.222. Click **OK**.

Boundary settings

(1) Make sure that **Conductive Media DC** is selected under the **Multiphysics** menu.
(2) Open the **Boundary Settings** dialog box from the **Physics** menu.
(3) Select Boundary 1 under **Boundary Selection**, change the **Boundary Condition** to Axial Symmetry.
(4) Select Boundaries 2 and 4 under **Boundary Selection**, change the **Boundary Condition** to Electric Potential and set V_0 as 25. We set the electric potential at the electrode surface to 25 V.
(5) Select Boundary 5 under **Boundary Selection**, change the **Boundary Condition** to Electric Insulation. Click **OK**.

Define expressions

The expressions for k and Q that were used in Subdomain Settings earlier are defined here.

(1) Open the **Expressions > Scalar Expressions** window from the **Options** menu.
(2) Define the following expressions:

Name	Expression
k	0.4925+0.001195*T
Q_blood	2000*(310.15−T)
Q	Q_blood+Q_dc

Note here that Q_dc is the heat source term due to radiofrequency ablation, calculated automatically by COMSOL.
(3) Click **OK**.

Solver settings

(1) Click the **Solver Parameters** button.
(2) Select **Time dependent** in the **Solver** list.
(3) In the **Timestepping** area, type 0:1:60 in the **Times** edit field.
(4) Click **OK**.

Solution

(1) Compute the final solution by clicking **Solve Problem** under **Solve** on the Main toolbar.

Postprocessing

We will look at the temperature at the top edge of the cylindrical probe used for RF ablation in contact with the tissue. This corresponds to the corner of the electrode for the axi-symmetric geometry. To obtain the temperature at this point as a function of time:

(1) Open the **Domain Plot Parameters** dialog box from the **Postprocessing** menu.
(2) On the **Point** page, select 4 under **Point Selection**.
(3) Select **Temperature** in the **Predefined quantities** area. Click **OK**.

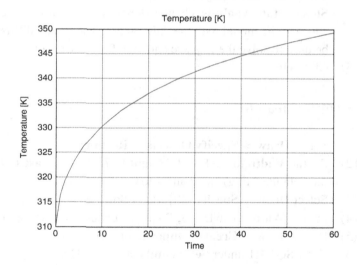

It is observed that the maximum temperature at the location is about 77 °C (350 K). The aim of RF ablation, as discussed earlier, is to obtain a temperature greater than 50 °C but less than 100 °C and it is seen that this is achieved at the particular location.

Alternative method

The electric potential equation (Equation 6.23) is basically a Laplace equation and can be solved in a more general form. In other words, the **Conductive Media DC** physics (in COMSOL) used earlier is not needed and the problem can be solved using the steady-state diffusion equation only. We will now solve the problem with this different approach.

Problem type specification

(1) Open COMSOL Multiphysics.
(2) In the **Model Navigator** select Axial Symmetry (2D) in the **Space dimensions** list.
(3) In the list of application modes select
COMSOL Multiphysics > **Heat Transfer** > **Conduction**> **Transient Analysis** and add it to the model with the **Add** button.
(4) Then select
COMSOL Multiphysics > **Convection and diffusion** > **Diffusion** > **Steady-state Analysis,** change **Dependent variable** name from **c** to V and add it to the model with the **Add** button. The steady diffusion equation will be used to solve the Laplace equation for electric potential.
(5) Click **OK**.

Geometry creation

(1) Select **Draw** > **Specify Objects** > **Rectangle**.
(2) Set the **Width** to $1.3\,E-3$, **Height** to $7.5\,E-3$, **Base** to **Corner** (which is the default setting) and **r** and **z** to 0. Click **OK**.
(3) Select **Draw** > **Specify Objects** > **Square**.
(4) Set the **Width** to $36\,E-3$, **Base** to **Corner** and **r** and **z** to 0. Click **OK**.
(5) Select **Draw** > **Create composite object**.
(6) Enter SQ1-R1 under **Set formula** and click **OK**.
(7) Click on **Zoom Extents** button on the top menu to fit the geometry in the screen.

Meshing

(1) Open the **Free Mesh Parameters** dialog box from the **Mesh** menu.
(2) Click the **Boundary** tab.

(3) Select 2 and 4 under **Boundary Selection** holding the Ctrl button, and enter 5E-5 as the **Maximum element size**. We need more elements near the electrode in the tissue since there is maximum temperature change in this region.

(4) Click on **Remesh**. Click **OK**.

Governing equations, source terms, I.C., B.C. – conduction

Subdomain settings

(1) Select **Heat Transfer by Conduction** under the **Multiphysics** menu.
(2) Open the **Subdomain Settings** dialog box from the **Physics** menu.
(3) Select subdomain 1 under **Subdomain selection** and enter the values of **k (isotropic)**, ρ, c_p and **Q** as shown:

Subdomain	1
k	k
ρ	1200
c_p	3200
Q	Q

(4) *Initial condition* Select subdomain 1, click on the **Init** tab and enter $T(t_0)$ as 310.15. Click **OK**.

Boundary settings

All boundaries for heat transfer have a zero flux condition due to insulation or symmetry. Since a zero flux boundary condition is the default in COMSOL, we will not change any settings.

Governing equations, source terms, I.C., B.C. – diffusion

Subdomain settings

(1) Select **Diffusion** under the **Multiphysics** menu.
(2) Open the **Subdomain Settings** dialog box from the **Physics** menu.
(3) Select subdomain 1, under **Subdomain Selection**.
(4) Check **D (isotropic)** and set **D (isotropic)** as 0.222. Click **OK**.

Boundary settings

(1) Make sure that **Diffusion** is selected under the **Multiphysics** menu.
(2) Open the **Boundary Settings** dialog box from the **Physics** menu.
(3) Select Boundary 1 under **Boundary Selection**, change the **Boundary Condition** to Axial Symmetry.
(4) Select Boundaries 2 and 4 under **Boundary Selection**, change the **Boundary Condition** to **Concentration** and set V_0 as 25. We set the electric potential at the electrode surface to 25 V. Note here that the concentration is treated as the electric potential in this case.
(5) Select Boundaries 3 and 6 under **Boundary Selection**, change the **Boundary Condition** to **Concentration** and set V_0 as 0. We set the electric potential at the outer surface of the model to 0 V. Note that in the previous method (i.e., using **Conductive Media DC** for modeling), 0 V is the default setting. For the diffusion equation, the default boundary condition is insulation (zero flux) and hence the boundary condition for boundary 5 is retained as default. Click **OK**.

Define expressions

The expressions for k and Q that were used in Subdomain Settings earlier are defined here.

(1) Open the **Expressions > Scalar Expressions** window from the **Options** menu.
(2) Define the following expressions:

Name	Expression
k	0.4925+0.001195*T
Q_blood	2000*(T − 310.15)
Q_dc	0.222*Vr*Vr+0.222*Vz*Vz
Q	Q_blood+Q_dc

Note here that Q_dc, the heat source term due to radiofrequency ablation, is calculated by specifying the explicit equation in COMSOL. Vr and Vz represent the differential of V with respect to r and z, respectively.

(3) Click **OK**.

Solver settings

(1) Click the **Solver Parameters** button.
(2) Select **Time dependent** in the **Solver** list.
(3) In the **Timestepping** area, type 0:1:60 in the **Times** edit field.
(4) Click **OK**.

Solution

(1) Compute the final solution by clicking **Solve Problem** under **Solve** on the Main toolbar.

Postprocessing

The results obtained from both the methods are the same hence, additional postprocessing is not shown here.

Adaptive mesh refinement

We now demonstrate how to proceed with adaptive mesh refinement in COMSOL. Since in COMSOL, adaptive mesh refinement is supported for steady-state problems only, we solve the radiofrequency ablation problem defined above to obtain the steady-state solution (by choosing the appropriate solver). We will use the COMSOL file created using the procedure described above. Please follow the additional steps below to obtain a solution for the problem using adaptive mesh refinement.

Open file

Open the file created in the last step.

Modify mesh

(1) Open the **Free Mesh Parameters** dialog box from the **Mesh** menu.
(2) Click the **Boundary** tab.

(3) Select 2 and 4 under **Boundary Selection** holding the Ctrl button, delete 5E-5 entered earlier in the **Maximum element size** box. We will now create a very coarse mesh in the domain and allow the solver to refine the mesh automatically through adaptive mesh refinement.
(4) In the **Global** tab, select **Extremely coarse** under **Predefined mesh size**.
(5) Click on **Remesh**. Click **OK**.

The total number of elements in the mesh just created is 65. The mesh is shown below. We will now obtain the steady-state solution to the problem using the coarse mesh first, so that we can compare it with the solution obtained after adaptive mesh refinement.

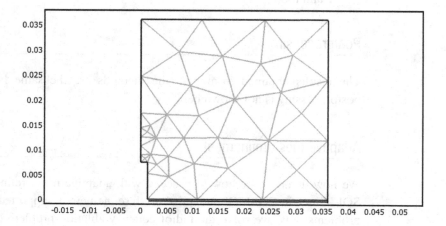

Change solver settings

(1) Click the **Solver Parameters** button.
(2) Select **Stationary** under **Analysis**.
(3) Click **OK**.

Solution

(1) Compute the final solution by clicking **Solve Problem** under **Solve** on the Main toolbar.

Postprocessing

We will plot the temperature contour plot so that we can compare the solution obtained from a coarse mesh and after adaptive mesh refinement more closely. To generate this plot:

(1) Open the **Plot Parameters** dialog box from the **Postprocessing** menu.
(2) On the **General** page, check **Contour**. Uncheck other plot type if already selected.
(3) On the **Contour** page, select Temperature under **Predefined quantities**, select °C as **Unit**.
(4) Click **OK**.

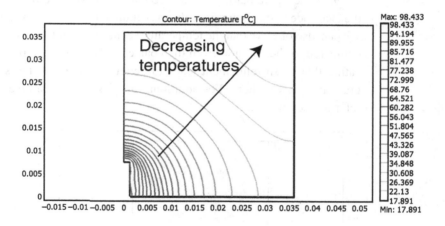

We now proceed with adaptive mesh refinement.

Change solver settings for adaptive mesh refinement

(1) Click the **Solver Parameters** button.
(2) Check.
(3) In the **Adaptive** tab, set **Maximum number of refinements** to 3 and **Maximum number of elements** to 20 000. This setting will limit the number of mesh refinements to 3 and the maximum number of elements in the mesh after mesh refinement to 20 000.
(4) Click **OK**.

Solution

(1) Compute the final solution by clicking **Solve Problem** under **Solve** on the Main toolbar.

Postprocessing

(1) Open the **Plot Parameters** dialog box from the **Postprocessing** menu.
(2) On the **General** page, check **Contour**. On the **Contour** page, select Temperature under **Predefined quantities**, select °C as **Unit**.
(3) Click **OK**.

The total number of elements obtained after 3 mesh refinements is found to be 1039. In the figure below, we can see that the mesh is automatically refined near the electrode in the tissue since the maximum temperature change is in this region. We can also observe that the temperature contours near the electrode are smoother compared to the solution from the coarse mesh above (see previous page), indicating that mesh refinement leads to better resolution of the steep temperature gradients present there. For additional discussion on adaptive mesh refinement refer to Section 5.3.2.

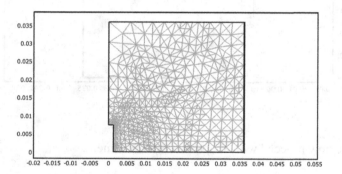

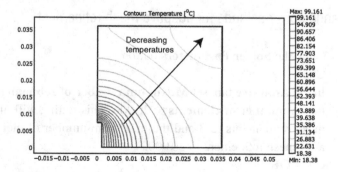

X Laser irradiation of human breast tumor

Laser irradiation is a treatment method for breast cancer. Hyperthermia during laser irradiation or other similar processes leads to change in blood flow and oxygen diffusion in the tissue and tumor regions (He *et al.*, 2006). The aim of this example is to model laser irradiation in a human breast tumor and determine the change in blood flow and oxygen diffusion. The model developed by He *et al.* (2006) and used by Sood *et al.* (2007) will be followed.

Problem formulation

Figure 6.15(a) shows the image obtained from mammography of a cancerous breast. Figure 6.15(b) shows a schematic of a breast tumor being irradiated by laser. The breast geometry includes subcutaneous fat and gland regions with the tumor embedded inside. The tumor is assumed to be spherical and placed symmetrically along the axis of the breast so that the problem can be reduced to axisymmetric. The laser heats the tumor and some part of the healthy region and thereby increases their average temperature. The change in temperature of the tumor changes the radii of the blood vessels within it. The blood flow and oxygen diffusion inside a single blood vessel is modeled using the schematic shown in Figure 6.16.

Governing equations

Heat transfer in the subcutaneous fat, gland and tumor regions is governed by the bioheat equation:

$$\rho C_p \frac{\partial T}{\partial t} = k \left[\frac{1}{r} \frac{\partial}{\partial r} \left(r \frac{\partial T}{\partial r} \right) + \frac{\partial^2 T}{\partial z^2} \right] + \rho_b C_{p,b} \dot{V}_b^v (T_a - T) + Q_m + Q \quad (6.25)$$

where Q_m represents the metabolic heat generation rate and Q is the heat due to laser irradiation, given by

$$Q = \alpha I_0 e^{-\alpha z} \quad (6.26)$$

Here, α is the specific absorption rate and I_0 is the laser power intensity at the surface of the breast.

Blood flow inside the capillary is governed by the continuity and Navier–Stokes equations:

$$\frac{\partial u_z}{\partial z} = 0 \quad (6.27)$$

(a)

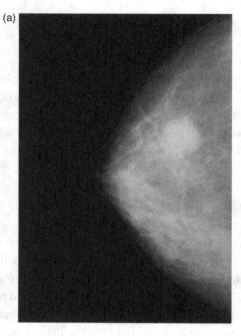

(b)
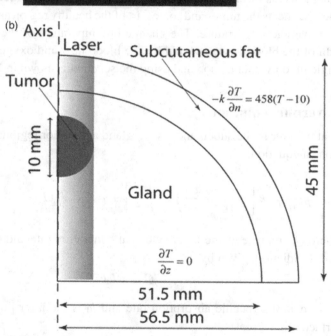

Figure 6.15

(a) Mammography image showing a cancerous breast (the cancerous part is the white circular area). Mammography is the process of examination of the breast using X-rays for early detection of tumors. (b) Assuming the breast to be axisymmetric and the tumor to be on the axis, the computational domain and boundary conditions for the heat transfer problem are as shown.

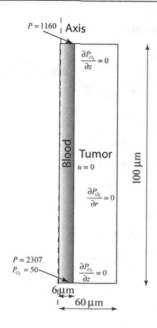

Figure 6.16

Schematic, boundary and initial conditions for the blood flow–oxygen diffusion problem.

$$\rho_b \left(\frac{\partial u_z}{\partial t} + u_z \frac{\partial u_z}{\partial z} \right) = -\frac{\partial p}{\partial z} + \mu \left[\frac{1}{r} \frac{\partial}{\partial r} \left(r \frac{\partial u_z}{\partial r} \right) + \frac{\partial^2 u_z}{\partial z^2} \right]. \quad (6.28)$$

Oxygen diffusion inside the capillary and tumor regions is given by the transient convection–diffusion equation:

$$\frac{\partial P_{O_2}}{\partial t} + u_z \frac{\partial P_{O_2}}{\partial r} = D \left[\frac{1}{r} \frac{\partial}{\partial r} \left(r \frac{\partial P_{O_2}}{\partial r} \right) + \frac{\partial^2 P_{O_2}}{\partial z^2} \right] - \frac{M}{a}, \quad (6.29)$$

where M is the metabolic rate of oxygen consumption in the tumor and a is the oxygen solubility. The flow only occurs in the z direction. There is no blood flow in the tumor region and therefore, the velocity is set to zero there.

The radius of the capillary at any time is a function of the average temperature of the tumor and is given by

$$r = r_0 \sqrt{e^{b(T-T_0)}} \quad (6.30)$$

where r_0 and T_0 are the capillary radius and temperature before heating, respectively.

The problem consists of two parts: heat transfer due to laser irradiation in the breast and blood flow and oxygen diffusion in a single capillary and adjoining tumor tissue. The two geometries used for the problem along with the boundary conditions are shown in Figures 6.15 and 6.16.

Table 6.8 Input parameters (all values are taken from He et al. (2006) except as shown).

Parameter	Value
Laser intensity, I_0	13000 W m^{-2}
Specific absorption rate, α	100 m^{-1}
Duration of heating, t	1200 s
Convective heat transfer coefficient, h	458 Wm^{-2}K^{-1}
Cooling water temperature, T_w	10 °C
Thermal conductivity, k	
Subcutaneous fat	0.22 Wm^{-1}K^{-1}
Gland	0.48 Wm^{-1}K^{-1}
Tumor	0.48 Wm^{-1}K^{-1}
Density, ρ	
Blood, ρ_b	1100 kgm^{-3}
Subcutaneous fat	930 kgm^{-3}
Gland	1050 kgm^{-3}
Tumor	1050 kgm^{-3}
Specific heat, C_p	
Subcutaneous fat	2770 Jkg^{-1}K^{-1}
Gland	3770 Jkg^{-1}K^{-1}
Tumor	3770 Jkg^{-1}K^{-1}
Metabolic heat generation, q_m	(Ng and Sudharsan, 2001)
Subcutaneous fat	400 W m^{-3}
Gland	700 W m^{-3}
Tumor	65400 W m^{-3}
Blood perfusion coefficient, $\rho_b C_{p,b} \dot{V}_b^v$	(Ng and Sudharsan, 2001)
Subcutaneous fat	800 Wm^{-3}K^{-1}
Gland	2400 Wm^{-3}K^{-1}
Tumor	48000 Wm^{-3}K^{-1}
Arterial blood temperature, T_a	37 °C

Table 6.8 (continued).

Parameter	Value
Initial tissue temperature, T_i	37 °C
Inlet pressure	2307 Pa (Jain, 1988)
Outlet pressure	1160 Pa (Jain, 1988)
Blood viscosity, μ	0.0044 Pa s
Oxygen diffusivity, D	1.5×10^{-9} m^2 s^{-1}
Oxygen solubility, a	3×10^{-5} cm^3 O$_2$ cm^{-3}/mmHg
Metabolic oxygen consumption rate, M	3.3×10^{-5} cm^3 O$_2$ cm^{-3}
Inlet oxygen concentration	50 mmHg

References

He, Y., Shirazaki, M., Liu, H., Himeno, R., and Sun, Z. G. (2006). A numerical coupling model to analyze the blood flow, temperature, and oxygen transport in human breast tumor under laser irradiation. *Computers in Biology and Medicine*, **36**, pp. 1336–1350.

Jain, R. K. (1988). Determinants of tumor blood-flow - a review, *Cancer Research*, **48**, pp. 2641–2658.

Ng, E. Y. K. and Sudharsan, N. M. (2001). An improved three-dimensional direct numerical modelling and thermal analysis of a female breast with tumour. Proceedings of the Institution of Mechanical Engineers Part H-*Journal of Engineering in Medicine*, **215**, pp. 25–37.

Sood, R., Nahlik, D., Nichols, W., and Graham E. (2007). Laser irradiation of tumors for the treatment of cancer: An analysis of blood flow, temperature and oxygen transport. On the web at http://ecommons.library.cornell.edu/handle/1813/7906

Implementation in COMSOL

The transient heat conduction equation in COMSOL is solved to obtain the temperatures inside the domain using the geometry shown in Figure 6.15. The average temperature of the tumor as a function of time is obtained during postprocessing.

The radius of the capillary is related to the average tumor temperature by Equation 6.30. The radius of the capillary as a function of time is thereby obtained using Equation 6.30 (knowing the average temperature as a function of time) in a spreadsheet application (Microsoft Excel 2003).

The Navier–Stokes and transient convection-diffusion equations are then solved in COMSOL using the geometry shown in Figure 6.16. The change of radius with time is incorporated as a mesh (representing the capillary region) moving with a specified velocity in COMSOL.

> **Box 6.29**
>
> **Steps : heat transfer problem**
>
> (1) **Problem type specification** We will setup the problem as a transient conduction problem to solve for temperatures in an axisymmetric setting.
> (2) **Geometry creation** We will draw the geometry shown in Figure 6.15.
> (3) **Meshing** We will create an unstructured mesh in the domain.
> (4) **Defining governing equations, source terms, I.C., B.C.** We will specify the thermal conductivity, specific heat, density and heat source terms in the different regions. The initial body temperature will be specified. We will also specify the boundary conditions as shown in Figure 6.15.
> (5) **Define postprocessing variables** We will define postprocessing variables to calculate the average temperature of the tumor as a function of time. This will be used to solve the next part of the problem.
> (6) **Solver setting and solution** We will solve the problem for 20 min.
> (7) **Postprocessing** We will plot the average tumor temperature as a function of time and save the data. We will then use the saved data to calculate the rate of change of capillary radius and create a COMSOL-compatible text input file to solve the flow and convection–diffusion problem.
>
> **Steps : blood flow problem**
>
> (1) **Problem type specification** We will use the Navier–Stokes and the transient convection–diffusion equation to solve the second part of the problem. Since mesh movement is involved in this problem, we will also select the Moving Mesh (ALE) module.
> (2) **Geometry creation** We will draw the geometry representing the capillary and tumor as shown in Figure 6.16.
> (3) **Meshing** We will create an unstructured mesh.
> (4) **Defining governing equations, source terms, I.C., B.C.** We will specify the mesh displacement and mesh velocity, density and viscosity of the blood, and diffusivity of oxygen in the blood stream and tumor, along with the reaction term. We will also specify the boundary conditions as shown in Figure 6.16.
> (5) **Define function** We will define a function that uses the data saved in the heat transfer problem (mesh velocity as a function of time) to interpolate the mesh velocity at different times.

(6) **Solver setting and solution** We will first obtain a steady-state solution as blood flow is already present at the initial time. We will then solve the problem for 20 min.

(7) **Postprocessing** We will plot the oxygen distribution in the capillary and tumor after 20 min.

Problem type specification

(1) Open COMSOL Multiphysics.
(2) In the **Model Navigator** select Axial Symmetry (2D) in the **Space dimensions** list.
(3) In the list of application modes select
    COMSOL Multiphysics > Heat Transfer > Conduction> Transient Analysis.
(4) Click **OK**.

Geometry creation

(1) Select **Draw > Specify Objects > Circle >** Enter **Radius** as $51.5\,E-3 >$ Click **OK**.
(2) Select **Draw > Specify Objects > Circle >** Enter **Radius** as $56.5\,E-3 >$ Click **OK**.
(3) Select **Draw > Specify Objects > Circle >** Enter **Radius** as $5\,E-3$, z as $-45.5\,E-3 >$ Click **OK**.
(4) Select **Draw > Specify Objects > Rectangle**.
(5) Set the **Width** to $113\,E-3$, **Height** to $68\,E-3$, **Base** to **Corner** and **r** and **z** to $-56.5\,E-3$ and $-11.5\,E-3$, respectively. Click **OK**.
(6) Select **Draw > Create Composite Object > Set Formula** as C1+C2+C3 > Click **Apply**.
(7) Now, in the **Create Composite Object** Window, **Set Formula** as CO1-R1. Click **OK**.
(8) Select **Draw > Specify Objects > Rectangle**.
(9) Set the **Width** to $56.5\,E-3$, **Height** to $56.5\,E-3$, **Base** to **Corner** and **r** and **z** to $-56.5\,E-3$ and $-56.5\,E-3$, respectively. Click **OK**.
(10) Select **Draw > Create Composite Object > Set Formula** as CO2-R1 > Click **OK**.

(11) Select **Draw** > **Modify** > **Move** > Enter z as 56.5 E−3. Make sure that the entire geometry is selected (shown in red) before it is moved. If it is not selected, hold Ctrl and click on the different subdomains to select each of them. Click on the **Zoom Extents** button on the top menu (in the second row) to fit the geometry in the screen.

Meshing

(1) Open the **Free Mesh Parameters** dialog box from the **Mesh** menu.
(2) Under **Predefined** mesh sizes select Finer.
(3) Click **Remesh**. Click **OK**.

Governing equations, source terms, I.C., B.C. – conduction

Subdomain settings

(1) Select **Conduction** under the **Multiphysics** menu.
(2) Open the **Subdomain Settings** dialog box from the **Physics** menu.
(3) Select the subdomains under **Subdomain selection** and enter the values of **k (isotropic)**, ρ, **c**$_p$ and **Q** as shown:

Subdomain	1	2	3
k	0.22	0.48	0.48
ρ	930	1050	1050
c$_p$	2770	3770	3770
Q	src_fat	src_gland	src_tumor

(4) *Initial condition* Select subdomains 1, 2 and 3 (holding Ctrl and selecting them under **Subdomain Selection**), click on the **Init** tab and enter $T(t_0)$ as 310.15. Click **OK**.

Boundary settings

(1) Make sure that **Conduction** is selected under the **Multiphysics** menu.
(2) Open the **Boundary Settings** dialog box from the **Physics** menu.

(3) Select **Boundary** 7 under **Boundary Selection**, change the **Boundary Condition** to Heat Flux and set **h** as 458 and T_{inf} as 283.15.
(4) Click **OK**. Note: All other boundaries have heat flux as zero due to insulation or symmetry. Since the default boundary condition of the solver is zero flux, we retain the default settings.

Define expressions

The expressions for the heat source term, Q, that were used in *Subdomain Settings* earlier are defined here.

(1) Open the **Expressions > Scalar Expressions** window from the **Options** menu.
(2) Define the following expressions:

Name	Expression
src_fat	800*(310.15−T)+400+100*1.3E4*exp(−100*z)*(r<=0.005)
src_gland	2400*(310.15−T)+700+100*1.3E4*exp(−100*z)*(r<=0.005)
src_tumor	48000*(310.15−T)+65400+100*1.3E4*exp(−100*z)

Note here that all the heat source terms have three components: blood perfusion, metabolic heat generation and laser heating. The laser does not act beyond a radius of 0.005 m in the subcutaneous fat and gland layer and hence, the logical expression (r<=0.005) is used in the equation with the laser heating term. The expression, r<=0.005, returns 1 if it is true, i.e., if r is less than or equal to 0.005, and 0 otherwise.

(3) Click **OK**.

Define postprocessing variables

The average temperature of the tumor as a function of time needs to be calculated as the radius of the capillary (which will be used in the second part of the problem) depends on it. We define the variable to calculate the average temperature of the tumor here, so that it can be used later.

(1) On the **Options** menu, point to **Integration Coupling Variables** and then click **Subdomain Variables**.

(2) Select Subdomain 3 under **Subdomain selection**.
(3) Type avg_tum_temp under **Name** and 2*pi*r*T/(4/3*pi*0.005^3) under **Expression**. 2*pi()*r is multiplied to T (temperature) as the problem is axisymmetric and therefore, we want to calculate the volume integral. The value of the integral is divided by the volume of the tumor (4/3*pi*0.005^3), represented as a sphere.
(4) Click **OK**.

Solver settings

(1) Click the **Solver Parameters** button.
(2) Select **Time dependent** in the **Solver** list.
(3) In the **Timestepping** area under the **General** tab, type 0:1:1200 in the **Times** edit field.
(4) Under the **Time Stepping** tab, select **Time steps taken by solver** to Strict.
(5) Click **OK**.

Solution

(1) Compute the final solution by clicking **Solve Problem** under **Solve** on the Main toolbar.

Postprocessing

The average tumor temperature as a function of time is needed to solve the subsequent problem, where the radius depends on the average temperature. Therefore, we will plot the average temperature in the tumor as a function of time and then save the data in a text file for further calculations.

(1) Open the **Global Variables Plot** dialog box from the **Postprocessing** menu.
(2) Type avg_tum_temp under **Expression** and click on the arrow (>) next to it to move it to the **Quantities to plot** list on the right.
(3) Select avg_tum_temp in the **Quantities to plot** list. Click **OK**. The Figure window opens up and displays the plot as shown below.
(4) Click on the **Save Current Plot in ASCII file** button in the **Figure** window (fourth button on the top denoted by **ASC=**).
(5) Specify the file name and save the file in a desired folder. We will use this file for calculating the radius as a function of time.

X Laser irradiation of human breast tumor

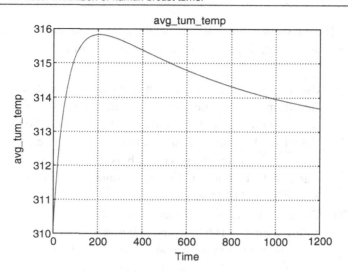

Calculations for subsequent analysis

The calculations shown here are performed using Microsoft Office Excel 2003. Any other spreadsheet application or code can also be used.

(1) Open the file in WordPad. It should look like this:

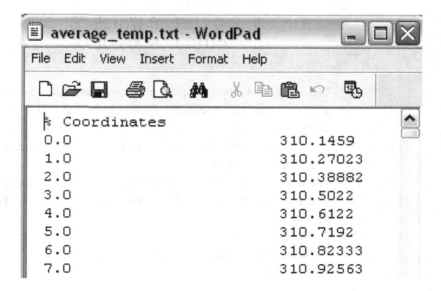

The numbers on the left indicate the time (these should go up to 1200) and the ones on the right are the average tumor temperatures.

(1) Use a spreadsheet application such as Microsoft Excel and import the data from the text file so that the times are listed in one column and the corresponding temperatures in another column to the right.
(2) Calculate the radius as a function of time given by the equation: $r = r_0 \sqrt{e^{b(T-T_0)}}$, using r_0 as 6 E−6, b as 0.1 if the average temperature is less than 315.15 K, b as −0.1 if the average temperature is greater than 315.15 K, and T_0 as 310.15 K.
(3) Calculate the change in radius for consecutive times. For example, if the radius at time 0 s is 6 E−6 and radius at time 1s is 6.04 E−6, then the change in radius for the first interval is (6.04 − 6) E − 6 = 0.4 E−6. This can be done very easily in a spreadsheet application. *Note here that the change in radius is also equal to the velocity with which the radius changes (since the time interval is 1s).*
(4) The calculation steps should look like this (shown in Microsoft Excel):

	A	B	C	D
1	0	310.1459	5.99877E-06	0
2	1	310.2702	6.03618E-06	3.74075E-08
3	2	310.3888	6.07206E-06	3.58796E-08
4	3	310.502	6.10655E-06	3.44897E-08
5	4	310.612	6.14021E-06	3.36663E-08
6	5	310.7189	6.17312E-06	3.29073E-08
7	6	310.823	6.20533E-06	3.22056E-08

The column A shows the times, B indicates the average tumor temperatures, C has the calculated radii values (calculated using the formula, =0.000006*SQRT(EXP(0.1*(B1−310.15))), for cell C1, in Excel) and D has the change in radius in different time intervals (calculated using the formula, = C2−C1, for cell D2).

(5) The last step is to create a text file that contains the velocity of radius change as a function of time which can be used in COMSOL. The format of the file is:

%Time
Time values separated by spaces
%Velocity

Velocity values separated by spaces
It should be made sure that there are no line breaks between the data.

```
velocity.txt - WordPad
File  Edit  View  Insert  Format  Help

%time
0       1       2       3       4       5       6
        26      27      28      29      30      31
        51      52      53      54      55      56
        214     215     216     217     218     219
        239     240     241     242     243     244
                401             402     403     404     405
        425     426     427     428     429     430
        450     451     452     453     454     455
        611     612     613     614     615     616
        636     637     638     639     640     641
        661     662     663             801             802
        822     823     824     825     826     827
        847     848     849     850     851     852
        1008    1009    1010    1011    1012    10:
        1033    1034    1035    1036    1037    10:
        1058    1059    1060    1061    1062    10(
%velocity
0       3.74E-08        3.59E-08        3.45E-08
08      2.79E-08        2.75E-08        2.7E-08
08      2.27E-08        2.23E-08        2.2E-08
08      1.85E-08        1.82E-08        1.8E-08
08      1.5E-08         1.48E-08        1.45E-08
11      -4.1E-11        -2.7E-11        -1.4E-11
11      1.11E-10        1.24E-10        1.31E-10
```

(6) The above format can be created using Excel: copy the data (as shown in (5) above) in a particular column > **Paste special** > Select **Paste Values** and **Transpose**. The procedure should be repeated as Excel (2003) provides a maximum of 256 columns and the data used here needs 1201 columns.

(7) Finally, copy the values from Excel into an empty WordPad file and create the file as shown below by typing %time and %velocity at the appropriate places. Remember to remove any line breaks between data manually in WordPad; otherwise the file cannot be used in COMSOL. To make sure there are no line

breaks, remove the *Word Wrap* feature (under View >> Options) in WordPad and paste the different row values from Excel on the same line. Save the file in a desired folder. The filename, velocity.txt, is used here and saved in the same folder as the COMSOL file. Also, make sure that you select file type as *Text Document* under *Save as type* in WordPad and *not* any other format such as *Rich Text Format (RTF)*.

As discussed earlier, we will now solve the Navier–Stokes and transient convection–diffusion equations in COMSOL to model blood flow and oxygen diffusion inside a single blood vessel, using the geometry shown in Figure 6.16. The radius of the capillary changes with time as obtained from the previous analysis and this is also incorporated in the model.

Problem type specification

(1) Open COMSOL Multiphysics.
(2) In the **Model Navigator** select Axial Symmetry (2D) in the **Space dimensions** list, then click the **Multiphysics** button.
(3) In the list of application modes select
COMSOL Multiphysics > Deformed Mesh > Moving Mesh (ALE) > Transient Analysis and add it to the model with the **Add** button.
(4) Then select
COMSOL Multiphysics > Fluid Dynamics > Incompressible Navier-Stokes > Transient Analysis and add it to the model with the **Add** button.
(5) Then select
COMSOL Multiphysics > Convection and diffusion > Convection and diffusion > Transient Analysis and add it to the model with the **Add** button.
(6) Click **OK**.

Geometry creation

(1) Select **Draw > Specify Objects > Rectangle**.
(2) Set the **Width** to 6 E−6, **Height** to 100 E−6, **Base** to **Corner** and **r** and **z** to 0. Click **OK**.
(3) Similarly, create 1 more rectangle with these specifications: Set the **Width** to 54 E−6, **Height** to 100 E−6, **Base** to **Corner**, **r** to 6 E−6 and **z** to 0. Click **OK**.
(4) Click on the **Zoom Extents** button on the top menu (in the second row) to fit the geometry in the screen.

Meshing

(1) Open the **Mapped Mesh Parameters** dialog box from the **Mesh** menu.
(2) Click the **Boundary** tab.
(3) Select Boundaries 1, 4 and 7 under **Boundary Selection** holding the Ctrl button, check **Constrained edge element distribution** and set **Number of edge elements** to 20.
(4) Similarly, select Boundaries 2 and 3 under **Boundary Selection**, check **Constrained edge element distribution** and set **Number of edge elements** to 10.
(5) Finally, select Boundaries 5 and 6 under **Boundary Selection**, check **Constrained edge element distribution** and set **Number of edge elements** to 60.
(6) Click on **Remesh**. Click **OK**.

Governing equations, source terms, I.C., B.C. : Moving mesh

Subdomain settings

The default values of **Subdomain settings** are retained for the **Moving Mesh (ALE)** application mode and only the boundary settings are changed.

Boundary settings

(1) Select **Moving Mesh (ALE)** under the **Multiphysics** menu.
(2) Open the **Boundary Settings** dialog box from the **Physics** menu.
(3) Select Boundaries 1 and 7 under **Boundary Selection**, select **Mesh Displacement**, check **dr** and **dz** and set them to zero (default value). These values make sure that the mesh does not move in any direction.
(4) Similarly, select Boundaries 2, 3, 5 and 6 under **Boundary Selection**, select **Mesh Displacement**, check **dz** and set it to zero. Do not check **dr**. This is done so that the mesh is free to move in the r direction.
(5) Check **Interior boundaries**. Boundary 4 under Boundary Selection becomes active. It is on this boundary that we will specify the mesh velocity calculated earlier and saved in file, velocity.txt.
(6) Select Boundary 4 under **Boundary Selection**, select **Mesh velocity**, check **vr** and enter vel(t) in the text edit field. Then, check **vz** and set it to zero. We will define vel(t) later. 't' is used to denote that velocity depends on time.
(7) Click **OK**.

Governing equations, source terms, I.C., B.C.: Incompressible Navier–Stokes

Subdomain settings

(1) Select **Incompressible Navier-Stokes (ns)** under the **Multiphysics** menu.
(2) Open the **Subdomain Settings** dialog box from the **Physics** menu.
(3) Select subdomain 1 under **Subdomain Selection**, set ρ as 1100 and η as 0.0044.
(4) Select subdomain 2 under **Subdomain Selection**, uncheck **Active in this domain**. Navier–Stokes equations are not solved in Subdomain 2 as it represents the tissue where no blood flow occurs.
(5) Click **OK**.

Boundary settings

(1) Make sure that **Incompressible Navier-Stokes (ns)** is selected under the **Multiphysics** menu.
(2) Open the **Boundary Settings** dialog box from the **Physics** menu.
(3) Select Boundary 1 under **Boundary Selection**, change the **Boundary Condition** to Axial Symmetry.
(4) Select Boundary 2 under **Boundary Selection**, change the **Boundary Condition** to Pressure and set P_0 as 2307.
(5) Select Boundary 3 under **Boundary Selection**, change the **Boundary Condition** to Pressure and set P_0 as 1160.
(6) Select Boundary 4 under **Boundary Selection**, change the **Boundary Condition** to No slip. Click **OK**.

Governing equations, source terms, I.C., B.C. : Convection and diffusion

Subdomain settings

(1) Select **Convection and Diffusion (cd)** under the **Multiphysics** menu.
(2) Open the **Subdomain Settings** dialog box from the **Physics** menu.
(3) Select subdomain 1, under **Subdomain Selection**, set **D isotropic** as $1.5\,\text{E}-9$ and **R** as 0.
(4) Select subdomain 2, under **Subdomain Selection**, set **D isotropic** as $1.5\,\text{E}-9$ and **R** as $-1.1*(c>0)+0*(c<=0)$. The expression for R denotes

that the reaction term is equal to -1.1 for concentration (c) less than zero and is 0 otherwise.
(5) Click **OK**.

Boundary settings

(1) Make sure that **Convection and Diffusion (cd)** is selected under the **Multiphysics** menu.
(2) Open the **Boundary Settings** dialog box from the **Physics** menu.
(3) Select Boundary 1 under **Boundary Selection**, change the **Boundary Condition** to Axial Symmetry.
(4) Select Boundary 2 under **Boundary Selection**, change the **Boundary Condition** to Concentration and set c_0 as 50.
(5) Select Boundary 3 under **Boundary Selection**, change the **Boundary Condition** to Convective flux. Note: all other boundaries have mass flux equal to zero as the boundary condition. Since, the default boundary condition of the solver is zero flux, we retain the default settings. Click **OK**.

Define function

The function, vel(t), used earlier needs to be defined before the problem is solved.

(1) Open the **Functions** window from the **Options** menu.
(2) Click on **New**, enter Function name as vel, select **Interpolation**, and select **File** in the drop down menu below **Use data from**.
(3) Click on **Browse** and select the file, velocity.txt (or your filename), created earlier. Click **OK**.
(4) Select **Interpolation Function** in the drop down menu next to **Extrapolation method** under Function Definition. Click **OK**.

Solver settings and solution

The problem is time-dependent. However, it first solved as a steady-state problem since blood flow is already present at the initial time. We define these settings now.

(1) Open the **Solver Parameters** dialog box from the **Solve** menu.
(2) Select **Static** under **Analysis** in the **General** tab. Click **OK**.
(3) Open the **Solver Manager** dialog box from the **Solve** menu.

(4) Click on the **Solve for** tab. All the three application modes, **Moving Mesh (ALE)**, **Incompressible Navier-Stokes (ns)** and **Convection and Diffusion (cd)** are selected initially.
(5) Hold Ctrl key and click on **Moving Mesh (ALE)**, to deselect it. We do not solve for the **Moving Mesh** in the steady-state case.
(6) Click **Solve**.

Once a steady-state solution is obtained, we will solve the actual transient problem.

(1) Open **Solver Parameters** dialog box from the **Solve** menu.
(2) Select **Transient** under **Analysis** in the **General** tab. The **Solver** should automatically be selected as **Time dependent**.
(3) In the **Timestepping** area in the **General** tab, type 0:1:1200 in the **Times** edit field.
(4) Under the **Time Stepping** tab, select **Time steps taken by solver** to Strict. Click **OK**.
(5) Open the **Solver Manager** dialog box from the **Solve** menu.
(6) Click on the **Solve for** tab. Only the two application modes, **Incompressible Navier-Stokes (ns)** and **Convection and Diffusion (cd)** are selected initially.
(7) Hold Ctrl key and click on **Moving Mesh (ALE)**, to select it.
(8) Under the **Initial value** tab, select **Current solution** under **Initial value**. This is done so that the initial conditions for the **Incompressible Navier-Stokes (ns)** and **Convection and Diffusion (cd)** application modes are taken from the steady-state solution.
(9) Click **Solve**.

Note that it will take around 10–15 min to solve the problem.

Postprocessing

We will plot the oxygen distribution in the capillary and tumor at 1200 s (end time) using a contours plot. To generate this plot:

(1) Open the **Plot Parameters** dialog box from the **Postprocessing** menu.
(2) On the **General** page, check **Contour**. Uncheck other items if they are already selected.
(3) On the **Contour** page, select **Concentration** under **Predefined quantities**.
(4) Under **Contour Levels**, select **Levels**, and type 27:3:48 under **Number of levels**. Click **OK**. This will plot contours starting from a value of 27 with increments of 3 to 48. The contour plot is shown below.

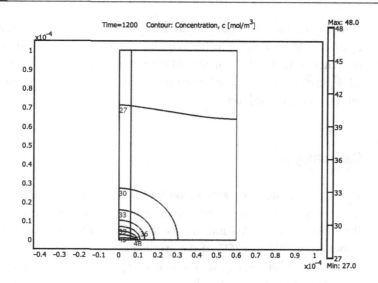

6.4 Problems

Supplementary case study questions *For these problems, complete the respective case study in the chapter and then perform the analysis mentioned below.*

6.4.1 Case study I

Perform a mesh convergence analysis. See Section 5.3 for an example of mesh convergence. Solve the same problem using two meshes with fewer elements and determine if the results are independent of the mesh.

6.4.2 Case study II

Include the blood perfusion term in the heat transfer equation for the normal tissue region. Solve the problem again and compare the results.

6.4.3 Case study III

In this case study, the drug supplied from the patch to the skin surface is assumed to be a constant flux of 8.849×10^{-7} g/m^{2}s. To model the problem more accurately, assume that the patch has a finite thickness of 0.3 mm. The radius of the patch is 24 mm (the same as in the original problem). The diffusivity of the drug in the patch is 1.11×10^{-14} m^2/s and the initial concentration of drug in the patch is 21.4 g/m^3. Solve the problem with the new formulation and compare the results with the original solution.

6.4.4 Case study IV

Perform a Monte Carlo simulation for the problem. Determine how the change in diffusivity of the drug in the contact lens affects the amount of drug delivered to the aqueous humor. Assume that the diffusivity of the drug in the contact lens layer varies from 75% to 125% of the original diffusivity value (D). Assume a normal distribution of diffusivity values in this range and determine the distribution of the amount of drug delivered to the aqueous humor region. Refer to Case study III for an example of Monte Carlo simulation.

6.4.5 Case study V

Perform adaptive mesh refinement for the problem. Refer to Case study IX for an example of adaptive mesh refinement.

6.4.6 Case study VI

Solve a highly simplified version of the problem by considering the geometry in 2D. Create the 2D geometry by modeling flow in only the cross-section along the axis. Compare the results with the original problem. Is 2D a good assumption in this case? What assumptions are we making when we model the geometry as 2D?

6.4.7 Case study VII

The initial concentration of antigen in the tumor was taken as 76 000 nM in the case study. Consider three different values of antigen concentration: 7600, 760 and 76 nM and solve the problem for each case. Plot the free antibody, antigen and complex concentrations as functions of time for the different initial antigen concentrations and comment on the results.

6.4.8 Case study VIII

The temperature of the heated disc was taken to be 90 °C in the case study. If the temperature of the heated disc is increased to 150 °C, solve the problem to determine the degree of burn injury in the tissue.

6.4.9 Case study IX

The thermal conductivity in the tissue is a function of temperature. Determine how the results would change if the thermal conductivity had a constant value of 0.55 W/mK.

6.4.10 Case study X

Solve the problem for a new location of the tumor and compare the results with the original solution if the location of the tumor inside the breast changes. The new location of the center of the tumor is 10 mm displaced toward the center of the breast.

6.4.11 Case study X (2)

Using the form of objective function (Eqn. 5.6) given in Problem 5.7.11, optimize the duration of laser heating. Refer to Case study II for an example of a COMSOL implementation of an optimization problem.

Complete problem formulations *The problems below are comprehensive, requiring knowledge of most of the chapters in this book. It is recommended that especially chapters 7–9 are studied before attempting these problems.*

6.4.12 Laser surgery

Laser-assisted reshaping of cartilage is a new surgical procedure designed to allow *in situ* treatment of deformities in the head and neck with less morbidity than traditional approaches. We would like to study the temperature distribution in a slab of porcine nasal cartilage during laser irradiation. The geometry, with location of laser, is shown in Figure 6.17(a). In a published work using the finite element model, the geometry and mesh are as shown in Figure 6.17(b). (1) Comment

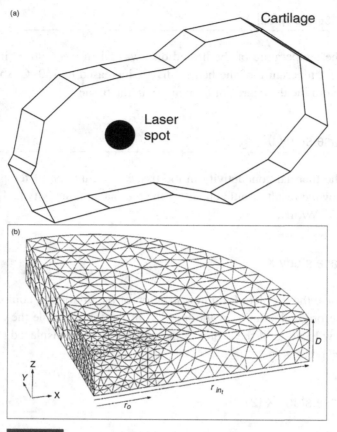

Figure 6.17

(a) Cartilage showing laser spot; (b) computational geometry and mesh. From Diaz et al., 2001 (©2001 IEEE). Reproduced with permission.

on whether assuming a circular disc of thickness D, with symmetry along the two planes (allowing the use of quarter-disk) is appropriate. (2) Is the variation in mesh density, i.e., more elements in some locations and fewer in the rest, appropriate? Explain. (3) What properties and parameters are needed to solve this problem? Comment on the expected level of difficulty in finding them. (4) The spatial distribution of laser beam intensity is given below. How would you incorporate this in a software such as COMSOL? Describe in words. (5) The predicted surface temperature history along the radial distance is given by Figure 6.18. Explain qualitatively, using your understanding of heat transfer, the variation of temperature with time and with location. (6) The authors determined the accuracy of predictions by comparing with experimental results, as shown in Figure 6.19. (a) How good is the comparison? Give reasons for your answer; (b) if experimental data

6.4 Problems

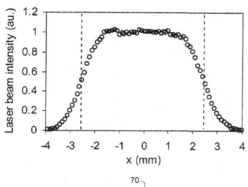

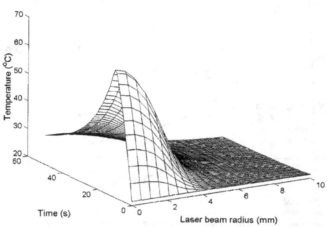

Figure 6.18
Spatial distribution of laser beam intensity. From Diaz et al., 2001 (©2001 IEEE). Reproduced with permission.

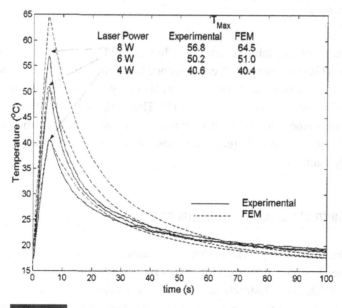

Figure 6.19
Computed surface temperature distribution (top figure) in the cartilage and comparison of predicted versus measured surface temperature (bottom figure). From Diaz et al., 2001 (©2001 IEEE). Reproduced with permission.

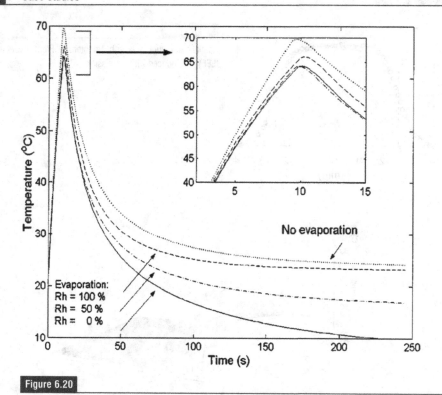

Figure 6.20

Temperatures as affected by relative humidity around the surface. From Diaz et al., 2001 (©2001 IEEE). Reproduced with permission.

were not available to compare, how would you check the accuracy of calculation? Remember: mesh convergence is not a choice since mesh convergence does not really provide information on accuracy. (7) What variable(s) would you consider for sensitivity analysis? (8) The authors performed a sensitivity analysis with respect to the relative humidity of the region surrounding the laser spot. This is shown in Figure 6.20. How sensitive is the process to relative humidity? Explain.

6.4.13 Thermal balloon endometrial ablation

One of the ways Menorrhagia, a condition associated with excessive menstrual bleeding that affects up to 19% of reproductive women worldwide, can be treated is by thermal destruction of the uterus interior surface wall. A latex balloon filled with hot water is placed in the uterus. Heat is transferred from the water in the balloon to the inside surface of the uterus, resulting in a burn injury.

6.4 Problems

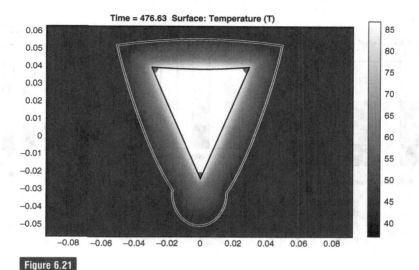

Figure 6.21

Temperature contours for the initial 2D simulation. From Baldwin *et al.* (2001). With kind permission from Springer Science+Business Media.

A model is used to investigate the effects of balloon surface temperature and treatment time on the depth of tissue injury and temperature distribution through the uterus wall, in an effort to better optimize the procedure. (1) Provide the complete problem formulation following Section 1.12.3. Figure 6.21 shows the temperature contours for the initial 2D simulation from the work of Baldwin *et al.* (2001). To simplify the geometry, we would like to consider a 1D situation. (2) Comment on whether a 1D assumption is a good idea in this case. (3) Provide details of how you would model the extent of thermal burn injury. Figure 6.22 summarizes some results of sensitivity analysis. (4) Summarize the effect of varying metabolic heat generation on tissue damage in one sentence and provide the probable reasoning behind your observation. (5) Summarize the effect of blood perfusion rate in one sentence and provide the probable reasoning behind your observation. (6) Summarize the effect of various temperature-time combinations for the fluid in the balloon in one sentence and provide the probable reasoning behind your observation.

Implementation in COMSOL *The problems below include the problem formulation steps and all information necessary for solution in COMSOL or any other software. Implement each of these problems in COMSOL.*

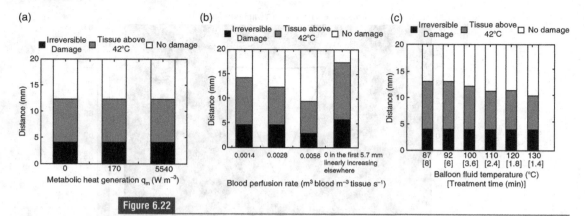

Figure 6.22

Effect of varying (a) metabolic heat generation; (b) blood perfusion rate; (c) temperature–time combinations. From Baldwin *et al.* (2001). With kind permission from Springer Science+Business Media.

6.4.14 Transdermal scopolamine patch

Motion sickness is a common ailment afflicting travelers. A transdermal patch is available on the market which prevents motion sickness. The small circular patch is applied to the hairless area behind the ear and it gradually delivers the drug scopolamine over a three-day period. A schematic of the patch is shown in the Figure 6.23. The objective is to model the diffusion of scopolamine through the patch into the skin and look at the efficacy of the patch (based on the work of Hung *et al.*, 2005).

Problem formulation Since the patch is circular, we can use axisymmetry to reduce the problem. We restrict our computational domain in the skin to a symmetric region around the patch, so that the problem remains axisymmteric. The computational domain with dimensions is shown in Figure 6.23. The patch consists of three layers: drug reservoir, microporous membrane and adhesive layers, as shown in Figure 6.23.

Governing equation

$$\frac{\partial c}{\partial t} = D \left[\frac{1}{r}\frac{\partial}{\partial r}\left(r\frac{\partial c}{\partial r}\right) + \frac{\partial^2 c}{\partial z^2} \right]$$

where c is the concentration of the drug scopolamine and D is its diffusivity in the different layers of the patch and the skin. The diffusivity values are given in Table 6.9.

6.4 Problems

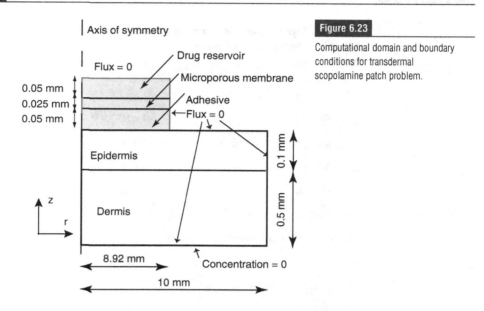

Figure 6.23

Computational domain and boundary conditions for transdermal scopolamine patch problem.

Boundary and initial conditions The boundary conditions are shown in Figure 6.23. The top edge and right edges of the patch have zero flux boundary condition due to insulation. The right edge of the skin is considered to be far away (semi-infinite) and hence, again a zero flux boundary condition is used. It is assumed that the blood stream completely takes away the drug from the bottom of the dermis layer and therefore, concentration of the drug is zero at that boundary. The initial concentrations of the drug in the reservoir and adhesive layers are 0.08 and 0.04 mg/mm^3, respectively. The microporous membrane layer does not have any drug initially.

Input parameters All input parameters are listed in Table 6.9.

Questions

(1) Create a structured mesh and run the problem for 72 h.
(2) Perform a mesh convergence analysis.
(3) Plot the concentration profile in the domain after 72 h.
(4) Change the diffusivity of the drug in the microporous membrane layer to 1×10^{-8} cm^2/s and solve the problem again. Compare the concentration profile obtained with the new diffusivity to the original solution and comment.

Table 6.9 Input parameters for the transdermal scopolamine patch problem.

Parameter	Value
Diffusivity, in cm^2/s	
Reservoir	1×10^{-7}
Microporous membrane	1×10^{-10}
Adhesive	1×10^{-7}
Epidermis	1.45×10^{-9}
Dermis	5.8×10^{-7}
Initial drug concentration, in mg/mm^3	
Reservoir	0.08
Microporous membrane	0
Adhesive	0.04
Application time, in h	72

6.4.15 Cold therapy

Cold therapy using ice packs is one of the most inexpensive and convenient treatment methods to reduce inflammation in sore and injured muscles (Chin *et al.*, 2004). The goal is to reduce muscle temperatures to provide comfort while avoiding temperatures that would damage tissue.

Problem formulation As shown in Figure 6.24, the ice pack is applied to the surface of the skin, separated by a thin layer of plastic. Three layers of tissues are considered: skin, fat and plastic. The ice pack is assumed to be cylindrical so that the problem is axi-symmetric. A small symmetric region of the tissue layers around the ice is included in the computational domain (shown in Figure 6.24) to look at the effect of cooling at the edges. It is also assumed that no volume change occurs in the ice layer as it melts to form water, and all the water that is formed remains in the ice pack. One way to model melting of ice is to assume that ice melts over a small range of temperatures (say from -1 to $1°C$) rather than exactly at $0°C$. Using this assumption, we can incorporate the latent heat of fusion of ice in the specific heat term (see Figure 9.9), thereby obtaining a parameter known as the apparent specific heat. For this problem, assume that melting of ice takes place between -1 to $1°C$ and obtain the apparent specific heat of ice as a function of

6.4 Problems

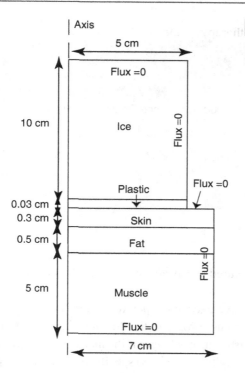

Figure 6.24
Computational domain and boundary conditions for the cold therapy problem.

temperature (as shown in Figure 9.9). Use this value of apparent specific heat to solve the problem.

Governing equations The heat transfer in the tissue is given by the bioheat equation:

$$\rho C_p \frac{\partial T}{\partial t} = k \left[\frac{1}{r} \frac{\partial}{\partial r} \left(r \frac{\partial T}{\partial r} \right) + \frac{\partial^2 T}{\partial z^2} \right] + \rho_b C_{p,b} w_b (T_a - T)$$

In the ice layer, the blood perfusion term is zero.

Boundary conditions The boundary conditions are shown in Figure 6.24. The top and right edges of the ice pack and the top edge of the skin are insulated. The right edges of the different tissue layers and the bottom edge are considered to be far away (semi-infinite) and therefore, have a zero flux boundary condition.

Input parameters The arterial blood temperature, T_a, is 37 °C and the time of application of the ice pack is 20 min. Other input parameters are shown in Table 6.10.

Table 6.10 Input parameters for the cold therapy problem.

Layer	Initial temperature (°C)	Thermal conductivity, k (W/m K)	Density, ρ (kg/m^3)	Specific heat, C_p (J/kg K)	Blood perfusion coefficient, $\rho_b C_{p,b} \dot{V}_b^v$ (W/m^{3}K)
ice	−10	0.567	1000	see Figure 9.9	0
plastic	−10	0.33	920	1900	0
skin	37	0.63	1030	3790	800
fat	37	0.20	938	2430	800
muscle	37	0.43	1044	3550	800

Questions

(1) Create a structured mesh and run the problem for 20 min. Verify mesh convergence.
(2) Determine the average ice temperature after 20 min.
(3) Plot the temperature profile in the domain after 20 min. Determine how the temperature distribution changes if the material of the plastic is changed. The thermal conductivity, density and specific heat of the new plastic material are 0.46 W/m K, 950 kg/m^3 and 1900 J/kg K, respectively.
(4) Implement a 20 min. on, 20 min. off cold therapy cycle, where the ice pack is applied for 20 min. and then removed and monitor the temperature distribution in the tissue for a period of 1 h.

References

Baldwin, S. A., A. Pelman, and J. L. Bert (2001). A heat transfer model of thermal balloon endometrial ablation. *Annals of Biomedical Engineering*, **29**(11):1009–1018.

Chin, J., Pham, H., Steck, A., Sterman, S. and Sun, M. (2004). Assessing the effects of icing the body for 20 minutes. On the web at http://ecommons.library.cornell.edu/handle/1813/143.

Diaz, S. H., Aguilar, G., Lavernia, E. J., and Wong, B. J. F. (2001). Modeling the thermal response of porcine cartilage to laser irradiation. *IEEE Journal of Selected Topics in Quantum Electronics*, **7**(6):944–951.

Hung, R., M. Tandon and Y. Sheido (2005). When motion sickness can't wait. On the web at http://ecommons.library.cornell.edu/handle/1813/2613.

III Background theory

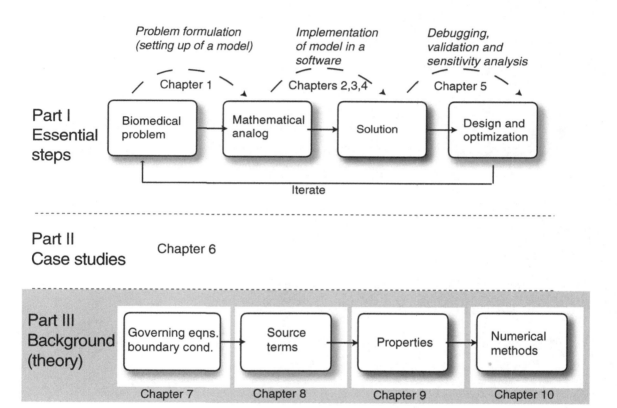

7 Governing equations and boundary conditions

As the reader is probably already aware, governing equations and boundary conditions are equations that describe physical processes, such as heat transfer or fluid flow. Together, these equations make up the mathematical analog of a physical process and allow us to simulate the physical process on a computer. Governing equations are simply universal laws, such as conservation of energy, stated mathematically.

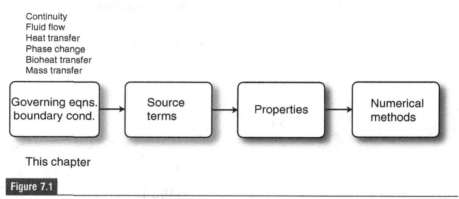

Figure 7.1

This chapter develops the governing equations and boundary conditions as part of the theoretical framework for modeling.

The physics of interest in this book are primarily fluid flow, heat transfer and mass transfer. We are also interested in the physics of electromagnetics and acoustics, as they relate to thermal therapy. However, the last two physics are appropriately covered in Chapter 8. Thus, this chapter will focus on the governing equations of our primary interest, as was stated in Section 1.7:

(1) Conservation of total mass (continuity equation)
(2) Conservation of a mass species (mass transfer equation)
(3) Conservation of momentum (fluid flow or Navier–Stokes equations)
(4) Conservation of thermal energy (heat transfer equation)

Governing equations and boundary conditions

This chapter provides a quick derivation of these governing equations (see Figure 7.1) showing where the various terms in the equations come from (and therefore what they represent). This will allow the user to decide when to drop (or keep) a particular term, with increased confidence and is one more step toward discouraging the use of the software as a blackbox.

7.1 Conservation of mass: the continuity equation

The total mass in a system is conserved – it cannot be created or destroyed. This can be contrasted in Section 7.6 later where individual mass species can be created or destroyed. For now, we are referring to the total mass in a system, not individual species. Consider an elemental control volume as shown in Figure 7.2. Conservation of mass requires that any increase in mass within the control volume is equal to the difference between mass coming into the control volume and the mass leaving the control volume. A word equation for this can be written as

$$\text{Mass in} - \text{Mass out} = \text{Mass stored} \tag{7.1}$$

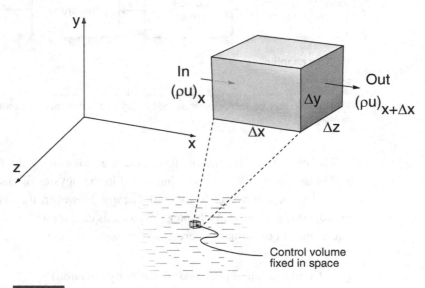

Figure 7.2

Control volume in fluid for setting up mass conservation. The quantity ρu is the rate at which total mass flows in and out of the control volume per unit area normal to the flow.

7.1 Conservation of mass: the continuity equation

Let ρ be the total mass density or total mass concentration in the system at any point x and time t. Let u be the x component of the velocity at any point x. Mass can enter and leave the control volume through bulk motion of the fluid. For simplicity, let us consider only flow in the x direction. The terms in Eqn. 7.1 can be written as

$$\text{Mass in} = (\rho u)|_x \Delta y \Delta z \Delta t,$$
$$\text{Mass out} = (\rho u)|_{x+\Delta x} \Delta y \Delta z \Delta t,$$
$$\text{Mass stored} = \Delta \rho \Delta x \Delta y \Delta z$$

Substituting these into Eqn. 7.1,

$$\left[(\rho u)_x - (\rho u)|_{x+\Delta x}\right] \Delta y \Delta z \Delta t = \Delta \rho \Delta x \Delta y \Delta z$$

Dividing throughout by $\Delta x \Delta y \Delta z \Delta t$ and rearranging the left-hand side,

$$-\frac{(\rho u)_{x+\Delta x} - (\rho u)|_x}{\Delta x} = \frac{\Delta \rho}{\Delta t}$$

Taking the limit, as Δx, Δt go to zero, we get

$$\frac{\partial \rho}{\partial t} + \frac{\partial}{\partial x}(u\rho) = 0 \tag{7.2}$$

At steady state, Eqn. 7.2 is written as

$$\frac{\partial}{\partial x}(u\rho) = 0$$

However, if the total density, ρ, can be assumed constant, as in an incompressible flow, Eqn. 7.2 is approximated for steady or unsteady flow as

$$\frac{\partial u}{\partial x} = 0 \tag{7.3}$$

Equation 7.2, which is in one dimension, can be generalized to multiple dimensions as

$$\frac{\partial \rho}{\partial t} + \frac{\partial}{\partial x}(u\rho) + \frac{\partial}{\partial y}(v\rho) + \frac{\partial}{\partial z}(w\rho) = 0 \tag{7.4}$$

Box 7.1 Software implementation

The continuity equation is always valid and therefore, for incompressible flow, the user typically does not explicitly choose to solve it – the software generally includes it in the system of equations to solve.

7.2 Conservation of momentum: governing equation for fluid flow

Our purpose here is to develop a general equation that describes fluid flow in many different physical situations. Many excellent books exist on fluid mechanics and our intent is not to repeat such information at length, but to have a simple derivation of the equation that will allow us to get a feel for where the equations come from and what the individual terms represent – making simulation a little less of a blackbox. We will derive the equations by applying Newton's law of motion to an elemental volume of fluid, as shown in Figure 7.3. Using Newton's law of motion on a fluid of mass Δm at velocity $\vec{V}$ on which a force of magnitude $\vec{F}$ is acting, we can write

$$\vec{F} = \Delta m \frac{D\vec{V}}{Dt} \tag{7.5}$$

which can be rewritten as

$$\vec{f} = \rho \frac{D\vec{V}}{Dt} \tag{7.6}$$

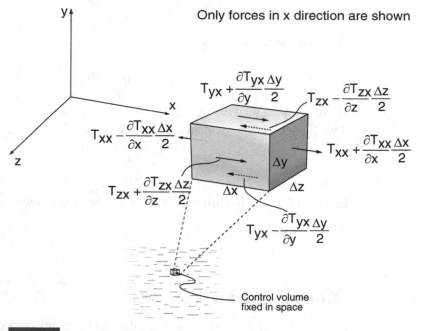

Figure 7.3

Control volume in a fluid showing various forces acting on it. Newton's law of motion is applied to the control volume to derive the governing equations of fluid flow.

7.2 Conservation of momentum: governing equation for fluid flow

where

$\vec{f}$ = total force acting on the fluid element per unit volume,
= body force per unit volume (f_b) + surface forces per unit volume (f_s)

Let T_{ij} be the stress acting on a surface (see Figure 7.3), with the index i denoting the direction normal to the surface on which stress is considered, and j denoting the direction in which stress acts. Thus, T_{xy} stands for stress in the y-direction over the surface oriented in the x-direction. The stress quantities T_{xx}, T_{yx}, T_{zx} are acting at the center of the control volume $\Delta x \Delta y \Delta z$. The net surface force on the control volume in the x-direction can be written as

$$f_s \Delta x \Delta y \Delta z = \left[\left(T_{xx} + \frac{\partial T_{xx}}{\partial x}\frac{\Delta x}{2}\right) - \left(T_{xx} - \frac{\partial T_{xx}}{\partial x}\frac{\Delta x}{2}\right)\right]\Delta y \Delta z$$
$$+ \left[\left(T_{yx} + \frac{\partial T_{yx}}{\partial y}\frac{\Delta y}{2}\right) - \left(T_{yx} - \frac{\partial T_{yx}}{\partial y}\frac{\Delta y}{2}\right)\right]\Delta x \Delta z$$
$$+ \left[\left(T_{zx} + \frac{\partial T_{zx}}{\partial x}\frac{\Delta z}{2}\right) - \left(T_{zx} - \frac{\partial T_{zx}}{\partial z}\frac{\Delta z}{2}\right)\right]\Delta x \Delta y$$

Simplifying, we get

$$f_s \Delta x \Delta y \Delta z = \left(\frac{\partial T_{xx}}{\partial x} + \frac{\partial T_{yx}}{\partial y} + \frac{\partial T_{zx}}{\partial z}\right)\Delta x \Delta y \Delta z \qquad (7.7)$$

Dividing throughout by $\Delta x \Delta y \Delta z$, we get the surface force per unit volume, f_s, as

$$f_s = \frac{\partial T_{xx}}{\partial x} + \frac{\partial T_{yx}}{\partial y} + \frac{\partial T_{zx}}{\partial z} \qquad (7.8)$$

If Δm is the mass of the element, $\Delta V = \Delta x \Delta y \Delta z$ its volume, and g_x is the component of gravity in the x-direction, body force per unit volume in the x-direction, f_b, is given by

$$f_b = \frac{\Delta m g_x}{\Delta V}$$
$$= \rho g_x \qquad (7.9)$$

Substituting into Eqn. 7.6, we get for the x-direction

$$\rho\frac{Du}{Dt} = \rho g_x + \frac{\partial T_{xx}}{\partial x} + \frac{\partial T_{yx}}{\partial y} + \frac{\partial T_{zx}}{\partial z} \qquad (7.10)$$

Similarly, for the y and z directions,

$$\rho \frac{Dv}{Dt} = \rho g_y + \frac{\partial T_{xy}}{\partial x} + \frac{\partial T_{yy}}{\partial y} + \frac{\partial T_{zy}}{\partial z} \quad (7.11)$$

$$\rho \frac{Dw}{Dt} = \rho g_z + \frac{\partial T_{xz}}{\partial x} + \frac{\partial T_{yz}}{\partial y} + \frac{\partial T_{zz}}{\partial z} \quad (7.12)$$

The surface stresses, T, can be related to shear and normal stresses, τ, and pressure, P, as (see upper level textbooks on fluid mechanics, such as Kundu and Cohen, 2004)

$$\begin{aligned}
\tau_{xx} &= T_{xx} - P & \tau_{xy} &= T_{xy} & T_{yx} &= T_{xy} \\
\tau_{yy} &= T_{yy} - P & \tau_{yz} &= T_{yz} & T_{zy} &= T_{yz} \\
\tau_{zz} &= T_{zz} - P & \tau_{zx} &= T_{zx} & T_{zx} &= T_{xz}
\end{aligned}$$

Substituting for T in terms of τ in Eqns. 7.10–7.12, we get

$$\begin{aligned}
\rho \frac{Du}{Dt} &= \rho g_x + \frac{\partial \tau_{xx}}{\partial x} + \frac{\partial \tau_{yx}}{\partial y} + \frac{\partial \tau_{zx}}{\partial z} - \frac{\partial P}{\partial x} \\
\rho \frac{Dv}{Dt} &= \rho g_y + \frac{\partial \tau_{xy}}{\partial x} + \frac{\partial \tau_{yy}}{\partial y} + \frac{\partial \tau_{zy}}{\partial z} - \frac{\partial P}{\partial y} \\
\rho \frac{Dw}{Dt} &= \rho g_z + \frac{\partial \tau_{xz}}{\partial x} + \frac{\partial \tau_{yz}}{\partial y} + \frac{\partial \tau_{zz}}{\partial z} - \frac{\partial P}{\partial z}
\end{aligned} \quad (7.13)$$

For a Newtonian fluid the viscous stress is proportional to the rate of shearing strain (angular deformation rate). The viscous stresses may be expressed in terms of viscosity, μ, and velocity gradients (here u, v and w are the velocities in the x, y and z-directions, respectively) as

$$\tau_{xx} = 2\mu \frac{\partial u}{\partial x}$$

$$\tau_{yy} = 2\mu \frac{\partial v}{\partial y}$$

$$\tau_{zz} = 2\mu \frac{\partial w}{\partial z}$$

$$\tau_{yx} = \tau_{xy} = \mu \left[\frac{\partial u}{\partial y} + \frac{\partial v}{\partial x} \right]$$

$$\tau_{zx} = \tau_{xz} = \mu \left[\frac{\partial u}{\partial z} + \frac{\partial w}{\partial x} \right]$$

$$\tau_{zy} = \tau_{yz} = \mu \left[\frac{\partial v}{\partial z} + \frac{\partial w}{\partial y} \right]$$

7.2 Conservation of momentum: governing equation for fluid flow

Therefore, Eqn. 7.13, upon substitution, becomes

$$\rho \frac{Du}{Dt} = \rho g_x + 2\mu \frac{\partial^2 u}{\partial x^2} + \mu \left[\frac{\partial^2 u}{\partial y^2} + \frac{\partial^2 v}{\partial y \partial x} \right]$$
$$+ \mu \left[\frac{\partial^2 u}{\partial z^2} + \frac{\partial^2 w}{\partial z \partial x} \right] - \frac{\partial P}{\partial x}$$
$$= \rho g_x + \mu \left[\frac{\partial^2 u}{\partial x^2} + \frac{\partial^2 u}{\partial y^2} + \frac{\partial^2 u}{\partial z^2} \right]$$
$$+ \mu \frac{\partial}{\partial x} \underbrace{\left[\frac{\partial u}{\partial x} + \frac{\partial v}{\partial y} + \frac{\partial w}{\partial z} \right]}_{= 0 \text{ from continuity}} - \frac{\partial P}{\partial x}$$

Substituting for Du/Dt, we get

$$\rho \left[\frac{\partial u}{\partial t} + u \frac{\partial u}{\partial x} + v \frac{\partial u}{\partial y} + w \frac{\partial u}{\partial z} \right] = \rho g_x + \mu \left[\frac{\partial^2 u}{\partial x^2} + \frac{\partial^2 u}{\partial y^2} + \frac{\partial^2 u}{\partial z^2} \right] - \frac{\partial P}{\partial x}$$

The complete set of governing equations for an incompressible, Newtonian fluid is

$$\rho \left[\frac{\partial u}{\partial t} + u \frac{\partial u}{\partial x} + v \frac{\partial u}{\partial y} + w \frac{\partial u}{\partial z} \right] = \rho g_x + \mu \left[\frac{\partial^2 u}{\partial x^2} + \frac{\partial^2 u}{\partial y^2} + \frac{\partial^2 u}{\partial z^2} \right] - \frac{\partial P}{\partial x} \quad (7.14)$$

$$\rho \left[\frac{\partial v}{\partial t} + u \frac{\partial v}{\partial x} + v \frac{\partial v}{\partial y} + w \frac{\partial v}{\partial z} \right] = \rho g_y + \mu \left[\frac{\partial^2 v}{\partial x^2} + \frac{\partial^2 v}{\partial y^2} + \frac{\partial^2 v}{\partial z^2} \right] - \frac{\partial P}{\partial y} \quad (7.15)$$

$$\rho \left[\frac{\partial w}{\partial t} + u \frac{\partial w}{\partial x} + v \frac{\partial w}{\partial y} + w \frac{\partial w}{\partial z} \right] = \rho g_z + \mu \left[\frac{\partial^2 w}{\partial x^2} + \frac{\partial^2 w}{\partial y^2} + \frac{\partial^2 w}{\partial z^2} \right] - \frac{\partial P}{\partial z} \quad (7.16)$$

Section 7.14 provides the governing equations in cylindrical coordinates. For non-Newtonian fluids, we start from Eqn. 7.13 and use the non-Newtonian relationships between stress and velocity gradients. Commercial CFD software typically solves a fairly general version of the governing equations for fluid flow.

Box 7.2 Software implementation

Implementation in COMSOL of the governing equations for fluid flow (Navier–Stokes equations for Newtonian fluids) is discussed in Section 2.8 and in Case studies (Table 6.1). One way to implement non-Newtonian fluids in COMSOL is to use its Chemical Engineering module (add-on software that extends the capabilities of COMSOL).

7.3 Conservation of thermal energy: governing equation for heat transfer

As for fluid flow in the previous section, our purpose here is to develop a general equation that describes heat transfer in many different physical situations. This general equation will provide the framework that will allow us to model temperatures in many real situations, always starting from the same equation. As in fluid flow, we will derive the general equation for heat transfer by considering an elemental control volume, as shown in Figure 7.4, where heat fluxes (rate of heat flow per unit area) due to conduction and flow are considered in a particular direction. The total heat flux at any location is made up of conductive flux and flux due to bulk flow. Heat flux due to bulk flow arises simply due to bulk movement of the fluid as it carries heat (thermal energy) with it:

$$\begin{aligned}\text{Rate at which heat}\\ \text{is carried by a flowing fluid}\end{aligned} = \dot{m} C_p (T - T_R)$$
$$= u A \rho C_p (T - T_R) \qquad (7.17)$$

where u is the fluid velocity in a particular direction, A is the area perpendicular to the flow direction, ρ is the density of the fluid, $\dot{m} = uA\rho$ is the mass flow rate in the flow direction, T is the fluid temperature and T_R is some reference temperature.

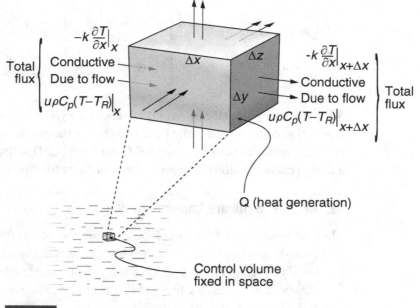

Figure 7.4

Control volume showing energy inflow and outflow by conduction (diffusion) and convection. From Datta (2002), with permission from Taylor and Francis Group.

7.3 Conservation of thermal energy: governing equation for heat transfer

Thus, the heat flux due to bulk flow is given by

$$\text{Heat flux due to bulk flow in a moving fluid} = u\rho C_p(T - T_R) \qquad (7.18)$$

This heat flux due to bulk flow in the x direction is noted in Figure 7.4 at the two faces of the control volume.

Within the control volume in Figure 7.4, heat is generated at the rate of Q per unit volume. An increase in the stored heat would manifest itself in an increase in temperature of the control volume. Conversely, if the stored heat in the control volume decreases, its temperature drops. Generation of heat is not to be confused with storage of heat. Generation is the transformation of energy from one form into heat. In the tissue, for example, heat can be generated through metabolism or due to absorption of laser energy Other examples of generation are described later in this section. A change in storage of energy (as manifested in a rise or fall in temperature) is the effect, whereas generation and conduction of energy are the causes.

Using the first law of thermodynamics for the control volume in Figure 7.4, which states that the energy in the control volume is conserved, we can write:

$$\begin{array}{c} \text{Energy} \\ \text{in} \end{array} - \begin{array}{c} \text{Energy} \\ \text{out} \end{array} + \begin{array}{c} \text{Energy} \\ \text{generated} \end{array} = \begin{array}{c} \text{Change in} \\ \text{energy stored} \end{array} \qquad (7.19)$$

Note that the balance in Eqn. 7.19 is for thermal energy or heat. For simplicity in deriving the heat transfer equation, let us consider heat transport only in the x-direction (the relevant area is $\Delta y \Delta z$), as illustrated in Figure 7.4. Let heat flux due to conduction at a location x be denoted by q_x'', which would be the flux over the area $\Delta y \Delta z$. We already have the expression for heat flux due to bulk flow, given by $u\rho C_p(T - T_R)$. The various quantities in Eqn. 7.19 can be written as:

$$\begin{array}{c} \text{Energy in} \\ \text{during time } \Delta t \end{array} = \left(q_x'' \Delta y \Delta z + [u \Delta y \Delta z \rho C_p (T - T_R)]_x \right) \Delta t$$

$$\begin{array}{c} \text{Energy out} \\ \text{during time } \Delta t \end{array} = \left(q_{x+\Delta x}'' \Delta y \Delta z + [u \Delta y \Delta z \rho C_p (T - T_R)]_{x+\Delta x} \right) \Delta t$$

$$\begin{array}{c} \text{Energy generated} \\ \text{during time } \Delta t \end{array} = Q \Delta x \Delta y \Delta z \Delta t$$

$$\begin{array}{c} \text{Change in energy stored} \\ \text{during time } \Delta t \end{array} = \Delta x \Delta y \Delta z \rho C_p \Delta T$$

Substituting in Equation 7.19, we get:

$$\Delta t\left(q''_x \Delta y \Delta z - q''_{x+\Delta x} \Delta y \Delta z + \rho C_p \Delta y \Delta z \left[u(T-T_R)_x - u(T-T_R)_{x+\Delta x}\right]\right.$$
$$\left. + Q \Delta x \Delta y \Delta z\right) = \rho C_p \Delta x \Delta y \Delta z \Delta T$$

Dividing throughout by $\Delta x \Delta y \Delta z \Delta t$ and rearranging:

$$-\frac{q''_{x+\Delta x} - q''_x}{\Delta x} - \rho C_p \frac{(u(T-T_R)_{x+\Delta x} - u(T-T_R)_x)}{\Delta x} + Q = \rho C_p \frac{\Delta T}{\Delta t}$$

Making Δx and Δt go to zero and using the definition of a derivative:

$$-\frac{\partial q''_x}{\partial x} - \rho C_p \frac{\partial}{\partial x}(uT) + Q = \rho C_p \frac{\partial T}{\partial t}$$

Note that T_R has been dropped since it is a constant. Using Fourier's law for heat conduction to substitute for the heat flux, $q''_x = -k\partial T/\partial x$, where k is the thermal conductivity of the material:

$$-\frac{\partial}{\partial x}\left(-k\frac{\partial T}{\partial x}\right) - \rho C_p \frac{\partial}{\partial x}(uT) + Q = \rho C_p \frac{\partial T}{\partial t}$$

If k can be assumed constant, this is simplified to:

$$\underbrace{\frac{\partial T}{\partial t}}_{\text{storage}} + \underbrace{\frac{\partial(uT)}{\partial x}}_{\substack{\text{flow or}\\\text{convection}}} = \underbrace{\frac{k}{\rho C_p}\frac{\partial^2 T}{\partial x^2}}_{\text{conduction}} + \underbrace{\frac{Q}{\rho C_p}}_{\text{generation}} \qquad (7.20)$$

Equation 7.20 is the general governing equation for energy transfer in a one-dimensional Cartesian coordinate system with constant thermal properties. This equation is also known as the energy equation or the heat equation. Often, the equation of continuity or mass conservation (see Eqn. 7.3)

$$\frac{\partial u}{\partial x} = 0 \qquad (7.21)$$

is used to write an alternative form of Eqn. 7.20. Using Eqn. 7.21, the second (convection) term in Eqn. 7.20 can be simplified as

$$\frac{\partial(uT)}{\partial x} = u\frac{\partial T}{\partial x} + T\frac{\partial u}{\partial x}$$
$$= u\frac{\partial T}{\partial x}$$

7.3 Conservation of thermal energy: governing equation for heat transfer

Using this, Eqn. 7.20 can be written as

$$\underbrace{\frac{\partial T}{\partial t}}_{\text{storage}} + \underbrace{u\frac{\partial T}{\partial x}}_{\substack{\text{flow or} \\ \text{convection}}} = \underbrace{\frac{k}{\rho C_p}\frac{\partial^2 T}{\partial x^2}}_{\text{conduction}} + \underbrace{\frac{Q}{\rho C_p}}_{\text{generation}} \quad (7.22)$$

Note that the convective term has been simplified. The heat generation term, Q, for various types of heating is the subject of Chapter 8.

Equation 7.22 is a simplified form of the heat equation in one dimension. For practical problems and since we have a numerical solution method accessible to us, we can deal with far more complex situations than given by this equation. For example, for a 3D problem, the governing equation can be developed following the same procedure as

$$\rho C_p \left(\frac{\partial T}{\partial t} + u\frac{\partial T}{\partial x} + v\frac{\partial T}{\partial y} + w\frac{\partial T}{\partial z} \right) = k \left[\frac{\partial^2 T}{\partial x^2} + \frac{\partial^2 T}{\partial y^2} + \frac{\partial^2 T}{\partial z^2} \right] + Q$$

We can develop more general forms of Eqn. 7.22 in other coordinate systems and in multiple dimensions. These are tabulated later in this chapter (Section 7.14) and the reader is referred to the well-known textbook (Bird et al., 1960) in this area for further details. To apply these general equations to a specific problem, the equations would need to be simplified by dropping the appropriate terms, as discussed at length in Section 1.7.1. As an example, we present here two common situations: the governing equation for heat conduction with constant properties in axisymmetric coordinates is given by:

$$\rho C_p \frac{\partial T}{\partial t} = k \left[\frac{1}{r}\frac{\partial}{\partial r}\left(r\frac{\partial T}{\partial r}\right) + \frac{\partial^2 T}{\partial z^2} \right] + Q \quad (7.23)$$

The governing equation for heat conduction with constant properties in spherically symmetric coordinates is given by:

$$\rho C_p \frac{\partial T}{\partial t} = k \frac{1}{r^2}\frac{\partial}{\partial r}\left(r^2 \frac{\partial T}{\partial r}\right) + Q \quad (7.24)$$

Box 7.3 Software implementation

The governing equations shown in Section 7.14 are less general than what is implemented in many major softwares. For example, equations in Section 7.14 are for constant properties but commercial software routinely allows property variations with temperature and other solution variables.

> Implementation of the governing equation for heat transfer in the software COMSOL has been discussed in Section 2.8 and in Case studies (see Table 6.1). Since numerical methods can accommodate arbitrary geometry, the computer code for solution can be done in any coordinate system. In the case of COMSOL, the program is written in the Cartesian coordinate system.

7.3.1 Lumped parameter analysis: special case of the governing equation for heat transfer

For some simplified heat transfer situations, the lumped parameter equation,

$$\frac{dT}{dt} = \frac{hA}{mC_p}(T - T_\infty) \tag{7.25}$$

which is a special adaptation of the general heat transfer equation (with only transient and conduction terms) can be valid. Here the spatial temperature variation is ignored, which leads to less than 5% error if the condition $h(V/A)/k < 0.1$ is satisfied. Here V is the volume of the solid and A is its surface area.

> ### Box 7.4 Software implementation
>
> Equation 7.25 has an analytical solution for constant surrounding fluid temperature, T_∞, and therefore does not require numerical solution. However, if numerical solution is desired, as it would be, for example, in the case when T_∞ changes with time, it can be accomplished in one of two ways. In the first approach, nothing special is needed since Eqn. 7.25 is only a special case of the energy equation (Eqn. 7.22, for example, without the convection and the generation terms). A large value of thermal conductivity can be used for the domain with a convective boundary condition having an appropriate h value. The other approach is to solve Eqn. 7.25 by itself using perhaps a simple finite differencing method (see Section 10.2) and programming into MATLAB or Excel.

7.4 Governing equation for heat conduction with change of phase

The governing energy equation (Eqn. 7.22), derived earlier does not include changing of phase, such as in freezing or evaporation, where a latent heat is involved. Heat transfer problems involving phase change can get considerably more complex.

7.4.1 Freezing/melting

Dissolved solutes in the water within a tissue depress the freezing point of the solution below the freezing point of pure water. This leads to gradual freezing of a tissue over a temperature range. This process of gradual freezing of a solution is shown as a schematic for a hypothetical system in Figure 7.5.

The various regions in the pie chart illustrate as examples the relative amounts of various components at a location in a solution. As the temperature is lowered, the amount of ice is zero initially, until the initial freezing point of the solution is reached, depending on the initial concentration of the solution. As ice begins to form, the rest of the solution concentrates since the ice formed reduces available liquid water, while the amount of solute remains the same. As the temperature is continually lowered, this process involving freezing of pure water and an increase in concentration of the remaining solution continues until all the water that can be crystallized as ice freezes, leaving only the solute and its water of hydration. At this solute concentration, the solution is termed eutectic. Eutectic temperature is defined as the lowest temperature at which the solution remains liquid. Further reduction in temperature results in the solidification of the remaining solution as a whole, i.e., without formation of pure ice.

Mathematically, change in enthalpy of a tissue with change in temperature, as ice forms or melts, can be written as follows. Let m_u, m_i and m_s be the masses of unfrozen water, ice and solute, respectively, in a small sample at temperature T

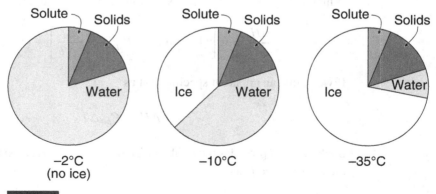

Figure 7.5

Schematic illustrating various components in gradual freezing of water in a biological tissue. From Datta (2002), with permission from Taylor and Francis Group.

(i.e., referring to any one temperature in Figure 7.5). Let the enthalpy change of that entire mass (water+ice+solute) be ΔH for a small change of temperature ΔT. Writing the enthalpy of the total mass in terms of the enthalpy of its components,

$$(m_u + m_i + m_s)\Delta H = m_u C_{p_u} \Delta T + m_i C_{p_i} \Delta T + m_s C_{p_s} \Delta T + \Delta m_i \lambda_f \quad (7.26)$$

which is rearranged as

$$\frac{\Delta H}{\Delta T} = \frac{m_u}{m_u + m_i + m_s} C_{p_u} + \frac{m_i}{m_u + m_i + m_s} C_{p_i}$$
$$+ \frac{m_s}{m_u + m_i + m_s} C_{p_s} + \frac{\Delta m_i / \Delta T}{m_u + m_i + m_s} \lambda_f \quad (7.27)$$

Let $f(T)$ be the fraction of total water frozen at temperature T, Thus

$$f(T) = \frac{m_i}{m_u + m_i} \quad (7.28)$$

Likewise, let w be the initial water content as a fraction of total weight, so that

$$w = \frac{m_u + m_i}{m_u + m_i + m_s} \quad (7.29)$$

Rewriting Eqn. 7.27 as

$$\frac{\Delta H}{\Delta T} = \frac{m_u + m_i}{m_u + m_i + m_s}\left[\frac{m_u}{m_u + m_i}C_{p_u} + \frac{m_i}{m_u + m_i}C_{p_i} + \frac{\Delta(m_i/(m_u + m_i))}{\Delta T}\lambda_f\right]$$
$$+ \frac{m_s}{m_u + m_i + m_s}C_{p_s} \quad (7.30)$$

Which is now rewritten in terms of water content, w, and frozen fraction, f, as

$$\frac{\Delta H}{\Delta T} = w\left[(1-f)C_{p_u} + fC_{p_i} + \lambda_f \frac{\Delta f}{\Delta T}\right] + (1-w)C_{p_s} \quad (7.31)$$

If we define an apparent specific heat as

$$\Delta H = C_{pa}\Delta T \quad (7.32)$$

Such that C_{pa} provides the total change in enthalpy, ΔH, then C_{pa} would relate to the ice fraction, f, as

$$C_{pa} = w\left[(1-f)C_{p_u} + fC_{p_i} + \lambda_f \frac{\partial f}{\partial T}\right] + (1-w)C_{p_s} \quad (7.33)$$

Note that C_{pa} relates to $\partial f/\partial T$ which is the rate of ice formation and is the latent heat contribution to C_{pa}. Since the type and amount of solutes dissolved in a tissue vary, so does $\partial f/\partial T$. There are two ways to obtain C_{pa}:

- Obtain from experimental measurement of enthalpy, H, as a function of temperature, T; typically using an instrument such as differential scanning calorimetry (DSC). From the measured $H(T)$, calculate C_{pa} as

$$C_{pa} = \frac{\Delta H}{\Delta T} \qquad (7.34)$$

- Predict apparent specific heat, C_{pa}, using Eqn. 7.33, when $f(T)$, the fraction of ice at any temperature, is available.

Computer implementation of C_{pa} is discussed in Section 9.8.2.

7.4.2 Evaporation

Evaporation occurs at all temperatures, but increases very significantly at the higher temperatures. Inclusion of significant evaporation in a tissue can be very complex, as evaporation is spatially distributed and also significant evaporation can generate pressures contributing to pressure-driven flow, outside the scope of the equations developed so far in this chapter. An extremely simple approach to include evaporation in a heat transfer situation is to assume that all the evaporation occurs at the surface and consider it as a boundary condition of negative heat flux:

$$-k\frac{\partial T}{\partial x} = \lambda_{\text{vap}} n_{A,s} \qquad (7.35)$$

where $n_{A,s}$ is the amount of moisture lost per unit area per unit time or the mass flux at the surface. This mass flux can be experimentally measured (for example, from simple weight loss measurements) or estimated using a surface convective mass transfer coefficient (see Eqn. 7.88).

7.5 The bioheat transfer equation for mammalian tissue

Although the general governing equation (Eqn. 7.20) and its more general versions are valid for most physical systems, it would be hard to apply the equations for a mammalian tissue that has blood vessels of varying sizes and varying blood flows. In such a system, thermal conductivity, density and specific heat (the thermal properties) would vary significantly over small distances, among other complications.

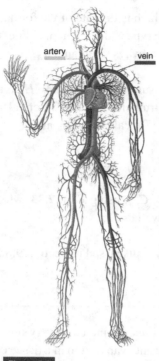

Figure 7.6

Arteries and veins of the circulatory system.

Researchers have attempted to develop a bioheat transfer equation that is simpler than applying the governing equations of fluid flow and heat transfer directly, but still captures the essence of the heat transfer process in a mammalian tissue with blood vessels (see Diller, 1992; Diller *et al.*, 1999; Shitzer and Eberhart, 1985).

The circulatory system in our body comprises two sets of blood vessels – arteries and veins (Figure 7.6) which carry blood from the heart and back. By the pumping action of the heart, blood flows through larger arteries to progressively smaller arteries to arterioles and capillaries to small veins to larger veins and eventually back to heart. Figure 7.7 shows a schematic of the arterial system with differently sized vessels. Figure 7.8 shows a schematic of temperature equilibration between the blood and the solid tissue, as the blood traverses different sized vessels in the systemic circulation. As blood leaves the heart and travels in the large arteries, its temperature remains essentially constant. This is the arterial blood temperature, T_a. Most of the temperature equilibration occurs as the blood passes through vessels whose diameter is between that of the arterial branch and that of the arteriole. As the blood reaches the latter, blood temperature becomes essentially that of the solid tissue (warmer or colder, as shown). Beyond this point, blood temperature

7.5 The bioheat transfer equation for mammalian tissue

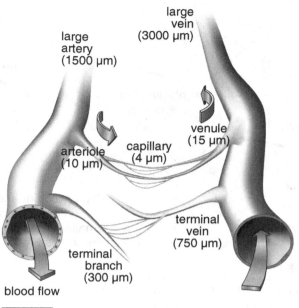

Figure 7.7

Variation of blood vessel sizes. From Datta (2002), with permission from Taylor and Francis Group.

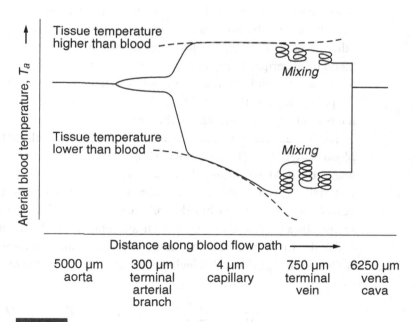

Figure 7.8

Variation of blood temperature in the blood vessels. From Datta (2002), with permission from Taylor and Francis Group.

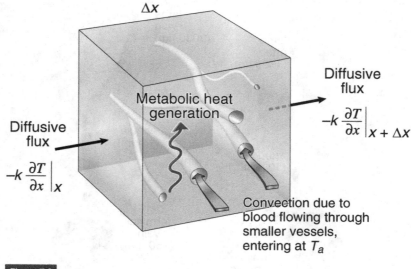

Figure 7.9

Idealized heat transfer in a tissue showing metabolic heat generation Q and convective heat transfer due to the passage of blood. From Datta (2002), with permission from Taylor and Francis Group.

follows the solid tissue temperature through its spatial and time variations until blood reaches the terminal veins. At this point the blood temperature ceases to equilibrate with the tissue, and remains virtually constant, except as it mixes with other blood of different temperatures at venous confluences. Finally, the cooler blood from peripheral regions and warmer blood from internal organs mix within the vena cavae and the right atrium and ventricle. Following thermal exchange in the pulmonary circulation and remixing in the left heart, the blood attains the same temperature it had at the start of the circuit.

The bioheat transfer equation can be derived for an idealized tissue system with blood vessels through it, as shown in Figure 7.9 in the same way as Equation 7.20 was derived. Many assumptions that have to be made include: (1) homogeneous material with isotropic (same in all directions) thermal properties; (2) large blood vessels are ignored; (3) blood capillaries are isotropic; (4) blood is at arterial temperature but quickly reaches the tissue temperature by the time it reaches the end of the artery system. Using these assumptions, the governing *bioheat equation* for a mammalian tissue that includes the heat carried by the blood vessels is

$$\underbrace{\rho c \frac{\partial T}{\partial t}}_{\substack{\text{change} \\ \text{in storage}}} = \underbrace{k \nabla^2 T}_{\text{conduction}} + \underbrace{\rho_b c_b \dot{V}_b^v (T_a - T)}_{\substack{\text{convection} \\ \text{due to blood flow}}} + \underbrace{Q}_{\substack{\text{metabolic} \\ \text{heat generation}}} \quad (7.36)$$

where $\dot{V}_b^v$ is the flow rate of blood in m³ of blood / m³ of tissue per second, k, ρ, c are the effective thermal properties of the tissue, ρ_b and c_b are the thermal properties of blood, T_a is the arterial blood temperature. Note that the energy carried in the blood is like an additional heat source term of magnitude $\rho_b c_b \dot{V}_b^v (T_a - T)$. This way, Equation 7.36 can be derived from Equation 7.20 by rewriting the source term with the energy carried by the blood fluid. Bioheat transfer is an active research area and alternative equations for bioheat transfer are being formulated. Equation 7.36 is the most widely used bioheat transfer equation today.

The bioheat equation is used generally to solve for tissue temperature distributions. While such considerations as tissue geometry, thermal property values, inhomogeneities and boundary conditions are very important, this equation can be used to predict the effects of frost bite, determine the depth of damage in a burn victim and find the amount of heat lost through various parts of the body.

> **Box 7.5 Software implementation**
>
> Equation 7.36 is easily included in a software since the second and the third term on the right-hand side can be treated together as a source term, the second term being temperature-dependent. An example of COMSOL implementation can be seen in Case studies VIII, IX and X.

7.6 Conservation of a mass species: governing equation for mass transfer

Following the development of the general equations for fluid flow and heat transfer, our purpose here is to develop a general equation that describes species mass transfer in many different physical situations. This general equation will provide the framework that will allow us to model species concentrations in many real situations, always starting from the same equation.

We will follow a procedure completely analogous to the one for heat transfer. Consider an elemental control volume as shown in Figure 7.10, fixed in space with mass flux into and out of the control volume. The total mass flux includes diffusive mass flux arising from the diffusion of the mass species A, and convective mass flux due to the bulk movement of the fluid as a whole. The diffusive mass flux of a species A in the x direction, $j_{A,x}$, is given by $j_{A,x} = -D_{AB} \partial c_A / \partial x$, which is Fick's law. Here D_{AB} is the diffusivity of the species A in some other species B, in m²/s. The mass flux due to bulk flow, which is completely analogous to Eqn. 7.18 for heat transfer, is given by

$$\begin{array}{c} \text{Flux of species A due to bulk flow} \\ \text{in a moving fluid in the } x \text{ direction} \end{array} = u c_A \qquad (7.37)$$

354 Governing equations and boundary conditions

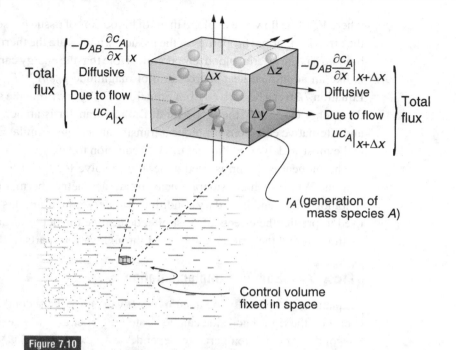

Figure 7.10

Control volume showing inflow and outflow of a mass species by diffusion and convection. From Datta (2002), with permission from Taylor and Francis Group.

where u is the velocity in the x direction. Thus, the total mass flux in the x direction, $n_{A,x}$, is written as

$$n_{A,x} = \underbrace{-D_{AB}\frac{\partial c_A}{\partial x}}_{\text{Diffusive flux}} + \underbrace{c_A u}_{\substack{\text{Flux due to} \\ \text{bulk flow}}} \tag{7.38}$$

where c_A is the concentration of mass species A in mass of A per unit volume. As the stored mass of species A increases in the control volume, it is reflected in an increased value of c_A. Conversely, if the stored amount decreases, the concentration decreases. Within the control volume, mass species A is generated at the rate of r_A in mass per unit volume. As in heat transfer, generation refers to transformation of other forms of mass into mass species A, as in chemical reactions. For example, O_2 in the fluid (mass species) can be used up due to biochemical reactions in the fluid, in which case r_{O_2} has a negative value. Using conservation of mass species A (see Eqn. 7.1)

$$\begin{array}{cccc} \text{Mass in} & - & \text{Mass out} & + & \text{Mass generated} & = & \text{Change in mass stored} \\ \text{(of species } A\text{)} & & \text{(of species } A\text{)} & & \text{(of species } A\text{)} & & \text{(of species } A\text{)} \end{array}$$

$$\tag{7.39}$$

7.6 Conservation of a mass species: governing equation for mass transfer

and referring to Figure 7.10, we can write the various quantities in Eqn. 7.39 as below. Here the relevant area over which diffusion and flow occurs is $\Delta y \Delta z$.

$$\text{Amount of } A \text{ in during time } \Delta t = n_{A,x} \Delta y \Delta z \Delta t$$

$$\text{Amount of } A \text{ out during time } \Delta t = n_{A,x+\Delta x} \Delta y \Delta z \Delta t$$

$$\text{Amount of } A \text{ generated during time } \Delta t = r_A \Delta x \Delta y \Delta z \Delta t$$

$$\text{Amount of } A \text{ stored during time } \Delta t = \Delta c_A \Delta x \Delta y \Delta z$$

Substituting in Eqn. 7.39 for mass balance on the control volume,

$$\left((n_{A,x} - n_{A,x+\Delta x}) \Delta y \Delta z + r_A \Delta x \Delta y \Delta z \right) \Delta t = \Delta c_A \Delta x \Delta y \Delta z$$

$$\frac{n_{A,x} - n_{A,x+\Delta x}}{\Delta x} + r_A = \frac{\Delta c_A}{\Delta t}$$

$$-\frac{\partial}{\partial x}(n_{A,x}) + r_A = \frac{\partial c_A}{\partial t} \tag{7.40}$$

which is the differential equation for mass balance for species A. Substituting Equation 7.38 for $n_{A,x}$ in Equation 7.40:

$$-\frac{\partial}{\partial x}\left(-D_{AB}\frac{\partial c_A}{\partial x} + c_A u\right) + r_A = \frac{\partial c_A}{\partial t}$$

$$-\frac{\partial}{\partial x}\left(-D_{AB}\frac{\partial c_A}{\partial x}\right) - \frac{\partial}{\partial x}(c_A u) + r_A = \frac{\partial c_A}{\partial t}$$

If the diffusivity D_{AB} can be considered a constant

$$D_{AB}\frac{\partial^2 c_A}{\partial x^2} - \frac{\partial}{\partial x}(c_A u) + r_A = \frac{\partial c_A}{\partial t}$$

Rearranging

$$\frac{\partial c_A}{\partial t} + \frac{\partial}{\partial x}(c_A u) = D_{AB}\frac{\partial^2 c_A}{\partial x^2} + r_A \tag{7.41}$$

Although Eqn. 7.41 is for conservation of mass species A, the total mass of all species together is also conserved. Using the continuity equation (Eqn. 7.3),

$$\frac{\partial u}{\partial x} = 0$$

we get

$$\underbrace{\frac{\partial c_A}{\partial t}}_{\text{storage}} + \underbrace{u\frac{\partial c_A}{\partial x}}_{\substack{\text{flow or} \\ \text{convection}}} = \underbrace{D_{AB}\frac{\partial^2 c_A}{\partial x^2}}_{\text{diffusion}} + \underbrace{r_A}_{\text{generation}} \quad (7.42)$$

Equation 7.42 is the general governing equation for mass transfer in a one-dimensional Cartesian coordinate system with constant diffusivity. Compare this equation with Eqn. 7.22 for heat transfer. Note that each term in Eqn. 7.42 has the dimensions of mass per unit volume per unit time or kg/m^3·s. Compare this with Eqn. 7.22, where, upon multiplying throughout by ρC_p, the terms have dimensions of energy per unit volume per unit time or J/m^3·s, again showing the analogy between the heat and mass equations. The reaction rate, r_A, is discussed in Chapter 8 starting in Section 8.10. As in heat transfer, we can develop more general forms of Eqn. 7.42 in other coordinate systems and in multiple dimensions, some of which are shown later in this chapter (Section 7.14).

As an example, we present here the governing equation for species mass diffusion for two common situations – for constant diffusivity in axisymmetric coordinates:

$$\frac{\partial c_A}{\partial t} = D_{AB}\left[\frac{1}{r}\frac{\partial}{\partial r}\left(r\frac{\partial c_A}{\partial r}\right) + \frac{\partial^2 c_A}{\partial z^2}\right] + r_A \quad (7.43)$$

and for constant diffusivity in spherically symmetric coordinates:

$$\frac{\partial c_A}{\partial t} = D_{AB}\frac{1}{r^2}\frac{\partial}{\partial r}\left(r^2\frac{\partial c_A}{\partial r}\right) + r_A \quad (7.44)$$

> **Box 7.6 Software implementation**
>
> See the general discussion at the end of the heat transfer section in this chapter. Implementation of the governing equations in the software COMSOL has been discussed in Section 2.8. The mass source or the reaction term has been discussed in more detail in Section 8.10, including its implementation.

7.6.1 Multiple species in the same problem and their coupling

Unlike the energy balance which is one equation, several mass species may be involved in the same problem which will lead to several species equations being solved simultaneously. See, for example, Case study VII, where three different species are involved. Obviously, we need as many species equations, i.e., as many occurrences of Eqn. 7.42, as the number of unknown species. Any reaction of the kind

$$\frac{\partial c_A}{\partial t} = k c_A c_B \tag{7.45}$$

can be thought in terms of a species equation with zero diffusivity as

$$\frac{\partial c_A}{\partial t} = (0)\frac{\partial^2 c_A}{\partial x^2} + k c_A c_B \tag{7.46}$$

and therefore the species equation can be used to solve for the concentration resulting from the reaction. When multiple species equations are involved, if the reaction term in one equation involves another species, like c_B in reaction term $k c_A c_B$ in the above equation, the reaction term is programmed in a routine manner since c_B is accessible in the coupled set of equations.

> **Box 7.7 Software implementation**
>
> It is typical for a computational software of this kind to allow many species equations and have species concentrations available to each other. In COMSOL, multiple species equations are allowed. For an example of the coupling of the reaction terms, see Case study VII.

7.7 Non-dimensionalization of the governing equations

Non-dimensionalization of the governing equations is often necessary for a number of reasons – ease of numerical solution, improved understanding of the physics and reduction of the number of parameters in sensitivity analysis. For example, in a diffusion problem where the diffusivities are very small, typically the computational software has difficulty handling such small diffusivity values.

7.7.1 Heat transfer

The heat transfer equation

$$\frac{\partial T}{\partial t} + u\frac{\partial T}{\partial x} = \alpha\frac{\partial^2 T}{\partial x^2} + \frac{Q}{\rho C_p}$$

can be non-dimensionalized as follows. Using a characteristic length L (for a slab, this is the half thickness), initial temperature, T_i and surrounding temperature, T_∞, we can rewrite the above equation by dividing throughout by α and multiplying by L^2 as

$$\frac{\partial T}{\partial \left(\frac{\alpha t}{L^2}\right)} + \frac{uL}{\alpha} \frac{\partial T}{\partial \left(\frac{x}{L}\right)} = \frac{\partial^2 T}{\partial \left(\frac{x}{L}\right)^2} + \frac{QL^2}{k} \tag{7.47}$$

Since T_i is a constant, we can replace T by $T - T_i$ in the above equation. Simultaneously dividing throughout by $T_\infty - T_i$ (also a constant), we get

$$\frac{\partial \left(\frac{T-T_i}{T_\infty-T_i}\right)}{\partial \left(\frac{\alpha t}{L^2}\right)} + \frac{uL}{\alpha} \frac{\partial \left(\frac{T-T_i}{T_\infty-T_i}\right)}{\partial \left(\frac{x}{L}\right)} = \frac{\partial^2 \left(\frac{T-T_i}{T_\infty-T_i}\right)}{\partial \left(\frac{x}{L}\right)^2} + \frac{QL^2}{k(T_\infty - T_i)} \tag{7.48}$$

Defining the following non-dimensional quantities

$$\theta = \frac{T - T_i}{T_\infty - T_i} \tag{7.49}$$

$$Fo = \frac{\alpha t}{L^2} \tag{7.50}$$

$$x^* = \frac{x}{L} \tag{7.51}$$

$$Re = \frac{uL\rho}{\mu} \tag{7.52}$$

$$Pr = \frac{\mu C_p}{k} \tag{7.53}$$

$$Pe = Re \cdot Pr = \left(\frac{uL\rho}{\mu}\right)\left(\frac{\mu C_p}{k}\right) = \frac{uL}{\alpha} \tag{7.54}$$

$$Q^* = \frac{QL^2}{k(T_\infty - T_i)} \tag{7.55}$$

we can rewrite the heat transfer equation as

$$\frac{\partial \theta}{\partial Fo} + RePr \frac{\partial \theta}{\partial x^*} = \frac{\partial^2 \theta}{\partial x^{*2}} + Q^* \tag{7.56}$$

where θ is the non-dimensional temperature, Fo is the non-dimensional time, x^* is the non-dimensional position, Q^* is the non-dimensional heat source term, Re is the Reynolds number (dimensionless), Pr is the Prandtl number (dimensionless) and Pe is the Peclet number (dimensionless).

7.7.2 Species mass transfer

The species mass transfer equation

$$\frac{\partial c}{\partial t} + u\frac{\partial c}{\partial x} = D_{AB}\frac{\partial^2 c}{\partial x^2} + r_A$$

can be non-dimensionalized in the same manner as the heat transfer equation. Multiplying both sides by L^2, dividing by D_{AB}, and rearranging,

$$\frac{\partial c}{\partial \left(\frac{D_{AB}t}{L^2}\right)} + \frac{uL}{D_{AB}}\frac{\partial c}{\partial \left(\frac{x}{L}\right)} = \frac{\partial^2 c}{\partial \left(\frac{x}{L}\right)^2} + \frac{r_A L^2}{D_{AB}} \quad (7.57)$$

For an initial concentration, c_i, and a surrounding concentration (in the bulk fluid) as c_∞, both assumed constants, we can write

$$\frac{\partial \left(\frac{c-c_i}{c_\infty - c_i}\right)}{\partial Fo} + \frac{uL}{D_{AB}} \cdot \frac{\partial \left(\frac{c-c_i}{c_\infty - c_i}\right)}{\partial x^*} = \frac{\partial^2 \left(\frac{c-c_i}{c_\infty - c_i}\right)}{\partial x^{*2}} + \frac{r_A L^2}{D_{AB}(c_\infty - c_i)} \quad (7.58)$$

which leads to the non-dimensional species mass transfer equation as

$$\frac{\partial C}{\partial Fo} + ReSc\frac{\partial C}{\partial x^*} = \frac{\partial^2 C}{\partial x^{*2}} + r_A^* \quad (7.59)$$

where

$$C = \frac{c - c_i}{c_\infty - c_i} \quad (7.60)$$

$$Fo = \frac{D_{AB}t}{L^2} \quad (7.61)$$

$$x^* = \frac{x}{L} \quad (7.62)$$

$$Re = \frac{uL\rho}{\mu} \quad (7.63)$$

$$Sc = \frac{\mu}{\rho D_{AB}} \quad (7.64)$$

$$Re \cdot Sc = \left(\frac{uL\rho}{\mu}\right)\left(\frac{\mu}{\rho D_{AB}}\right) = \frac{uL}{D_{AB}} \quad (7.65)$$

$$r_A^* = r_A L^2 / D_{AB}(c_\infty - c_i) \quad (7.66)$$

Here C is the non-dimensional concentration, Fo is the non-dimensional time, x^* is the non-dimensional position, r_A^* is the non-dimensional mass source term, Re is the Reynolds number, and Sc is the Schmidt number (dimensionless).

7.7.3 Momentum equation

For simplicity, let us consider the momentum equation in one dimension

$$\rho \left[\frac{\partial u}{\partial t} + u \frac{\partial u}{\partial x} \right] = -\frac{\partial P}{\partial x} + \mu \frac{\partial^2 u}{\partial x^2} + \rho g_x$$

which can be rewritten and non-dimensionalized as follows. Using a characteristic length, L, and a characteristic velocity (typically the bulk or free stream velocity), u_∞, we can write

$$\frac{\partial \left(\frac{u}{u_\infty} \right)}{\partial \left(\frac{tu_\infty}{L} \right)} + \frac{u}{u_\infty} \frac{\partial (u/u_\infty)}{\partial (x/L)} = -\frac{\partial \left(\frac{P}{\rho u_\infty^2} \right)}{\partial (x/L)} + \frac{\mu}{\rho u_\infty L} \frac{\partial^2 (u/u_\infty)}{\partial (x/L)^2} + \frac{g_x L}{u_\infty^2} \quad (7.67)$$

which can be written using dimensionless quantities as

$$\frac{\partial u^*}{\partial t^*} + u^* \frac{\partial u^*}{\partial x^*} = -\frac{\partial P^*}{\partial x^*} + \frac{1}{Re} \frac{\partial^2 u^*}{\partial x^{*2}} + \frac{g_x L}{u_\infty^2} \quad (7.68)$$

where $u^* = u/u_\infty$ is the non-dimensional velocity, $t^* = tu_\infty/L$ is the non-dimensional time, and $P^* = P/\rho u_\infty^2$ is the non-dimensional pressure. In the heat transfer equation, time was scaled as $Fo = \alpha t/L^2$. If we use the same scaling of time instead, we get

$$\frac{1}{RePr} \cdot \frac{\partial u^*}{\partial Fo} + u^* \frac{\partial u^*}{\partial x^*} = -\frac{\partial P^*}{\partial x^*} + \frac{1}{Re} \frac{\partial^2 u^*}{\partial x^{*2}} + \frac{g_x L}{u_\infty^2}$$

> **Box 7.8 Software implementation**
>
> In implementing the non-dimensional governing equations in a software, the same procedure as described in Section 2.8 is followed, with substitution of values for variables in the input data, such as the dimensions, boundary conditions, initial conditions, heat or mass source terms, according to their non-dimensional counterpart, listed in Table 7.1.

7.8 Coupling of governing equations

In a given physical situation, we often have two or more different physics that occur over the entire domain or, different physics occurring over different regions or subdomains of the same problem. For example, in Case study V, we have two domains with one solid, while the other is fluid. Thus, fluid equations are applied only over the fluid domain while the species diffusion equation is applied

Table 7.1 Non-dimensional quantities to be substituted in a solution process for their dimensional counterparts.

Quantity	Variable for dimensional solver	Variable for non-dimensional solver
Temperature	T	$\dfrac{T - T_i}{T_\infty - T_i}$
Concentration	c	$\dfrac{c - c_i}{c_\infty - c_i}$
Position	x	$\dfrac{x}{L}$
Time	t	$\dfrac{\alpha t}{L^2}$
Velocity (in heat equation)	u	$\dfrac{uL}{\alpha}$ or $RePr$
Velocity (in mass equation)	u	$\dfrac{uL}{D_{AB}}$ or $ReSc$
Heat generation	Q	$\dfrac{QL^2}{k(T_\infty - T_i)}$
Mass generation	r_A	$\dfrac{r_A L^2}{D_{AB}(c_\infty - c_i)}$
Thermal diffusivity	α	1
Mass diffusivity	D_{AB}	1

over the entire domain. Most softwares allow different governing equations to be applied over different regions. The coupling is generally made possible by solving the equations for the two physics either together or sequentially, with the variables from one equation available to the other equation, as needed. In another example, in electromagnetic heating, the energy equation is solved together with the Maxwell's equations for electromagnetics, sequentially, and the coupling term

Box 7.9 Software implementation

Implementation of coupled problems in COMSOL can be seen in the following examples:

- Heat transfer and species transport: Case study VIII
- Fluid flow and heat transfer: Case study X
- Fluid flow and species transport: Case studies V, X
- Electromagnetics and heat transfer: Case study IX

is the heat generation, $Q = 2\pi f \epsilon_0 \epsilon''_{\text{eff}} |E|^2$, where E, the electric field, comes from solving the Maxwell's equations. Although it is possible to compute a particular physics in one software and take the information to a different software to couple with another physics, it is generally harder to do this, compared to when both of the physics are included in the same software.

7.9 Summary: governing equations

Depending on the physics of interest, governing equations of fluid flow, heat transfer and mass transfer, and their combinations are chosen. These equations are mathematical statements of conservation of momentum, energy and mass species, and are used to solve for velocity (or pressure), temperature and concentration, respectively. A fairly general set of governing equations is provided in Section 7.14, from Bird *et al.* (1960). However, even these equations have many limitations in terms of the physics. For example, they are for incompressible flow and for constant property. Most general purpose CFD softwares, however, implement quite general versions of the equations.

Two different versions of a governing equation can be applied to two subregions of the same geometry. When two or more governing equations are chosen to represent more than one physics in a given application, the equations can be coupled to different extents – the exact implementation of the coupling depends on the software.

7.10 Boundary conditions: general comments

The description of the physical problem in mathematical terms is incomplete without the boundary conditions. The need for boundary conditions has also been discussed in Section 1.8. The boundary conditions are the conditions on the boundary enforced by the physics of the problem.

7.11 Boundary conditions: fluid mechanics

To solve the equations for fluid mechanics (Eqns. 7.14–7.16, together with Eqn. 7.4, the continuity equation), boundary conditions are needed for velocity and pressure. However, since pressure appears in the equations as a gradient, pressure boundary conditions are not needed unless we are specifically solving for pressure.

7.11.1 Velocity at the boundary is specified

No slip on a solid boundary Fluid in contact with a solid wall will have the same velocity as the wall. Often the wall is not moving, so the fluid velocity is zero at that boundary (wall).

$$u\big|_{\text{boundary}} = 0 \tag{7.69}$$

This is illustrated at the wall for a tube flow in Figure 7.11. If the boundary is moving, the fluid velocity is equal to the velocity at the boundary wall:

$$u\big|_{\text{boundary}} = u_s$$

where u_s is the velocity of the boundary wall.

Velocity continuity at a fluid boundary When another fluid forms one of the boundaries of the flow, the velocity is assumed continuous from one fluid to the other:

$$u(\text{fluid 1})\big|_{\substack{\text{fluid}\\\text{boundary}}} = u(\text{fluid 2})\big|_{\substack{\text{fluid}\\\text{boundary}}}$$

Symmetry condition Flow situations can arise where the geometry and the boundary conditions are symmetric. An example of this is the tube flow shown in Figure 7.11 where the centerline of the tube is the line of symmetry. The velocity

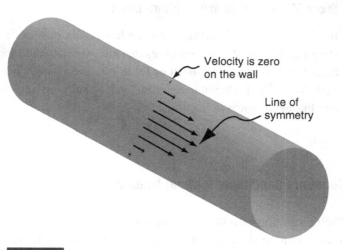

Figure 7.11

Example of boundary conditions for fluid flow in a tube.

is the same at equal distance from the centerline on either side of the plane of symmetry, thus having a zero slope at the centerline. This is expressed as

$$\left.\frac{\partial u}{\partial x}\right|_{\text{line of symmetry}} = 0 \qquad (7.70)$$

Velocity at an inlet is specified Quite often, at an inlet, the velocity is specified. For example, at the inlet of a circular tube where the flow is already fully developed, the velocity profile can be specified as

$$u_{\text{inlet}} = u_{\text{max}}\left(1 - \left(\frac{r}{R}\right)^2\right) \qquad (7.71)$$

where u_{max} is the velocity at the centerline, R is the radius of the tube and r is the radial location.

7.11.2 Pressure at a boundary is specified

Pressure values at the boundaries (typically the outlet) can be specified:

$$p_{\text{outlet}} = p_s \qquad (7.72)$$

where p_s is the pressure at the outlet. Using this value of pressure at the outlet, all other pressure values can be found explicitly.

> **Box 7.10 Software implementation**
>
> Implementation in COMSOL of boundary conditions for fluid mechanics is shown in Section 2.9. Case studies that show implementation of some of the boundary conditions are listed in Table 1.3. Specifying a parabolic inlet velocity profile is shown in Section 2.9.3. Specifying time-varying boundary conditions is shown in Section 2.9.2.

7.12 Boundary conditions for heat transfer

The description of a heat transfer problem in a system is not complete without information on the thermal conditions on the bounding surfaces of the system. The following are the three most common types of thermal conditions that can occur on the boundary of a system in a heat transfer situation:

7.12 Boundary conditions for heat transfer

7.12.1 Surface temperature is specified

One of the simplest thermal conditions that can occur on a surface is a specified temperature. For a one-dimensional heat transfer, this boundary condition is expressed as

$$T\Big|_{x=0} = T_s \tag{7.73}$$

The temperature at the surface, T_s, can be specified as a constant or a function of time. For examples of physical situations that lead to this type of boundary condition, look under various case studies, as noted in Table 1.3.

7.12.2 Surface heat flux is specified

Sometimes it is possible to know and specify the rate of heat transfer or *heat flux* on a surface. For a one-dimensional heat transfer, this boundary condition is expressed as

$$-k\frac{dT}{dx}\Big|_{x=0} = q_s'' \tag{7.74}$$

where the surface heat flux, q_s'', can be specified as a constant or a function of time. There are two important special cases of specified surface heat flux:

Special case: insulated condition When a surface is highly insulated, the heat flux through the surface is very small and can be approximated as zero. This boundary condition is expressed as

$$-k\frac{dT}{dx}\Big|_{x=0} = 0 \tag{7.75}$$

Special case: symmetry condition Another common situation arises in a heating or cooling process when the geometry and the boundary conditions are symmetric. The resulting temperature profile in the slab will also be symmetric, having a zero slope at the line of symmetry. This is expressed as

$$-k\frac{dT}{dx}\Big|_{\text{Line of symmetry}} = 0 \tag{7.76}$$

Note that this symmetry condition resembles the insulated condition mentioned above. To maintain symmetry, heat flux has to be zero at the line of symmetry.

7.12.3 Convection at the surface

Perhaps the most common type of thermal condition that can occur on a boundary is convection of a fluid over it. Thus heat conducted out of the boundary is convected away by the fluid. This condition is written as simply a heat balance at the boundary:

$$\underbrace{-k \left.\frac{dT}{dx}\right|_{x=0}}_{\text{heat conduction}} = \underbrace{h(T|_{x=0} - T_\infty)}_{\text{heat convection}} \qquad (7.77)$$

Here h is the heat transfer coefficient. Note that the surface temperature $T(x=0)$ is not known and will come out of the solution. This is in contrast with Eqn. 7.73, where the surface temperature was known. As a special case of Eqn. 7.77, when $h \to \infty$, it is approximated to

$$T|_{x=0} = T_\infty$$

which is the temperature-specified (first type of) boundary condition mentioned above.

7.12.4 Evaporation at surface

For a wet biomaterial, depending on the temperature, there can be significant evaporation at the surface. This was partly discussed in Section 7.4.2. The evaporative boundary condition at a surface can be included analogous to the convective boundary condition at the surface given by Eqn. 7.77 as

$$-k \left.\frac{\partial T}{\partial x}\right|_{x=0} = \lambda n_{A,s} \qquad (7.78)$$

where $n_{A,s}$ is the amount of moisture lost per unit area per unit time or the mass flux at the surface.

7.12.5 Radiation at surface

Radiative boundary conditions can be important when the temperature difference between the surface and the surroundings is large. For a simple situation, radiative flux at a surface can be included analogous to the convective boundary condition at the surface given by Eqn. 7.77 as

$$\underbrace{-k \left.\frac{dT}{dx}\right|_{x=0}}_{\text{heat conduction}} = \underbrace{\sigma(T^4|_{x=0} - T^4_{\text{wall}})}_{\text{heat radiation}} \qquad (7.79)$$

7.12.6 Combined convection, radiation and evaporation

Convection, radiation and evaporation at the surface can happen simultaneously and therefore can be combined as

$$\underbrace{-k \left.\frac{dT}{dx}\right|_{x=0}}_{\text{heat conduction}} = \underbrace{h(T|_{x=0} - T_\infty)}_{\text{heat convection}} + \underbrace{\sigma(T^4|_{x=0} - T^4_{\text{wall}})}_{\text{heat radiation}} + \underbrace{\lambda n_{A,s}}_{\text{evaporative loss}} \qquad (7.80)$$

> **Box 7.11 Software implementation**
>
> Implementation in COMSOL of boundary conditions for heat transfer is shown in Section 2.9. Case studies that show implementation of some of the boundary conditions are listed in Table 1.3. Specifying time-varying boundary conditions is shown in Section 2.9.2.

7.13 Boundary conditions for mass transfer

This section parallels Section 7.12 on heat transfer. Like the description of a heat transfer problem is not complete without specifying the thermal conditions at the boundary, concentration conditions at the boundary are necessary for the full description of a mass transfer problem. Mass transfer boundary conditions parallel their counterpart in heat transfer for the first three situations. However, before getting to the boundary conditions, it is important to note the complications that arise in mass transfer due to the presence of more than one phase.

7.13.1 Interphase equilibrium as it relates to boundary conditions

Unlike in heat transfer, when two different materials meet, there is often a discontinuity in concentration at the boundary. We consider three separate cases: (1) gas and liquid; (2) liquid and solid; and (3) solid and gas.

Gas and liquid When a gas phase is in contact with a liquid phase, and we are considering a species that is in both the gas and the liquid phase, the concentrations of the species in the liquid and gas phases are related by Henry's law

$$x_A = \frac{p_A}{H} \qquad (7.81)$$

where x_A is the the concentration of the species in the liquid phase and p_A is partial pressure (a measure of concentration) of the same species in the gas phase that is in contact with the liquid phase. Here H is the Henry's law constant, values for which are fairly easily available in the literature. An example of such an equilibrium would be water in contact with air, the species being O_2 that is in both gas and water. If the computation is being done on the liquid phase, x_A, calculated from Eqn. 7.81 with known p_A, is set as the boundary condition. Conversely, if the computations are being done for the gas phase, its boundary condition p_A is set from the known liquid concentration, x_A.

Liquid and solid When a liquid phase is in contact with a solid phase, the ratio of the concentration of a species in the solid phase to that in the liquid phase where they are in contact is termed a partition or distribution coefficient. For example, consider a dialyzing fluid in contact with a dialysis membrane, as shown in the schematic of Figure 7.12. The concentrations of urea (the transporting species) in the liquid and in the solid membrane where they contact are not the same. The ratio of these two concentrations is the partition coefficient for urea (the species), between the membrane (the solid phase) and the dialyzate solution (the liquid phase). Thus, referring to Figure 7.12, partition coefficient, K is defined as

$$K = \frac{c_1^*}{c_1} = \frac{c_2^*}{c_2} \tag{7.82}$$

where c_1^* is the concentration in the solid phase and c_1 is the concentration in the liquid phase. To set the concentration boundary condition in the solid phase when the concentration in the liquid phase in contact is known, we use $c_1^* = Kc_1$ and, when the concentration in the solid phase is known, we use $c_1 = c_1^*/K$ for setting the concentration boundary condition in the liquid phase. See also Section 9.17.

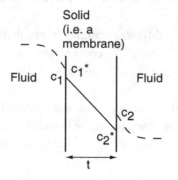

Figure 7.12

Illustration of partition coefficient showing how the concentration of a species can be different in the two phases where the phases are in contact.

Solid and gas As in the case of a solid-liquid boundary, in a solid-gas boundary, the concentrations in the solid phase and gas phase are related through a similar partition or distribution coefficient, K, as in Eqn. 7.82, with c_1^* being concentration in the solid phase and c_1 being the concentration in the gas phase. For example, consider a contact lens surface in touch with air. The oxygen concentration in the contact lens will not be the same as that in the air. The ratio of these two concentrations is the partition coefficient for oxygen between air and the contact lens. To set the concentration boundary condition in the solid phase when the concentration in the gas phase in contact is known, we use $c_1^* = Kc_1$ and, when the concentration in the solid phase is known, we use $c_1 = c_1^*/K$ for setting the concentration boundary condition in the gas phase. The values for K come from experiment. When the diffusing species is moisture, the relationship between concentration in the solid and that in the gas phase at equilibrium is known as a moisture isotherm.

We now return to the discussion on the types of mass transfer boundary conditions. Like heat transfer, there are three most common types of boundary conditions present in a mass transfer situation. These are described in the following sections.

7.13.2 Surface concentration is specified

Perhaps the simplest situation occurs when the concentration can be specified at a surface. For a one-dimensional mass transfer, this boundary condition can be expressed as

$$c_A\bigg|_{x=0} = c_{A,s} \qquad (7.83)$$

where $c_{A,s}$, the concentration at the surface, can be specified as a constant or a function of time. At a solid-fluid interface, if we are setting up the mass transfer problem for the solid, we need $c_{A,s}$ for the solid. Sometimes, at such an interface, the boundary condition on the fluid side is provided instead. In such a case, $c_{A,s}$ for the solid is obtained from the boundary condition on the fluid side and equilibrium relations. On the other hand, if we are setting up a mass transfer problem for a fluid (i.e., we need to calculate concentrations in the fluid), we need the boundary condition for the fluid. An example is the evaporation from the surface of free water or a very wet solid into air. Here the concentration of vapor in air at the boundary will be given by the vapor pressure of water,

$$c_{\text{vapor}}\bigg|_{x=0} = \frac{p_{\text{vapor}}}{RT} \qquad (7.84)$$

where p_{vapor} is the vapor pressure of water at temperature T. Another example of a concentration-specified boundary condition is the disappearance of a species at a surface, e.g., diffusing species O_2 is completely consumed at a surface, leading to $c_{O_2} = 0$ at the surface.

7.13.3 Surface mass flux is specified

Sometimes it is possible to know and specify the rate of mass transfer or mass flux on a surface. For a one-dimensional mass transfer, this boundary condition is expressed as

$$-D_{AB} \left.\frac{\partial c_A}{\partial x}\right|_{x=0} = n_{A,s} \tag{7.85}$$

Here $n_{A,s}$, the surface mass flux, can be specified as a constant or a function of time. Two important special cases of surface mass flux occur frequently in practice:

Special case: impermeable condition Surfaces are sometimes made impermeable in an attempt to stop a particular type of mass movement. When the surface is highly impermeable to a particular species, the mass flux of that species is very small and can be approximated as zero, This boundary condition is given by

$$-D_{AB} \left.\frac{\partial c_A}{\partial x}\right|_{x=0} = 0 \tag{7.86}$$

For example, when a biomaterial containing water is wrapped with aluminum foil to reduce moisture loss, the flux of water at the foil surface can be considered zero, leading to the above boundary condition.

Special case: symmetry condition Another common situation arises in a mass transfer process when the geometry, as well as the boundary conditions, is symmetric, as it would be in the drying of a slab of uniform thickness (making it symmetric about the centerline) as well as it is being dried symmetrically (same boundary conditions on both faces). The resulting concentration profile in the slab will also be symmetric about the centerline, having a zero slope at the centerline. Thus, the symmetry boundary condition at the centerline can be written as

$$\left.\frac{\partial c_A}{\partial x}\right|_{\text{line of symmetry}} = 0 \tag{7.87}$$

Note that this boundary condition resembles the impermeable condition mentioned above, since to maintain symmetry, mass flux has to be zero at the line of symmetry.

7.13.4 Convection at the surface

Another common mass transfer boundary condition occurs when a fluid is flowing over a surface. At the surface ($x = 0$), the amount of mass diffused through solid or liquid is equal to the amount of mass convected away. This boundary condition is written simply as a mass balance at the surface, which is:

$$\underbrace{-D_{AB} \frac{\partial c_A}{\partial x}\bigg|_{x=0}}_{\text{mass diffusion to the left}} = \underbrace{h_m(c_{A,x=0}^{\text{fluid}} - c_{A,\infty}^{\text{fluid}})}_{\text{mass convection to the right}} \qquad (7.88)$$

Here h_m is the mass transfer coefficient, defined analogous to the heat transfer coefficient, and is explained in later chapters. In Eqn. 7.88, it is important to note that the left-hand side uses concentration of species A in the solid, c_A, while the right-hand side uses concentration of species A in the fluid, c_A^{fluid}. A common example of this type of boundary condition is that at the surface of a wet solid with air flowing over it, after the solid has experienced some drying, i.e., when the surface is not very wet. (When the surface is very wet, the boundary condition is given by Eqn. 7.84.) In this case, the left-hand side of Eqn. 7.88 uses the concentration of water in the solid, c_{water}, while the right-hand side uses the concentration of water vapor in air, c_{vapor}. At the surface, the concentration of water in the solid, $c_{\text{water,surface}}$ and concentration of water vapor in air, $c_{\text{vapor,surface}}$ are not the same, but they are related by isotherms, as discussed in Section 9.14.

Also, analogous to the boundary conditions for heat transfer, in Section 7.12 when $h_m \to \infty$ (such as in high fluid velocity), Eqn. 7.88 leads to:

$$c_{A,\text{surface}}^{\text{fluid}} = c_{A,\infty}^{\text{fluid}}$$

or the concentration in the fluid at the surface becomes equal to the concentration in the bulk fluid. Thus, the boundary condition given by Eqn. 7.88 reverts to the first kind of boundary condition given by Eqn. 7.83, with the surface concentration in the solid being specified as that in equilibrium with the surface concentration in the fluid, $c_{A,\text{surface}}^{\text{fluid}}$.

7.13.5 Multiple materials and discontinuities at the boundary

In Section 7.13.1, it was mentioned that, at the interface of two phases, the concentration is different depending on which phase we are referring to, i.e., the concentration value is discontinuous at the interface of two materials. This causes no problem if the computations are restricted to one side of the boundary or the other, as discussed in that section, since the concentration in the other domain is not part of the computation. However, when the computational domain includes both materials, we run into computational difficulties. Some workaround is noted below.

> **Box 7.12 Software implementation**
>
> Implementation in COMSOL of boundary conditions for mass transfer is shown in Section 2.9. Case studies that show implementation of some of the boundary conditions are listed in Table 1.3. Specifying time-varying boundary conditions is shown in Section 2.9.2.
>
> In COMSOL, like most other softwares, the discontinuity in concentration where two materials meet, as given by the partition coefficient or its equivalent, cannot be specified directly. However, the partition coefficient can still be implemented using different species equations for different subdomains, as follows. Consider species equations for c_1 and c_2 in subdomains 1 and 2, respectively, as shown in Figure 7.13. An interface boundary condition is specified for subdomain 1 of the type $-D\partial c_1/\partial x = P(c_2 - Kc_1)$, where P is an arbitrary parameter whose value can be taken to be very large so that the exact value chosen has no effect on the solution. Similarly, for subdomain 2, a boundary condition at the same interface of $-D\partial c_2/\partial x = P(Kc_1 - c_2)$ is specified. This enforces the concentration discontinuity given by Eqn. 7.82 while automatically maintaining the flux continuity through the governing equations.

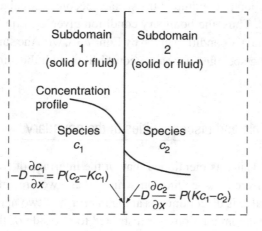

Figure 7.13

Schematic for implementation of concentration discontinuities (partition coefficient) at the boundary.

7.13.6 Summary of boundary conditions

Boundary conditions go with the governing equations in completing the mathematical description of a (biomedical) process. Three types of boundary conditions are most common – the value (temperature or concentration) is set, the flux (of temperature or concentration) is set or convective condition at the surface. In mass transfer, there are complexities at the boundary when two different materials are involved–additional equilibrium relationships are needed.

> **Box 7.13 Software implementation**
>
> COMSOL implementation of the boundary conditions is noted above under various types of boundary conditions.

7.14 Governing equations in various coordinate systems

Although general, these equations (adapted from Bird *et al.*, 1960) still have many assumptions. For example, most of them use constant properties. More general versions of these equations are available in advanced textbooks or in research publications. A note about the symbols used here. The velocity components noted below, v_x, v_y and v_z correspond to the velocities u, v and w, respectively, used earlier in this chapter.

7.14.1 The equation of continuity in rectangular, cylindrical and spherical coordinate systems

Rectangular coordinates (x, y, z):

$$\frac{\partial \rho}{\partial t} + \frac{\partial}{\partial x}(\rho v_x) + \frac{\partial}{\partial y}(\rho v_y) + \frac{\partial}{\partial z}(\rho v_z) = 0$$

Cylindrical coordinates (r, θ, z):

$$\frac{\partial \rho}{\partial t} + \frac{1}{r}\frac{\partial}{\partial r}(\rho r v_r) + \frac{1}{r}\frac{\partial}{\partial \theta}(\rho v_\theta) + \frac{\partial}{\partial z}(\rho v_z) = 0$$

Spherical coordinates (r, θ, ϕ):

$$\frac{\partial \rho}{\partial t} + \frac{1}{r^2}\frac{\partial}{\partial r}(\rho r^2 v_r) + \frac{1}{r \sin \theta}\frac{\partial}{\partial \theta}(\rho v_\theta \sin \theta) + \frac{1}{r \sin \theta}\frac{\partial}{\partial \phi}(\rho v_\phi) = 0$$

7.14.2 The governing equations for fluid flow in rectangular coordinates (x, y, z): in terms of shear stress τ

x-component $\quad \rho \left(\dfrac{\partial v_x}{\partial t} + v_x \dfrac{\partial v_x}{\partial x} + v_y \dfrac{\partial v_x}{\partial y} + v_z \dfrac{\partial v_x}{\partial z} \right) = -\dfrac{\partial P}{\partial x}$

$\quad - \left(\dfrac{\partial \tau_{xx}}{\partial x} + \dfrac{\partial \tau_{yx}}{\partial y} + \dfrac{\partial \tau_{zx}}{\partial z} \right) + \rho g_x$

y-component $\quad \rho \left(\dfrac{\partial v_y}{\partial t} + v_x \dfrac{\partial v_y}{\partial x} + v_y \dfrac{\partial v_y}{\partial y} + v_z \dfrac{\partial v_y}{\partial z} \right) = -\dfrac{\partial P}{\partial y}$

$\quad - \left(\dfrac{\partial \tau_{xy}}{\partial x} + \dfrac{\partial \tau_{yy}}{\partial y} + \dfrac{\partial \tau_{zy}}{\partial z} \right) + \rho g_y$

z-component $\quad \rho \left(\dfrac{\partial v_z}{\partial t} + v_x \dfrac{\partial v_z}{\partial x} + v_y \dfrac{\partial v_z}{\partial y} + v_z \dfrac{\partial v_z}{\partial z} \right) = -\dfrac{\partial p}{\partial z}$

$\quad - \left(\dfrac{\partial \tau_{xz}}{\partial x} + \dfrac{\partial \tau_{yz}}{\partial y} + \dfrac{\partial \tau_{zz}}{\partial z} \right) + \rho g_z$

7.14.3 The governing equations for fluid flow in rectangular coordinates (x, y, z): in terms of velocity gradients for a Newtonian fluid with constant ρ and μ

x-component $\quad \rho \left(\dfrac{\partial v_x}{\partial t} + v_x \dfrac{\partial v_x}{\partial x} + v_y \dfrac{\partial v_x}{\partial y} + v_z \dfrac{\partial v_x}{\partial z} \right) = -\dfrac{\partial P}{\partial x}$

$\quad + \mu \left(\dfrac{\partial^2 v_x}{\partial x^2} + \dfrac{\partial^2 v_x}{\partial y^2} + \dfrac{\partial^2 v_x}{\partial z^2} \right) + \rho g_x$

y-component $\quad \rho \left(\dfrac{\partial v_y}{\partial t} + v_x \dfrac{\partial v_y}{\partial x} + v_y \dfrac{\partial v_y}{\partial y} + v_z \dfrac{\partial v_y}{\partial z} \right) = -\dfrac{\partial P}{\partial y}$

$\quad + \mu \left(\dfrac{\partial^2 v_y}{\partial x^2} + \dfrac{\partial^2 v_y}{\partial y^2} + \dfrac{\partial^2 v_y}{\partial z^2} \right) + \rho g_y$

z-component $\quad \rho \left(\dfrac{\partial v_z}{\partial t} + v_x \dfrac{\partial v_z}{\partial x} + v_y \dfrac{\partial v_z}{\partial y} + v_z \dfrac{\partial v_z}{\partial z} \right) = -\dfrac{\partial P}{\partial z}$

$\quad + \mu \left(\dfrac{\partial^2 v_z}{\partial x^2} + \dfrac{\partial^2 v_z}{\partial y^2} + \dfrac{\partial^2 v_z}{\partial z^2} \right) + \rho g_z$

7.14.4 The governing equations for fluid flow in cylindrical coordinates (r, θ, z): in terms of shear stress τ

r-component
$$\rho\left(\frac{\partial v_r}{\partial t} + v_r\frac{\partial v_r}{\partial r} + \frac{v_\theta}{r}\frac{\partial v_r}{\partial \theta} - \frac{v_\theta^2}{r} + v_z\frac{\partial v_r}{\partial z}\right) = -\frac{\partial P}{\partial r}$$
$$- \left(\frac{1}{r}\frac{\partial}{\partial r}(r\tau_{rr}) + \frac{1}{r}\frac{\partial \tau_{r\theta}}{\partial \theta} - \frac{\tau_{\theta\theta}}{r} + \frac{\partial \tau_{rz}}{\partial z}\right) + \rho g_r$$

θ-component
$$\rho\left(\frac{\partial v_\theta}{\partial t} + v_r\frac{\partial v_\theta}{\partial r} + \frac{v_\theta}{r}\frac{\partial v_\theta}{\partial \theta} + \frac{v_r v_\theta}{r} + v_z\frac{\partial v_\theta}{\partial z}\right) = -\frac{1}{r}\frac{\partial P}{\partial \theta}$$
$$- \left(\frac{1}{r^2}\frac{\partial}{\partial r}(r^2\tau_{r\theta}) + \frac{1}{r}\frac{\partial \tau_{\theta\theta}}{\partial \theta} + \frac{\partial \tau_{\theta z}}{\partial z}\right) + \rho g_\theta$$

z-component
$$\rho\left(\frac{\partial v_z}{\partial t} + v_r\frac{\partial v_z}{\partial r} + \frac{v_\theta}{r}\frac{\partial v_z}{\partial \theta} + v_z\frac{\partial v_z}{\partial z}\right) = -\frac{\partial P}{\partial z}$$
$$- \left(\frac{1}{r}\frac{\partial}{\partial r}(r\tau_{rz}) + \frac{1}{r}\frac{\partial \tau_{\theta z}}{\partial \theta} + \frac{\partial \tau_{zz}}{\partial z}\right) + \rho g_z$$

7.14.5 The governing equations for fluid flow in cylindrical coordinates (r, θ, z): in terms of velocity gradients for a Newtonian fluid with constant ρ and μ

r-component
$$\rho\left(\frac{\partial v_r}{\partial t} + v_r\frac{\partial v_r}{\partial r} + \frac{v_\theta}{r}\frac{\partial v_r}{\partial \theta} - \frac{v_\theta^2}{r} + v_z\frac{\partial v_r}{\partial z}\right) = -\frac{\partial P}{\partial r}$$
$$+ \mu\left[\frac{\partial}{\partial r}\left(\frac{1}{r}\frac{\partial}{\partial r}(rv_r)\right) + \frac{1}{r^2}\frac{\partial^2 v_r}{\partial \theta^2} - \frac{2}{r^2}\frac{\partial v_\theta}{\partial \theta} + \frac{\partial^2 v_r}{\partial z^2}\right] + \rho g_r$$

θ-component
$$\rho\left(\frac{\partial v_\theta}{\partial t} + v_r\frac{\partial v_\theta}{\partial r} + \frac{v_\theta}{r}\frac{\partial v_\theta}{\partial \theta} + \frac{v_r v_\theta}{r} + v_z\frac{\partial v_\theta}{\partial z}\right) = -\frac{1}{r}\frac{\partial P}{\partial \theta}$$
$$+ \mu\left[\frac{\partial}{\partial r}\left(\frac{1}{r}\frac{\partial}{\partial r}(rv_\theta)\right) + \frac{1}{r^2}\frac{\partial^2 v_\theta}{\partial \theta^2} + \frac{2}{r^2}\frac{\partial v_r}{\partial \theta} + \frac{\partial^2 v_\theta}{\partial z^2}\right] + \rho g_\theta$$

z-component
$$\rho\left(\frac{\partial v_z}{\partial t} + v_r\frac{\partial v_z}{\partial r} + \frac{v_\theta}{r}\frac{\partial v_z}{\partial \theta} + v_z\frac{\partial v_z}{\partial z}\right) = -\frac{\partial P}{\partial z}$$
$$+ \mu\left[\frac{1}{r}\frac{\partial}{\partial r}\left(r\frac{\partial v_z}{\partial r}\right) + \frac{1}{r^2}\frac{\partial^2 v_z}{\partial \theta^2} + \frac{\partial^2 v_z}{\partial z^2}\right] + \rho g_z$$

7.14.6 Governing equation for heat transfer (*assuming Newtonian fluids of constant ρ and k*) in rectangular coordinates*

$$\rho C_p \left(\frac{\partial T}{\partial t} + v_x \frac{\partial T}{\partial x} + v_y \frac{\partial T}{\partial y} + v_z \frac{\partial T}{\partial z} \right) = k \left[\frac{\partial^2 T}{\partial x^2} + \frac{\partial^2 T}{\partial y^2} + \frac{\partial^2 T}{\partial z^2} \right] + Q$$

$$+ 2\mu \left\{ \left(\frac{\partial v_x}{\partial x} \right)^2 + \left(\frac{\partial v_y}{\partial y} \right)^2 + \left(\frac{\partial v_z}{\partial z} \right)^2 \right\} + \mu \left\{ \left(\frac{\partial v_x}{\partial y} + \frac{\partial v_y}{\partial x} \right)^2 \right.$$

$$\left. + \left(\frac{\partial v_x}{\partial z} + \frac{\partial v_z}{\partial x} \right)^2 + \left(\frac{\partial v_y}{\partial z} + \frac{\partial v_z}{\partial y} \right)^2 \right\} \qquad (7.89)$$

7.14.7 Governing equation for heat transfer (*assuming Newtonian fluids of constant ρ and k*) in cylindrical coordinates*

$$\rho C_p \left(\frac{\partial T}{\partial t} + v_r \frac{\partial T}{\partial r} + \frac{v_\theta}{r} \frac{\partial T}{\partial \theta} + v_z \frac{\partial T}{\partial z} \right) = k \left[\frac{1}{r} \frac{\partial}{\partial r} \left(r \frac{\partial T}{\partial r} \right) + \frac{1}{r^2} \frac{\partial^2 T}{\partial \theta^2} + \frac{\partial^2 T}{\partial z^2} \right] + Q$$

$$+ 2\mu \left\{ \left(\frac{\partial v_r}{\partial r} \right)^2 + \left[\frac{1}{r} \left(\frac{\partial v_\theta}{\partial \theta} + v_r \right) \right]^2 + \left(\frac{\partial v_z}{\partial z} \right)^2 \right\} + \mu \left\{ \left(\frac{\partial v_\theta}{\partial z} + \frac{1}{r} \frac{\partial v_z}{\partial \theta} \right)^2 \right.$$

$$\left. + \left(\frac{\partial v_z}{\partial r} + \frac{\partial v_r}{\partial z} \right)^2 + \left[\frac{1}{r} \frac{\partial v_r}{\partial \theta} + r \frac{\partial}{\partial r} \left(\frac{v_\theta}{r} \right) \right]^2 \right\} \qquad (7.90)$$

7.14.8 Governing equation for heat transfer (*assuming Newtonian fluids of constant ρ and k*) in spherical coordinates*

$$\rho C_p \left(\frac{\partial T}{\partial t} + v_r \frac{\partial T}{\partial r} + \frac{v_\theta}{r} \frac{\partial T}{\partial \theta} + \frac{v_\phi}{r \sin \theta} \frac{\partial T}{\partial \phi} \right) = k \left[\frac{1}{r^2} \frac{\partial}{\partial r} \left(r^2 \frac{\partial T}{\partial r} \right) \right.$$

$$\left. + \frac{1}{r^2 \sin \theta} \frac{\partial}{\partial \theta} \left(\sin \theta \frac{\partial T}{\partial \theta} \right) + \frac{1}{r^2 \sin^2 \theta} \frac{\partial^2 T}{\partial \phi^2} \right] + Q + 2\mu \left\{ \left(\frac{\partial v_r}{\partial r} \right)^2 \right.$$

$$\left. + \left(\frac{1}{r} \frac{\partial v_\theta}{\partial \theta} + \frac{v_r}{r} \right)^2 + \left(\frac{1}{r \sin \theta} \frac{\partial v_\phi}{\partial \phi} + \frac{v_r}{r} + \frac{v_\theta \cot \theta}{r} \right)^2 \right\}$$

$$+ \mu \left\{ \left[r \frac{\partial}{\partial r} \left(\frac{v_\theta}{r} \right) + \frac{1}{r} \frac{\partial v_r}{\partial \theta} \right]^2 + \left[\frac{1}{r \sin \theta} \frac{\partial v_r}{\partial \phi} + r \frac{\partial}{\partial r} \left(\frac{v_\phi}{r} \right) \right]^2 \right.$$

$$\left. + \left[\frac{\sin \theta}{r} \frac{\partial}{\partial \theta} \left(\frac{v_\phi}{\sin \theta} \right) + \frac{1}{r \sin \theta} \frac{\partial v_\theta}{\partial \phi} \right]^2 \right\} \qquad (7.91)$$

---

* Includes heat generation due to viscous dissipation.

7.14.9 Governing equation for species mass transfer in rectangular coordinates for constant ρ and D_{AB}

$$\frac{\partial c_A}{\partial t} + \left(v_x \frac{\partial c_A}{\partial x} + v_y \frac{\partial c_A}{\partial y} + v_z \frac{\partial c_A}{\partial z}\right) = D_{AB}\left(\frac{\partial^2 c_A}{\partial x^2} + \frac{\partial^2 c_A}{\partial y^2} + \frac{\partial^2 c_A}{\partial z^2}\right) + R_A$$

7.14.10 Governing equation for species mass transfer in cylindrical coordinates for constant ρ and D_{AB}

$$\frac{\partial c_A}{\partial t} + \left(v_r \frac{\partial c_A}{\partial r} + v_\theta \frac{1}{r}\frac{\partial c_A}{\partial \theta} + v_z \frac{\partial c_A}{\partial z}\right)$$
$$= D_{AB}\left(\frac{1}{r}\frac{\partial}{\partial r}\left(r\frac{\partial c_A}{\partial r}\right) + \frac{1}{r^2}\frac{\partial^2 c_A}{\partial \theta^2} + \frac{\partial^2 c_A}{\partial z^2}\right) + R_A$$

7.14.11 Governing equation for species mass transfer in spherical coordinates for constant ρ and D_{AB}

$$\frac{\partial c_A}{\partial t} + \left(v_r \frac{\partial c_A}{\partial r} + v_\theta \frac{1}{r}\frac{\partial c_A}{\partial \theta} + v_\phi \frac{1}{r\sin\theta}\frac{\partial c_A}{\partial \phi}\right)$$
$$= D_{AB}\left(\frac{1}{r^2}\frac{\partial}{\partial r}\left(r^2 \frac{\partial c_A}{\partial r}\right) + \frac{1}{r^2 \sin\theta}\frac{\partial}{\partial \theta}\left(\sin\theta \frac{\partial c_A}{\partial \theta}\right)\right.$$
$$\left. + \frac{1}{r^2 \sin^2\theta}\frac{\partial^2 c_A}{\partial \phi^2}\right) + R_A$$

References

Bird, R. B., W. E. Stewart and E. N. Lightfoot (1960). *Transport Phenomena*. John Wiley & Sons. Pages 83–91, 318–319, 555–562.

Datta, A. K. (2002). *Biological and Bioenvironmental Heat and Mass Transfer*, CRC Press, Boca Raton, Florida.

Diller, K. R. (1992). Modeling of bioheat transfer processes at high and low temperatures. *Advances in Heat Transfer*, **22**:157–357.

Diller, K. R., J. W. Valvano and J. A. Pearce (1999). Bioheat transfer. In *CRC Handbook of Thermal Engineering*, Edited by Frank Kreith. Pages 4–114 to 4–187.

Kundu, P. and I. M. Cohen (2004). *Fluid Mechanics*. UK, Elsevier Academic Press.

Shitzer, A. and R. C. Eberhart (1985). *Heat Transfer in Medicine and Biology*. Volumes 1 and 2. New York, Plenum Press.

7.15 Problems

7.15.1 Heat transfer with larger vessels

Consider the bioheat equation discussed in Section 7.5:

$$\rho C_p \frac{\partial T}{\partial t} = k \frac{\partial^2 T}{\partial x^2} + \rho_b c_b w_b (T_b - T) + Q$$

(1) Blood flows from only the smallest blood vessels are included in the above equation. Why? (2) Formulate the problem for heat transfer between tissue and the largest of the blood vessels (arteries and veins).

7.15.2 Governing equation for a freezing problem

Describe how you would formulate the governing equations for a freezing problem, as in cryosurgery, using only the standard heat conduction equation (i.e., without any term for latent heat).

7.15.3 Difficulties with modeling freezing using a sharp interface formulation

The computational work reported in Figure 5.19 used a two-region formulation of freezing. Why is formulating biomaterial freezing as a two-region problem (with a sharp freezing interface) not the best choice? How would a two-region problem be implemented in COMSOL?

8 Source terms

Heat source term The governing equation for heat transfer (Eqn. 7.22) is a statement of energy balance and is repeated here for convenience:

$$\rho C_p \left(\frac{\partial T}{\partial t} + u \frac{\partial T}{\partial x} \right) = k \frac{\partial^2 T}{\partial x^2} + \underbrace{Q}_{\text{heat source}} \tag{8.1}$$

This equation has in it a *heat source* term which represents generation of thermal energy in the material. This term is distinct from the terms due to conduction of heat (second term on the right) and flow (second term on the left). Such generation of energy can be due to, for example, electromagnetic and ultrasonic heating, where energy converts from another form to heat within the tissue (see, for example, Diederich, 2005). The principles behind these heating methods that lead to the formulation of the respective heat source terms will be introduced in this chapter (see Figure 8.1). We will also see how metabolic heat generation and the effect of blood flow can be treated as heat source terms. Source terms can also be negative (also called sink terms) – negative values signify thermal energy being converted into other forms of energy which is therefore 'lost' in the thermal energy balance represented by Eqn. 8.1.

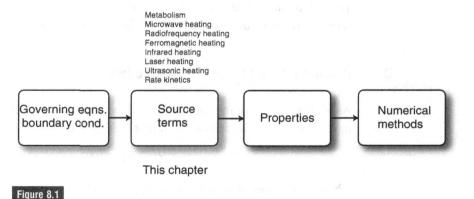

Figure 8.1
Source terms as part of the theoretical framework for modeling is the topic of this chapter.

Mass source term Likewise, the governing equation for mass transfer, which is a statement of species mass balance and shown in Eqn. 7.42, is repeated here:

$$\frac{\partial c_A}{\partial t} + u\frac{\partial c_A}{\partial x} = D_{AB}\frac{\partial^2 c_A}{\partial x^2} + \underbrace{r_A}_{\text{mass source}} \tag{8.2}$$

This equation has a *mass source* term for mass species generation or depletion. Such a term would represent various types of reactions that lead to generation or disappearance (elimination or depletion) of a quantity. For example, in modeling a drug delivery process, the clearance (removal) of the drug is considered as a negative source term in the mass balance equation for drug. Thus, a negative mass source means that the mass species is converting to another one and therefore 'disappearing.' This chapter will introduce the mass source terms that arise from some of the common reactions.

8.1 Heat source terms due to metabolism and blood flow

In the bioheat equation, described in Section 7.5, metabolic heat generation is considered distributed throughout the tissue and included in the equation as a heat source term. Data for metabolic heat generation for various situations are discussed in Section 9.5.4.

Blood perfusion is considered in the bioheat equation (Eqn. 7.36) as a source term:

$$\rho c \frac{\partial T}{\partial t} = k\nabla^2 T + \overbrace{\underbrace{\rho_b c_b \dot{V}_b^v (T_a - T)}_{\substack{\text{heat deposition}\\\text{due to blood flow}}} + \underbrace{Q_m}_{\substack{\text{metabolic}\\\text{heat generation}}}}^{\text{heat source term}} \tag{8.3}$$

The volumetric blood flow rate is $\dot{V}_b^v$, in units of m^3 of blood per m^3 of tissue per second. Available data for blood perfusion are discussed in Section 9.5.3.

8.2 A generic form for the heat source term

As will be shown in the following sections, *under certain simplified conditions*, the heat source term for many different heating situations can be approximated as

$$Q = Q_0 e^{-x/\delta} \tag{8.4}$$

where Q is the rate of volumetric heating at a location x, Q_0 is the rate of volumetric heating at the surface, $x = 0$, and δ is the penetration depth of the energy into a

8.2 A generic form for the heat source term

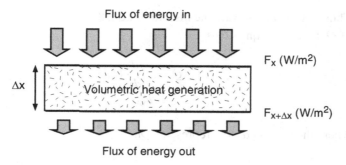

Figure 8.2

Relationship between flux and volumetric quantity.

material. Sometimes, however, propagation of energy into a material is described in terms of a flux, as opposed to the volumetric heating rate. Thus, in terms of a flux, F, the description of the energy deposition is provided as

$$F = F_0 e^{-\alpha x} \tag{8.5}$$

where F_0 is the flux at the surface. Here the unit of flux is W/m^2. Since the governing equation for heat transfer needs Q, the volumetric heat source which is the energy deposition per unit volume (e.g., in W/m^3), we need to convert F to Q. To do this, consider a 1D energy propagation (electromagnetic, ultrasonic etc.) in Figure 8.2 where flux at two different positions, x and $x + \Delta x$, are given by F_x and $F_{x+\Delta x}$, respectively. Assuming an area A perpendicular to the direction of the propagation of energy, the energy balance over a layer of thickness Δx for a particular type of energy such as electromagnetic or ultrasonic, is performed as follows

$$\text{In} - \text{Out} + \text{Generation} = 0$$
$$F_x A - F_{x+\Delta x} A - QA\Delta x = 0$$

where Q is the conversion of the particular type of energy into thermal energy (and therefore is a loss for that type of energy). Simplifying,

$$Q = -\frac{\Delta F}{\Delta x}$$

By making $\Delta x \to 0$, we get

$$Q = -\frac{dF}{dx} \tag{8.6}$$

382 Source terms

Thus, if Eqn. 8.5 is available instead of Eqn. 8.4, we can convert Eqn. 8.5 in terms of Q, by using Eqn. 8.6 as

$$Q = -\frac{d}{dx}\left(F_0 e^{-\alpha x}\right)$$
$$= F_0 \alpha e^{-\alpha x}$$

where the term $F_0\alpha$ now represents Q_0.

8.3 Heat source term for electromagnetic heating

Electromagnetic waves (see Figure 8.3) have an electric field and a magnetic field at right angles to each other, while the plane containing the electric and magnetic fields is at a right angle to the direction of travel of the waves, traveling at the speed of light. The waves interact with the material they are traveling through and generate heat volumetrically, whereby electromagnetic energy is converted to heat. In a heat transfer model, this heat generated is included as a source term. Mechanisms of interaction of electromagnetic waves with a material vary

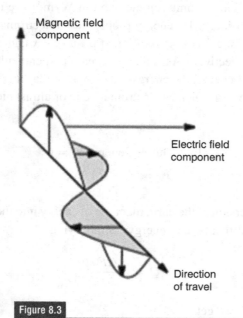

Figure 8.3

Schematic of a traveling electromagnetic wave. The electric and magnetic fields vary in directions perpendicular to each other and they are both perpendicular to the direction of travel.

8.3 Heat source term for electromagnetic heating

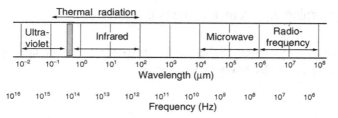

Figure 8.4

The electromagnetic spectrum.

Table 8.1 Industrial, Scientific and Medical (ISM) band frequencies that could be used for medical purposes. Frequencies in use for laser heating can be seen in Figure 8.12.

Band	Frequency range
Radiofrequency	13.56 ± 6.68 kHz
	27.12 ± 160.00 kHz
	40.68 ± 20.00 kHz
Microwaves	915 ± 25 MHz
	2450 ± 50 MHz
	5800 ± 75 MHz
	24225 ± 125 MHz

depending the frequency involved. Figure 8.4 shows the electromagnetic spectrum with various wavelengths and frequencies. The frequency and wavelength are related by

$$\lambda f = c \tag{8.7}$$

where λ is the wavelength in m, f is the frequency in Hz (= cycles/s), and c is the speed of light, in m/s. In a vacuum, the speed of light is approximately 3×10^8 m/s. Many different types of electromagnetic waves, for example microwaves, are useful for thermal therapy in biomedical applications. A partial list of allowed frequencies from the industrial, scientific and medical (ISM) band which includes the frequencies for medical use, is shown in Table 8.1. For general aspects of various electromagnetic heating, the reader is referred to excellent sources such as Metaxas (1996) for microwaves and radio frequencies and Niemz (2002) for lasers.

8.3.1 Governing equations and the heat source term

The electromagnetic fields that are responsible for the heating of the biomaterial are described by Maxwell's equations

$$\nabla \times \mathbf{E} = -\frac{\partial}{\partial t}(\mu \mathbf{H}) \tag{8.8}$$

$$\nabla \times \mathbf{H} = \frac{\partial}{\partial t}(\epsilon \epsilon_0 \mathbf{E}) + \epsilon_{\text{eff}}'' \epsilon_0 \omega \mathbf{E} \tag{8.9}$$

$$\nabla \cdot (\epsilon \mathbf{E}) = 0 \tag{8.10}$$

$$\nabla \cdot \mathbf{H} = 0 \tag{8.11}$$

where $\mathbf{E}$ and $\mathbf{H}$ are the electric and magnetic field vectors, respectively. The quantity ϵ_0 is called the permittivity of free space ($= 8.86 \times 10^{-12}$ F/m). The complex permittivity ϵ is given by

$$\epsilon = \epsilon' + j\epsilon_{\text{eff}}'' \tag{8.12}$$

The parameter ϵ' is called the dielectric constant and it represents the material's ability to store electromagnetic energy. The other parameter, ϵ_{eff}'', is called the effective dielectric loss factor and it represents the energy dissipation or heat generation ability of the material. In a similar manner as for dielectric heat generation, the other quantity, μ, represents magnetic heat generation. In biomaterials, heating is primarily through interaction of the electric field with water and ions. For a short discussion of Maxwell's equations and solutions, see Saltiel and Datta (1998).

The Maxwell's equations must be solved to obtain the electric field $\mathbf{E}$ as a function of position in the biomaterial and heating time. The rate of volumetric heat generation (that is the heat source term due to microwave or radiofrequency absorption in a tissue, for example) can be calculated from this electric field as

$$Q = \pi f \epsilon_0 \epsilon_{\text{eff}}'' |E|^2 \tag{8.13}$$

where $|E|$ is the magnitude of the peak value of electric field at any location. As the wave travels in the tissue, its energy is dissipated ($|E|$ drops) in the tissue and appears as heat.

8.4 Microwave heating and its heat source term

Microwave heating is part of electromagnetic heating. Microwave hyperthermia has been used generally in combination with chemotherapy and/or radiotherapy because hyperthermia enhances the tumor's response to radio/chemotherapy

(Laguna *et al.*, 2000; Rappaport, 2004). This has been particularly effective for superficial tumors. Also, the effects of unintended exposure to microwaves and associated heating is another area where microwave heating is important in a biomedical context.

8.4.1 Mechanism of heating

Two electrically active components in the biomaterial are the major sources of interaction with the electric field of the microwaves. These are water molecules due to their dipolar nature (centers of positive and negative charges are separated by a distance) and ions (carrying electrical charges). For a commonly used microwave frequency of 2450 MHz, the electric field changes directions 2450 million times a second, making the dipoles and ions move with it and generate heat.

Two electromagnetic properties characterize the biomaterial's interaction with the microwaves. These are the dielectric constant (represented by the symbol ϵ') which measures the ability to store energy (not convert to thermal) and the dielectric loss (represented by the symbol ϵ'') which measures the ability to absorb microwaves and generate heat. The dielectric loss parameter is also reported in terms of an equivalent conductivity, σ of the material which relates to the dielectric loss as

$$\epsilon'' = \frac{\sigma}{\epsilon_0 \omega} \qquad (8.14)$$

The term ϵ_0 is the permittivity of free space (8.86×10^{-12} F/m) and $\omega = 2\pi f$ is the angular frequency of the microwaves (f being the frequency). The SI unit of conductivity is in Siemens per meter (S/m). These properties depend on the temperature, moisture content and other composition and they have been measured for many biomaterials. Dielectric loss increases significantly with temperature when significant amounts of ions are present and also during the change from a frozen to a thawed state. Dielectric loss decreases as the moisture content of the biomaterial is reduced. Various components in a biomaterial affect the dielectric properties in a complex way and measurement is the only way to obtain data for a particular material.

8.4.2 Dependence of electromagnetic properties on frequency, temperature, composition and other factors

The dielectric properties of a biomaterial can be strong function of frequency, temperature and composition of the material. The mechanism of interaction changes

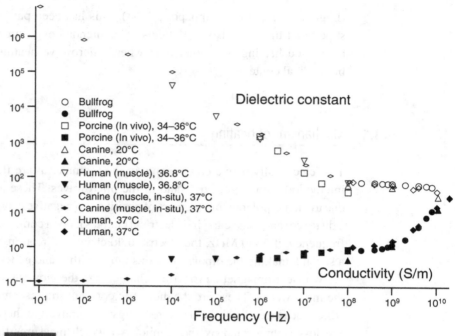

Figure 8.5

Dielectric properties of the heart for various sources (data from Gabriel et al., 1996ab; Data also available on the web at http://niremf.ifac.cnr.it/docs/DIELECTRIC/AppendixB2.html).

depending on the frequency of the wave. For example, in the gigahertz region (microwaves), polarization of water molecules plays a dominant role in determining the permittivity. In the hundreds of kilohertz region, the permittivity value mainly reflects the polarization of cellular membranes which acts as a barrier to the flow of ions between the intra and extracellular media. Protein and other organic macromolecules are also active in this region. An example of frequency dependence in heart tissues can be seen in Figure 8.5 (Gabriel *et al.*, 1996a,b; Gabriel *et al.*, 1998). At a given frequency, the dielectric properties also depend on temperature, as illustrated in Figure 8.6. The data in this figure resemble that for water with a small amount of added salt.

8.4.3 Clinical applicator example

An example of a microwave applicator can be seen in Figure 8.7. It is a helical microwave antenna and was designed to investigate its potential in the thermal therapy of Barrett's esophagus – a condition in which the normal squamous epithelium

8.4 Microwave heating and its heat source term

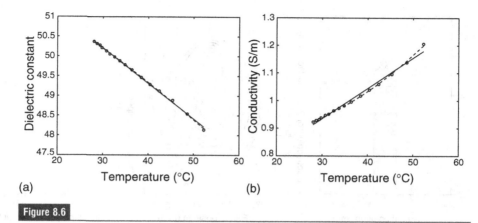

Figure 8.6

Example of the temperature dependence of the dielectric properties of liver tissue at 915 MHz. The circles represent raw measured data and the solid lines are the linear fits to the data. The data shown are for a cooling cycle and in general, data from a heating cycle differ from those during a cooling cycle. From Lazebnik *et al.* (2006). Reproduced with permission from IOP Publishing Limited.

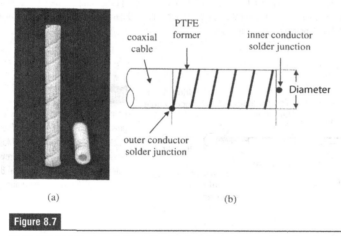

Figure 8.7

Photograph (a) and schematic illustration (b) of a helical antenna configuration of a microwave heating applicator. From Reeves *et al.* (2002). Reproduced with permission.

of the esophagus is replaced by abnormal columnar epithelium. Ablation of the abnormal mucosa is necessary to reverse this condition and allow regeneration of a normal epithelium. The microwave signal used was 915 MHz. An example of the type of temperature distribution obtained using such an antenna is shown in Figure 8.8. Here the antenna was placed between two muscle-equivalent phantoms and microwave power in the range 20–40W applied for 20 s, followed by separation of the phantoms, removal of the antenna and measurement of the cut surface temperature using an infrared camera. The temperatures are normalized

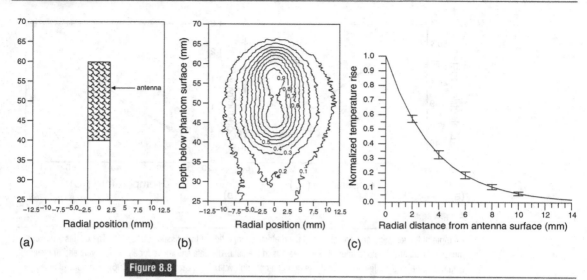

Figure 8.8

(a) Antenna; (b) measured temperature (normalized) distribution using an infrared camera for a 20 mm helical antenna ($S = 2.3$ mm), positioned from the edge as shown in (a); (c) radial temperature (normalized) profile for the 20 mm antenna ($S = 2.3$ mm). Vertical bars represent the standard deviation of ten independent trials. From Reeves *et al.* (2002). Reproduced with permission.

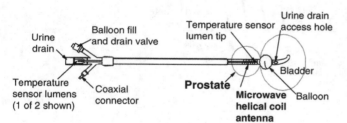

Figure 8.9

Example of a system for microwave thermotherapy of prostate: TMx-3000™ Office Thermo Therapy™ System Treatment Catheter. American Medical Systems, Minnetonka, Minnesota. A microwave helical coil antenna is discussed in Section 8.4.3. Image from TMx-3000™ manual. Reproduced with permission.

to the maximum temperature value, leading to a maximum temperature of 1, as shown in the figure.

Another example of an applicator is the TMx-3000™ BPH Thermotherapy system, illustrated in Figure 8.9. The microwave source here is also a helical coil antenna and is operated at 915 MHz.

8.4.4 Simplified heat generation term

In general, the heat source term will be given by Eqn. 8.13. Under some very restricted conditions (see Section 8.4.5 below), the spatial variation of the electric field can be approximated as an exponential decay from the incident surface. Thus, the heat source term for this restricted situation can be written as

$$Q = Q_0 e^{-x/\delta} \tag{8.15}$$

Here δ is the penetration depth (a measure of energy penetration into the material) given by

$$\delta = \frac{1}{2}\frac{\lambda_0}{2\pi}\left[\frac{\epsilon'}{2\epsilon_0}\left(\sqrt{1+(\epsilon''/\epsilon')^2}-1\right)\right]^{-1/2} \tag{8.16}$$

An example of a temperature profile that qualitatively follows Eqn. 8.16 is shown in Figure 8.8(c).

8.4.5 Complexities

Since the spatial variation in Eqn. 8.16 can only be proven for a plane wave traveling in a semi-infinite domain and constant properties, which are difficult to achieve in typical thermal therapy situations, complications arise very easily. In general, one needs to start from the governing equations of electromagnetics, given by Eqn. 8.8–8.11.

> **Box 8.1 Software implementation**
>
> Simulation of microwave heating will involve obtaining the heat generation term Q from simple considerations (Eqn. 8.15), or from a precise formulation using the complete governing equations of electromagnetics (Eqn. 8.8–8.11). As has been illustrated above, tissue dielectric properties change as temperature rises during a heating process, this will change the heating rate itself as heating progresses. Thus, the heat transfer equation and the electromagnetics (simple or more complex versions) need to be *coupled*. Two of the softwares where electromagnetics and heat transfer equations can be solved and coupled are ANSYS and COMSOL. An example of coupling of heat transfer and electromagnetics that combines the two different software, COMSOL and ANSYS, can be seen in Rakesh *et al.* (2008).

8.5 Radiofrequency heating and its heat source term

An example of radiofrequency heating is radiofrequency ablation (RFA), a minimally invasive method (through controlled heating) for destroying tumors that cannot be removed surgically. Radiofrequency ablation has been studied for treating cardiac arrythmia, and destruction of tumors in liver, kidney, lung, bone, prostate and breast (see, for example, Berjano, 2006; Tsuda *et al.*, 1996). Another popular method is RF-capacitive heating to destroy tumors.

8.5.1 Mechanism of heating

Of the two mechanisms of dipolar reorientation and ionic conduction, mentioned earlier in Section 8.4.1, the latter mechanism dominates at radiofrequencies (≈ 500 kHz for RFA). This is in contrast to microwave heating where the mechanism of dipolar reorientation is dominant.

8.5.2 Clinical applicator example

An example of a radiofrequency clinical applicator, a cluster electrode, is shown in Figure 8.10(a). The picture shows an array of three cluster electrodes (each of the three has multiple electrodes within them making up the cluster). The computed current density and the experimental lesions created in fresh bovine liver for a heating period of 12 minutes using the electrodes are shown in Figure 8.10(b).

8.5.3 Simplified heat generation term

When the wavelength of RF in tissues is much greater than the tissue dimensions involved, quasistatic electric field approximation can be applied (i.e., it can be treated like a DC voltage). One study, for example, assumed that cardiac tissue is purely resistive at frequencies from 10 kHz to 500 kHz and the reactive term is negligible, based on experimental data on the impedance of normal sheep myocardium. When the RF field can be treated as a DC voltage, the governing equation will be given by

$$\nabla \cdot (\nabla \phi) = 0 \tag{8.17}$$

where ϕ is electric potential (i.e., voltage). The electric field, E, is given by $E = -\nabla \phi$. The rate of heat generation from this electric field is given by

$$Q = \frac{1}{2}\sigma |E|^2 \tag{8.18}$$

Here σ is the electrical conductivity of the tissue at the particular frequency.

8.5 Radiofrequency heating and its heat source term

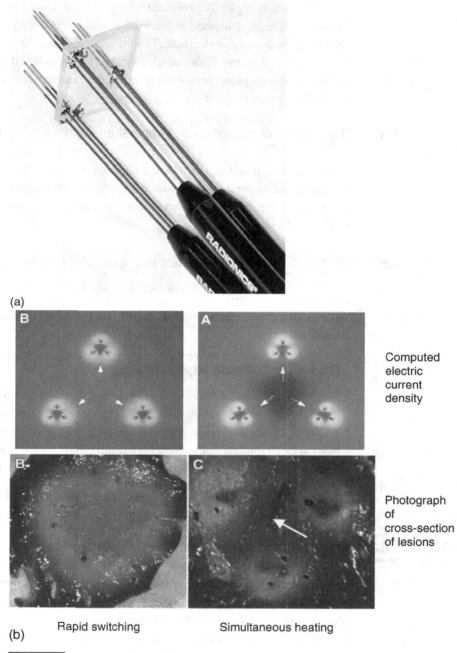

Figure 8.10

(a) Three cool tip cluster electrodes. Each electrode has an active tip that is water-cooled to prevent charring or overcooking. (b) Computed electric current density and experimental lesions after 12 minutes of heating. From Haemmerich et al. (2005). Reproduced with permission.

In RF ablation, typically a constant voltage is used. Thus, boundary conditions in RF ablation are implemented with constant voltage ($\neq 0$ V) at the active electrode and 0 V at the dispersive electrode. In a different scenario, in constant-temperature RF ablation, where the delivered electrical power is modulated by the RF generator to maintain a preset temperature value, the boundary condition on the active electrode is a function of time.

8.5.4 Dependence of properties on temperature, composition and other factors

Like dielectric properties in other frequency ranges, radiofrequency dielectric properties are dependent on tissue composition and temperature. For example, temperature-dependence for heart tissue can be seen in Figure 8.5; see also Schmid *et al.* (2003).

8.5.5 Complexities: when a simplified heat generation term is not enough

Literature has primarily used the heat generation term given by Eqn. 8.17, therefore, this equation can be considered adequate for many different frequency and material combinations.

> **Box 8.2 Software implementation**
>
> The form of Equation 8.17, which is Laplace's equation, can be obtained from the mass diffusion equation if the transient, convection and source terms are dropped. Thus, the electric potential can be solved as another species using a mass species transfer equation, with the diffusivity of the species replaced by the electrical conductivity. Details of this implementation in COMSOL are shown in Case study IX, involving RF cardiac ablation.

8.6 Ferromagnetic heating and its heat source term

Ferromagnetic hyperthermia utilizes cylindrically-shaped metallic alloys, called thermoseeds, that are placed surgically into tumours. See Figure 1.19 for an example of a thermoseed. A coil placed around the patient produces an electromagnetic field that causes eddy-current heating of the thermoseeds. Tissue in contact with thermoseeds is then heated via thermal conduction. This will not be discussed in further detail and the reader is referred to publications such as Tompkins *et al.* (1994). A simple computer model for this technique can be seen in a class project at Cornell University (Avissar *et al.*, 2001).

8.7 Infrared heating and its heat source term

Infrared heating is part of electromagnetic heating. One of the ways that infrared helps to heal and soothe pain is due to its effects on increasing blood flow. In the clinical setting, infrared heating typically refers to the use of an ordinary incandescent bulb to generate infrared rays that provide the heat. Such infrared heat may be preferred over hot packs when the patient is difficult to position or cannot tolerate pressure.

8.7.1 Mechanism of heating

In infrared heating, the electromagnetic energy of the infrared rays converts into heat energy as they are absorbed into the biomaterial.

8.7.2 Clinical applicator example

An illustration of an infrared heat lamp is shown in Figure 8.11. Typically a 250 W incandescent bulb is placed at 40–50 cm from the patient. Heating rates are controlled by adjusting the power of the lamp or the distance between the lamp and the patient.

8.7.3 Simplified heat generation term

When infrared rays are able to penetrate the tissue significantly, an exponentially decaying heat generation term representing the absorption of the infrared energy can be a good assumption:

$$Q = Q_0 e^{-x/\delta} \tag{8.19}$$

Here δ is the penetration depth of the infrared energy at its particular frequency. Data on penetration depth in various tissues is hard to locate. When penetration of energy is very small, a surface heat flux boundary condition can be used instead:

$$-k\frac{\partial T}{\partial x} = I_0 \tag{8.20}$$

Here I_0 is the incident energy flux (W/m^2) on the tissue surface. Note that if an energy flux boundary condition is used at the surface, the volumetric rate of heating (as in Eqn. 8.19) should be set to zero so that the energy absorption is not double counted.

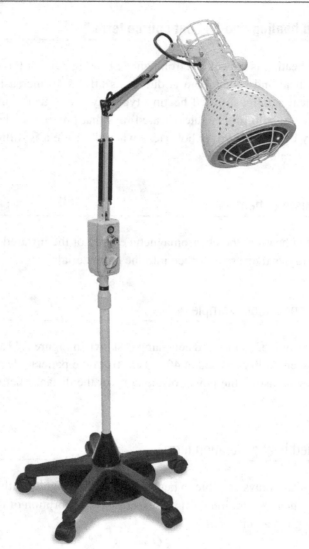

Figure 8.11

A typical infrared heat lamp (IR-Pro 250, Lhasa OMS, Inc.). Picture from the website http://www.lhasaoms.com/infrared_heat_therapy-43-1600-page.html. Reproduced with permission.

Box 8.3 Software implementation

Implementation of Eqn. 8.19 is straightforward when δ is constant. COMSOL implementation of Eqn. 8.19 has been discussed in Case study X under Define expressions.

8.8 Laser heating and its heat source term

Laser heating is also a type of electromagnetic heating. Laser heating of tissue can be painless or 'bloodless.' It is precise, heating only the target and has a reduced risk of infection. Examples of clinical uses of laser heating are in opthalmology, tonsillectomy, thermotherapy, skin rejuvenation and birthmark or tatoo removal.

8.8.1 Mechanism of heating

Laser stands for light amplification by stimulated emission of radiation. The characteristics of laser light are that it is of single wavelength, narrow, well-defined, polarized, a coherent (all the light is in phase) beam of light. Various types of lasers, such as the excimer laser, are illustrated in Figure 8.12. The mechanism of tissue disruption by laser energy can be *photothermal*, *photoablative*, *photodisruptive* (photomechanical or plasma induced) or *photochemical*. The dependence of various laser–tissue interaction mechanisms on power density and exposure time is illustrated in Figure 8.13.

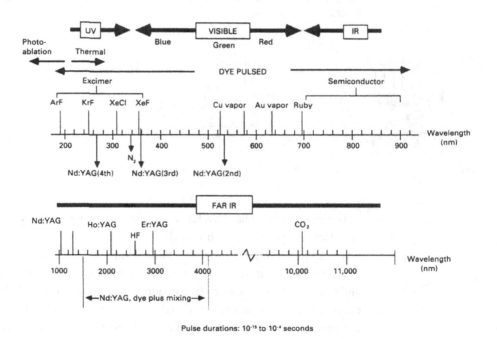

Figure 8.12

Pulsed lasers available for clinical and research applications. From Welch *et al.* (1989). Copyright ©1989 by the Texas Heart Institute.

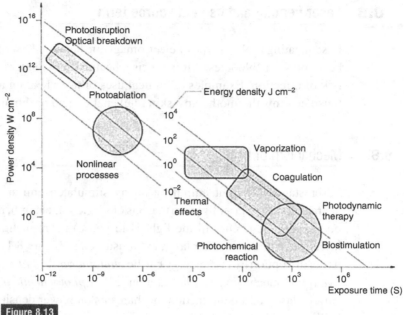

Figure 8.13

Different types of laser–tissue interactions depending on power density and exposure time. From Knappe et al. (2004). Reproduced with permission.

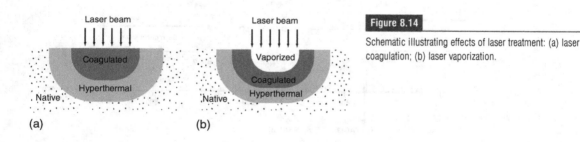

Figure 8.14

Schematic illustrating effects of laser treatment: (a) laser coagulation; (b) laser vaporization.

For most surgical laser systems the primary influence on the tissue is *photothermal*. Here the photons are absorbed into the tissue, converting to thermal energy and resulting in alteration of the tissue induced by temperature change. It is the principal mechanism in soft tissue surgery. Between 60–65 °C, proteins are denatured and tissue necrosis happens (the coagulated region in Figure 8.14(a)). At 100 °C, evaporation results in explosions (the vaporized region in Figure 8.14(b)). When temperatures reach 150 °C, proteins are broken down, leaving carbon behind in a layer known as char. *Photoablation* is the process of tissue decomposition from bond breaking that results from direct absorption of photons, without a thermal influence. This mechanism is induced by wavelengths in the UV range and is used commonly for corneal sculpting. *Photomechanical disruption* of tissue may be induced when high pulse energies are deposited into tissue in successive pulses of short duration, resulting in the generation of shock waves, cavitation or jet formation within the tissue. *Laser-induced plasma* formation is a type of photodisruption in which tissue breakdown results from an increase in local electrical field strength in the tissue, resulting in ionization of molecules and atoms (plasma formation). *Photochemical* mechanisms of tissue alteration are a result of biochemical reactions in tissue that are induced by the absorption of photons and are typical of the reactions induced in the application of photodynamic therapy. The thermal and non-thermal effects just discussed can also be roughly distinguished based on pulse duration. An approximate rule is that for pulse durations greater than $1\mu s$, thermal effects usually become measurable. Below $1\mu s$ pulse duration, thermal effects are usually negligible, if moderate repetition rates are chosen. *Modeling in this book is primarily heat transfer modeling and therefore is in the context of photothermal effects.*

8.8.2 Clinical applicator example

Many applicator types exist, depending on the particular laser application. For example, for localized tissue coagulation, a technique called *laser-induced interstitial thermotherapy*, or LITT, can be used to treat various types of tumors such as in the retina, brain, prostate and liver. A schematic of the LITT procedure is shown in Figure 8.15.

Laser radiation is applied to the tissue through an optical fiber. The fiber is placed inside the tissue by means of a specially designed, transparent catheter. Tissue necrosis occurs in selected, coagulated volumes only. For large lesions, the distance between adjacent puncturing canals needs to be small enough (<1.5 mm) to ensure an overlap of coagulated tissue volumes.

8.8.3 Simplified heat generation term

Laser power output is measured in Watts while the laser spot size is the laser beam cross-sectional area, measured in area units. *Irradiance* is defined as the laser power output per unit spot size, for example, in W/cm^2. Laser *fluence* is defined as the energy (not power) output of the laser per unit area of spot, i.e.,

$$\text{Fluence} = \frac{(\text{laser output}) \times (\text{pulse duration})}{\text{spot size}} \quad (8.21)$$

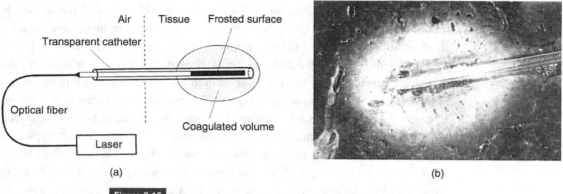

Figure 8.15

(a) A schematic for laser induced interstitial thermotherapy (LITT). Laser radiation is applied to the tissue through an optical fiber. The fiber is placed inside the tissue by means of a specially designed, transparent catheter. (b) Liver tissue after exposure to a Nd:YAG laser, using a standard LITT fiber applicator (laser power: 5.5 W, exposure duration: 10 min). The lighter region shows the coagulated volume where tissue necrosis occurs. From Niemz (2002). Reproduced with permission from Springer Science + Business Media.

Units of fluence can be in J/cm^2, for example. Between continuous wave or long-pulsed laser and short-pulsed laser, the latter is preferred for tumor heating because of its ability to produce highly localized heating at the desired location. This is due to a higher instantaneous peak power of the short pulsed lasers for the same energy input.

The decay of laser energy into a tissue can be approximated by an exponential decay given by

$$Q = Q_0 e^{-\alpha x} \tag{8.22}$$

where Q_0 is the volumetric heating at the surface which is related to the irradiance, I_0, of the laser (output before decay) as $Q_0 = I_0 \alpha$. The quantity α is the inverse of penetration depth of the laser of a specific frequency in a specific tissue (penetration depth is the distance at which the power level drops to $1/e$ of its value at the surface).

A more general expression for the heat generation term in a pulsed laser can be seen in Banerjee et al. (2005). For a two-dimensional, axisymmetric, cylindrical coordinate system, they use the heat generation term $Q(r, z)$:

$$Q = \left[(1 + \delta(t)) - (1 + \delta(t - t_p))\right] I_0 k_a \exp\left(\frac{-2r^2}{\sigma^2}\right) \exp(-zk_a) \tag{8.23}$$

Here $\delta(t)$ is the delta function, I_0 is the laser intensity at the surface, k_a is the absorption coefficient, σ is the spot radius, and r and z are distances in the radial

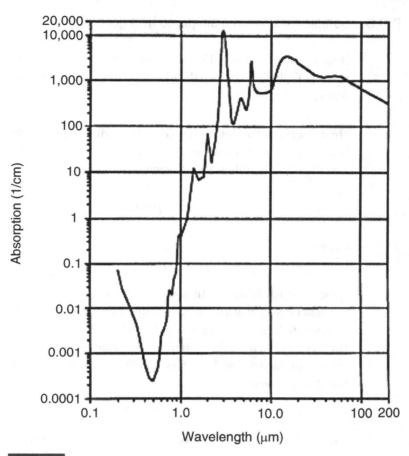

Figure 8.16

Absorption characteristics of water at various wavelengths, which is of relevance to laser use. From Welch *et al.* (1989). Copyright ©1989 by the Texas Heart Institute.

and axial directions, respectively. The laser source is approximated as a square pulse of width t_p.

8.8.4 Dependence of laser absorption on various factors

Laser absorption depends on its wavelength. Since data for tissues are hard to find and water is the major component of many tissues, perhaps one can try to understand this wavelength dependence by following the property data for water, as shown in Figure 8.16. This shows that absorption can vary by several orders of magnitude, depending on wavelength.

8.8.5 Complexities: when a simplified heat generation term is not enough

Both Eqns. 8.22 and 8.23 are simpler forms of a heat generation term in laser heating. Following are two detailed models for heat generation in laser heating of tissue.

Heat source term computed from light diffusion In this model, which has been used for skin tissue (see, for example, Zhang *et al.*, 2005), light transport is described by the following differential equation:

$$\frac{\partial \phi}{\partial t} - D\nabla^2 \phi + c_* \mu_a \phi = 0 \qquad (8.24)$$

where

$$D = c_* \left[3(\mu_a + (1-g)\mu_s)\right]^{-1} \qquad (8.25)$$

is the optical diffusion coefficient [m^2/s] of skin, ϕ is the fluence rate [W/m^2] of the incident laser light at a location, c_* is the speed of light [m/s] in the particular tissue component (epidermis, dermis or blood vessel), μ_a is the absorption coefficient [m^{-1}] of tissue, μ_s is the scattering coefficient [m^{-1}] of tissue and g is the optical anisotropy factor. The boundary condition on the tissue surface receiving laser radiation is given by

$$(1-r)P(t)c_0 = -D\frac{\partial \phi}{\partial n} \qquad (8.26)$$

where $P(t)$ is the laser irradiance [W/m^2], r is the ratio of reflected light to the laser power output, c_0 is the speed of light [m/s] in vacuum and n represents the normal direction. Other boundary conditions are discussed in Zhang *et al.* (2005). The heat source term is calculated from the fluence rate, ϕ, as

$$Q = \mu_a \phi \qquad (8.27)$$

Very short duration laser heating Research work has also shown that for very short durations of laser heating, prediction of temperatures using the classical heat equation (Eqn. 8.1) can be erroneous and a hyperbolic wave model of heat conduction is to be used instead, given by

$$\frac{\partial T}{\partial t} + \tau \frac{\partial^2 T}{\partial t^2} = \alpha \nabla^2 T + \frac{Q}{\rho C_p} \qquad (8.28)$$

where τ is the thermal relaxation time. More details can be seen in research papers such as Banerjee *et al.* (2005) and Jaunich *et al.* (2008).

> **Box 8.4 Software implementation**
>
> Heat generation terms as in Eqns. 8.22 and 8.23 can be readily implemented in a software as a special function or subroutine as long as there is access to the position variables r and z. In COMSOL, variables r and z can be accessed and therefore, Eqns. 8.22 and 8.23 can be entered under Define expressions, as shown in Case study X.

8.9 Ultrasonic heating

Ultrasound can be used in diagnostic imaging, therapeutic applications and tissue characterization. Ultrasonically induced heating is also studied because of its importance in evaluating ultrasonic safety. One of the novel techniques in ultrasound thermal therapy is the magnetic resonance imaging-guided high-intensity focused ultrasound (MRIgFUS), discussed below.

8.9.1 Mechanism of heating

Sound waves are mechanical pressure waves. They are longitudinal, i.e., the direction of particle motion is parallel to the direction of energy propagation. The range of human hearing is typically accepted to be 20 Hz to 20 kHz. Ultrasound refers to sound waves with a frequency above 20 kHz. Ultrasound is produced by transducers that are piezoelectric crystals in which a varying voltage causes a correspondingly varying strain. In imaging applications, such as in cardiology, obstetrics and gynaeclology, typically a frequency range of 1–6 MHz is used for deeper structures such as the liver and the kidney, whereas 7–15 MHz frequencies are used for superficial structures such as muscles, tendons, testes and breast. In therapeutic applications, such as in localized heating of tumors, lower frequencies (250 kHz–2000 kHz) are used at a significantly higher energy level. The physics of ultrasound waves are generally similar to those of waves in the audible range. In a tissue, ultrasound is attenuated due to absorption and scattering and the absorbed portion appears as heat. For a comprehensive review of mechanisms of interaction of ultrasound with biomaterial, see O'Brien (2007).

8.9.2 Clinical applicator example

An example of ultrasonic heating is the MRIgFUS mentioned earlier. This system, illustrated in Figure 8.17(a), has been reported to have the potential for precise

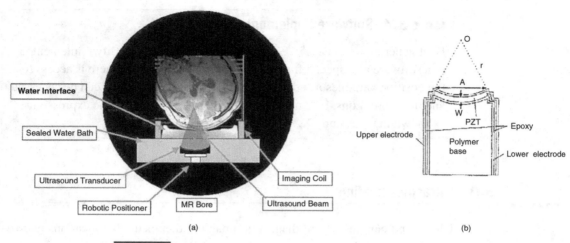

Figure 8.17

(a) An example of the use of focused ultrasound for brain treatments (ExAblate 2000 schematic), Ram et al. (2006); (b) schematic of a focused ultrasonic probe. Radius of curvature is r, focal point is at O, thickness of piezoelectric transducer is w and diameter of aperture is A. From Yasui et al. 2005 (©2005 IEEE).

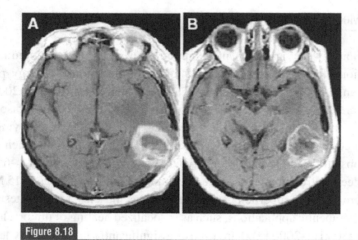

Figure 8.18

Enhanced T1-weighed MRI scans obtained from a patient before (A) and after (B) MRIgFUS treatment. Note the loss of enhancement at the treated site (Ram et al., 2006).

image-guided thermocoagulation of intracranial lesions. This particular system delivers small volumetric sonications from an ultrasound phased-array transmitter that focuses energy selectively to destroy the target, with verification by magnetic resonance imaging-generated thermal maps. An example of a focused ultrasound transducer is shown in Figure 8.17(b). MRI images obtained from the MRIgFUS procedure are shown in Figure 8.18.

8.9.3 Simplified heat generation term

For plane waves of a small signal level travelling in a homogeneous medium, the average volumetric rate of heat generation at a location is given by

$$Q = 2\alpha I \qquad (8.29)$$
$$= \alpha \rho c u_0^2$$

where I is the intensity of the wave, α is the absorption coefficient for the particular frequency and the medium, ρ is the tissue density, c is the speed of sound and u_0 is the particle velocity amplitude. Such a simplified equation would generally not apply to many practical situations of biomedical interest such as focused ultrasound, discussed earlier (see further discussion below).

8.9.4 Dependence of properties on temperature, composition and other factors

In the context of high temperature thermal therapies, data on dependence of ultrasonic absorption properties on the type of tissue and its temperature are generally hard to find. Available data are often in the context of diagnostic equipment. For high intensity focused ultrasound application, a polyacrylamide phantom containing egg white has been proposed as tissue-mimicking material.

Figure 8.19 shows the speed of sound and absorption properties for this phantom. Increase in the speed of sound and a general decrease in the attenuation coefficient with temperature can be seen. The dependence of attenuation coefficient, α, on frequency, f, is given by the expression $\alpha = \alpha_0 f^y$, where α_0 represents the attenuation coefficient at 1 MHz. Variation in y in Figure 8.19(c) shows that the frequency effect is not the same at all temperatures.

8.9.5 Complexities: when a simplified heat generation term is not enough

When amplitudes are not as small as they would be in focused ultrasound, propagation leads to distortion of the waveform and increased absorption. At a critical combination of distance and amplitude, a pressure discontinuity develops and there is, subsequently, a continuous process of shock formation and decay, with excess absorption associated uniquely with the shock front (Dalecki *et al.*, 1991). Such non-linear propagation of a wave is most commonly modeled using the following equation, also called the Khokholov–Zabolotskaya–Kuznetsov (KZK) equation. The approximation in this equation is valid for acoustic sources that are many

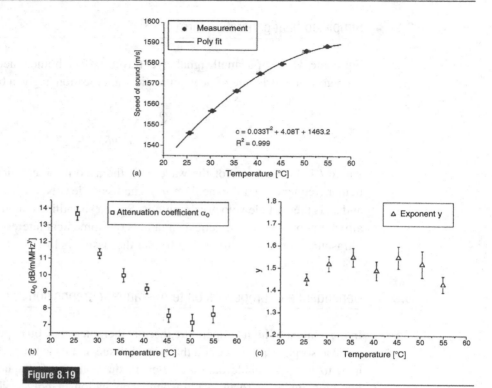

Figure 8.19

Temperature-dependence of acoustic parameters of an egg-white and acrylamide mixture phantom that is tissue-mimicking for a high intensity focused ultrasound application. The error bars show the standard deviation for four measurements. See text for definition of symbols. Reprinted from Divkovic *et al.* (2007). Copyright (2007), with permission from Elsevier.

wavelengths across, and for field points that are not too far from the beam axis or too near the source plane. This equation is written as (see also Hallaj and Cleveland, 1999; Humphrey, 2000; Nyborg, 1981)

$$\frac{\partial^2 p}{\partial z \partial \tau} = \frac{c_0}{2} \nabla_\perp^2 p + \frac{\delta}{2c_0^2} \frac{\partial^3 p}{\partial \tau^3} + \frac{\beta}{2\rho_0 c_0^3} \frac{\partial^2 p}{\partial \tau^2} \qquad (8.30)$$

Here p is the sound pressure, the z-axis is the direction of beam propagation, and the transducer lies in the (x, y) plane normal to the z-axis, c_0 is the small signal sound speed, δ is the sound diffusivity, β is the nonlinearity coefficient, ρ_0 is the ambient density and $\tau = t - z/c_0$ is the retarded time. The heat source term due to nonlinear wave propagation is calculated from the pressure (obtained by solving Eqn. 8.30) as

$$Q = \frac{2\alpha}{\rho_0 c_0 \omega^2} \left(\frac{\partial p}{\partial t} \right)^2 \qquad (8.31)$$

> **Box 8.5 Software implementation issues**
>
> Various softwares are available for modeling acoustics. In particular, the COMSOL software has a module that handles acoustics in fluids and solids in both 2D and 3D. Simulation in this module can be coupled with any other physics from COMSOL Multiphysics application modes.

8.10 Mass source terms

The source term r_A in Eqn. 8.2 denotes any generation or disappearance (negative source or sink) of the species A. For example, metabolic consumption of oxygen in the tissue is included in modeling as a negative value for r_A.

8.10.1 Zero-order reactions

A reaction is zeroth order when the rate of reaction is independent of concentration, i.e.,

$$r_A = k'' \tag{8.32}$$

where $r_A = -dc_A/dt$ stands for the rate of reaction and k'' is its constant value. The temperature dependency of the reaction rate constant, k'', is typically assumed to follow an Arrhenius type relationship:

$$k'' = A e^{-E/RT} \tag{8.33}$$

Here A is called the frequency factor and E is the activation energy for the reaction, T is the absolute temperature and R is the gas constant. Thermal injury, for example, has been modeled as a zeroth order reaction (see discussion in Section 8.10.4).

8.10.2 First-order reactions

A reaction is first order when the rate of reaction is linearly related to the concentration of the species:

$$r_A = k'' c_A \tag{8.34}$$

where $r_A = -dc_A/dt$ stands for the rate of reaction and k'' is the reaction rate constant. The temperature dependency of the reaction rate constant is typically assumed to follow the same Arrhenius type relationship as shown in Eqn. 8.33.

8.10.3 Various other reactions

Reactions of orders higher than zero or one are readily implemented following the same methodology. Another common type of reaction kinetics in a biological context is the Michaelis–Menten kinetics. For example, tissue oxygen consumption in some situations can be modeled as Michaelis–Menten kinetics. This kinetics is written as:

$$r_A = r_{A0} \frac{c}{c + c_m} \tag{8.35}$$

Here c_m is the value of the concentration c at which the rate of reaction is half of its maximum value r_{A0}.

When several reactions are to be included, for every reactant or product species of interest, we typically need a species transport equation. Computation time obviously increases as new equations are to be solved at every nodal point (as when using a finite element method). An example of software implementation of three simultaneous reactions in COMSOL can be seen in Case study VII.

8.10.4 Modeling of various reactions

Tissue oxygen consumption Tissue oxygen consumption is included in the species mass balance equation as a negative source term (also called a sink term). In tissue, the reaction in which oxygen is utilized has been modeled as an irreversible reaction, different from the cases of reversible chemical reactions between oxygen and hemoglobin and myoglobin. The rate of reaction for tissue oxygen consumption has been modeled as: (1) zeroth order; (2) first order; (3) mixed zeroth and first order; and (4) Michaelis–Menten kinetics (Popel, 1989). Which kinetics is appropriate depends on many factors, including the oxygen level in the tissue. An example of software implementation of tissue oxygen consumption as a zeroth-order reaction can be seen in Case study X.

Thermal injury of tissue Thermal injury or damage during either accidental burns or thermal therapy discussed under various heating methods has been modeled as a zeroth-order reaction (Sapareto and Dewey, 1984). Let the amount of injury be denoted by Ω. Following zeroth-order reaction, we can write

$$\frac{d\Omega}{dt} = A e^{-E/RT} \tag{8.36}$$

Integrating for the situation that temperature is not varying with time,

$$\Omega = A e^{-E/RT} t \tag{8.37}$$

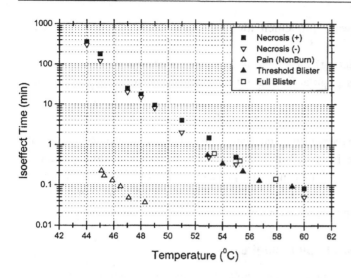

Figure 8.20

Time-temperature combinations that achieve varying thresholds of thermal damage to human skin. Data from Moritz and Henriques (1947) and Stoll and Greene (1959). Necrosis (+) refers to full thickness, or third-degree burn. Necrosis (−) refers to less serious injury or no injury observed at all. From Dewhirst et al. (2003). Copyright 2003 by Informa Healthcare - Journals. Reproduced with permission.

Table 8.2 Activation energies calculated from double exponential fits to the data in Figure 8.20 and similar data from elsewhere (Dewhirst et al., 2003).

Species	Temp range (°C)	Activation energy (kcal/mole)
Man	44–47	182.2
	47–60	95.78
Pig	44–47	150.75
	48–56	106.38
Mouse	41.5–42.5	273.89
	42.5–44.5	138.26

The parameters A and E have been obtained by curve fitting experimental data on thermal damage. The most commonly used data is that of Moritz and Henriques (1947). This and the data of Stoll and Greene (1959) are plotted in Figure 8.20. Note that the isoeffect curves are parallel for the various forms of injury. Also, the pain threshold is significantly lower than the threshold for significant injury, indicating that pain avoidance effectively minimizes significant injury to skin. The activation energies calculated from double exponential fits to data from Figure 8.20 are shown in Table 8.2. In one study, the values of Ω were estimated from the data of Moritz and Henriques (1947) as

$$\Omega = \begin{cases} 0.53, & \text{Irreversible epidermal injury (injury threshold)}; \\ 1.0, & \text{Complete transepidermal necrosis}. \end{cases} \quad (8.38)$$

An example of software implementation of burn injury is shown in Case study VIII.

Another formula relating time-temperature equivalence (which leads to the same thermal injury) that is popular in making thermal dose determination in cancer therapy is given by

$$CEM 43°C = tR^{(43-T)} \tag{8.39}$$

Here CEM43°C is the *cumulative number of equivalent minutes* of thermal injury at 43 °C, t is the time interval (min), T is average temperature during the time interval t, R is the ratio of the times that produce equal thermal injury when the temperature is increased by 1°C (t_{T+1}/t_T). Obviously, when temperature T is changing, CEM43°C needs to be (integrated) over the total time.

> **Box 8.6 Software implementation**
>
> Implementation of the mass source term is done by including the expression for r_A, as described above, for the source term in the software. In COMSOL, this is done by providing the appropriate expression (for zero, first order etc.) for r_A under Define expressions, as discussed in Section 2.8.3 and also in Case study VII, Case study VIII and Case study X.

8.11 Summary

The equation $Q = Q_0 \exp(-x/\delta)$ can describe (sometimes qualitatively) spatial variation of volumetric heating for a number of heating situations including microwave, RF, infrared, laser and ultrasonic heating; although important deviations from this equation may need to be considered. When a surface flux $I_0[\text{W/m}^2]$ is provided instead of surface heat generation, $Q_0[\text{W/m}^3]$, the volumetric heat generation would relate to this equation as $Q = (I_0/\delta) \exp(-x/\delta)$. Reactions are implemented as mass source terms. Heat or mass source terms can be functions of position, time or both.

References

General

Dewhirst, M. W., Viglianti, B. L., Lora-Michiels, M., Hanson, M., Hoopes, P. J. (2003). Basic principles of thermal dosimetry and thermal thresholds for tissue damage from hyperthermia. *International Journal of Hyperthermia*, **19**(3):267–294.

Diederich, C. J. (2005). Thermal ablation and high-temperature thermal therapy: Overview of technology and clinical implementation. *International Journal of Hyperthermia*, **21**(8):745–753

Metaxas, A. C. (1996). *Foundations of Electroheat*. Chichester, UK, John Wiley & Sons.

Moritz, A. R. and F. C. Henriques (1947) Studies of thermal injury 2. The relative importance of time and surface temperature in the causation of cutaneous burns. *American Journal of Pathology* **23**:695–720.

Sapareto, S. A. and W. C. Dewey (1984). Thermal dose determination in cancer therapy. *International Journal of Radiation Oncology Biology Physics*, **10**:787–800.

Stoll, A. M. and L. C. Greene (1959). Relationship between pain and tissue damage due to thermal radiation. *Journal of Applied Physiology* **14**:373–382.

Laser

Banerjee, A., A. A. Ogale, C. Das, K. Mitra, C. Subramanian (2005). Temperature distribution in different materials due to short pulse laser irradiation. *Heat Transfer Engineering*, **26**(8):41–49.

Jaunich, M., S. Raje, K. Kim, K. Mitra, Z. X. Guo (2008). Bio-heat transfer analysis during short pulse laser irradiation of tissues. *International Journal of Heat and Mass Transfer*, **51**(23–24): 5511–5521.

Knappe, V., F. Frank and E. Rohde (2004). Principles of lasers and biophotonic effects. *Photomedicine and Laser Surgery*, **22**(5):411–417.

Niemz, M. H. (2002). *Laser-Tissue Interactions. Fundamentals and Applications*. Berlin, Springer-Verlag.

Welch, A. J., Torres, J. H., Cheong, W. F. (1989). Laser physics and laser-tissue interaction. *Texas Heart Institute Journal*, **16**(3):141–149.

Zhang R, Verkruysse W, Aguilar G, Nelson JS. 2005. Comparison of diffusion approximation and Monte Carlo based finite element models for simulating thermal responses to laser irradiation in discrete vessels. Physics in Medicine and Biology, 50 (17): 4075-4086.

Microwave

Gabriel, C., Gabriel, S., Corthout, E. (1996a). The dielectric properties of biological tissues .1. Literature survey. *Physics in Medicine and Biology*, **41**(11):2231–2249.

Gabriel, S., Lau, R. W., Gabriel, C. (1996b). The dielectric properties of biological tissues .2. Measurements in the frequency range 10 Hz to 20 GHz. *Physics in Medicine and Biology*, **41**(11):2251–2269.

Gabriel, C., Gabriel, S., Grant, E. H., Halstead, B. S. J., Mingos D. M. P. (1998). Dielectric parameters relevant to microwave dielectric heating. *Chemical Society Reviews*, **27**(3):213–223.

Laguna, M. P., R. Muschter, F. M. J. Debruyne (2000). Microwave Thermotherapy: Historical Overview. *Journal of Endourology* **14**(8) (October 2000) 603–609.

Lazebnik, M., M. C. Converse, J. H. Booske, S. C. Hagness (2006). Ultrawideband temperature-dependent dielectric properties of animal liver tissue in the microwave frequency range. *Physics in Medicine and Biology* **51**:1941–1955.

Rakesh, V., A. K. Datta, M. H. G. Amin, L. D. Hall (2009). Heating uniformity and rates in a domestic microwave combination oven. *Journal of Food Process Engineering*: **32**(3):398–424.

Rappaport, C. (2004). Cardiac tissue ablation with catheter-based microwave heating. *Int. J. Hyperthermia* **20**(7) (November 2004) 769–780.

Reeves, J., M. Birch, K. Munro, R. Collier. (2000). Investigation into the thermal distribution of microwave helical antennas designed for the treatment of Barrett's oesophagus. *Phys. Med. Biol.* **47**, 3557–3564.

Saltiel, C. and A. K. Datta (1998). Heat and mass transfer in microwave processing. In *Advances in Heat Transfer*, **32**:1–94.

Radiofrequency

Berjano, E. J. (2006). Theoretical modeling for radiofrequency ablation: State-of-the-art and challenges for the future. Biomedical Engineering OnLine, 5:24, On the web at http://www.biomedical-engineering-online.com/content/5/1/24

Haemmerich D, Lee F. T., Schutt, D. J., Sampson, L. A., Webster, J. G., Fine, J. P., Mahvi, D. M. (2005). Large-volume radiofrequency ablation of ex vivo bovine liver with multiple cooled cluster electrodes. *Radiology*, **234**(2):563–568.

Schmid, G, Neubauer, G, Mazal, P. R. (2003). Dielectric properties of human brain tissue measured less than 10 h postmortem at frequencies from 800 to 2450 MHz. *Bioelectromagnetics*, **24**(6):423–430.

Tsuda, N., K. Kuroda and Y. Suzuki (1996). An inverse method to optimize heating conditions in RF-capacitive hyperthermia. *IEEE Transactions on Biomedical Engineering*, **43**(10):1029–1037.

Ferromagnetic

Avissar, M., Ishman, N., Lee, S. P., Patel, P. (2001). Ferromagnetic Thermal Ablation of Prostate Tumor. Class Project at Cornell University. On the web at http://ecommons.library.cornell.edu/handle/1813/259

Tompkins, D. T., R. Venderby, S. A. Klein, W. A. Beckman, R. A. Steeves, D. M. Frye and B. R. Paliwal. (1994). Temperature-dependent versus constant-rate blood perfusion modelling in ferromagnetic thermoseed hyperthermia: results with a model of the human prostate. *Int. J. Hyperthermia*, **10**(4):517–536.

Ultrasonics

Dalecki, D., E. L. Carstensen, K. J. Parker, D. R. Bacon (1991). Absorption of finite amplitude focused ultrasound, *J. Acoust. Soc. Am.* **89**, (May 1991) 2435–2447.

Divkovic, G. W., Liebler, M., Braun, K., Dreyer, T., Huber, P. E., Jenne, J. W. (2007). Thermal properties and changes of acoustic parameters in an egg white phantom during heating and coagulation by high intensity focused ultrasound. *Ultrasound in Medicine and Biology* **33**(6):981–986.

Hallaj, I. M., R. O. Cleveland (1999). FDTD simulation of finite-amplitude pressure and temperature fields for Biomedical ultrasound. *J. Acoust. Soc. Am.* **105**(5), (May 1999), L7–L12.

Humphrey, V. F. (2000). Nonlinear propagation in ultrasonic fields: measurements, modeling and harmonic imaging. *Ultrasonics*, **38**: 267–272.

Nyborg, W. L. (1981). Heat generation by ultrasound in a relaxing medium. *J. Acoust. Soc. Am.* **70** (2), (Aug. 1981) 310–312.

O'Brien Jr., W. D. (2007). Ultrasound-biophysics mechanisms. *Progress in Biophysics and Molecular Biology*, **93**:212–255.

Ram, Z., Z. R. Cohen, S. Harnof, S. Tal, M. Faibel, D. Nass, S. E. Maier, M. Hadani, Y. Mardor (2006). Magnetic resonance imaging-guided, high-intensity focused ultrasound for brain tumor therapy. *Neurosurgery*, **59**(5):949–956.

Yasui, A., Y. Haga, J. J. Chen, M. Esashi, H. Wada (2005). Focused ultrasonic transducer for localized sonodynamic therapy. Proceedings of the 13th International Conference on Solid-State Sensors, Actuators and Microsystems, Seoul, Korea, June 5–9.

Mass transfer

Popel, A. S. (1989). Theory of oxygen transport to tissue. *Critical Reviews in Biomedical Engineering*, **17**(3):257–321.

8.12 Problems

8.12.1 Penetration depth in the heat source term

The equation $Q = Q_0 e^{-x/\delta}$ can describe spatial variation of volumetric heating for microwave, radio frequency, infrared, laser and ultrasonic heating. (1) What is the interpretation of the variable δ in the above equation? (2) List the above heating applications in order of their typical δ values from large to small.

8.12.2 Variable penetration depth

Show that for the simple formulation of a heat source, given by $Q = Q_0 e^{-x/\delta}$, if δ is varying with position, a possible way to take this variation into account is given by

$$Q = Q_0 \exp\left(\int_0^x -\frac{dx}{\delta}\right) \tag{8.40}$$

where δ is a function of x.

8.12.3 Zero penetration depth

When penetration depth is zero, how would you implement $Q = Q_0 e^{-x/\delta}$? Hint: Implementation as a heat source term will not work (see Section 1.8.6).

8.12.4 Thermal dosimetry

Derive Eqn. 8.39 used in determining thermal dose in cancer therapy.

8.12.5 Heat source term for ferromagnetic heating

Provide details (no software-specific commands) of how you would set up the source term for ferromagnetic heating (see introduction to ferromagnetic heating in this chapter).

8.12.6 Heat source term for laser heating

For laser heating, three different formulations for the heat generation term have been discussed (Eqns. 8.22, 8.23 and 8.27). Provide examples of practical situations where each one would be appropriate, given that we always strive for the simplest formulation.

8.12.7 Coupling of equations: effect of temperature dependence of properties

The properties that describe electromagnetic heating or ultrasonic heating are generally temperature dependent. Thus, as heating progresses, these properties change and that, in turn, changes the rate of heat generation itself. This can couple the physics of electromagnetic or ultrasonic wave propagation with that of heat transfer in a strong way. Such coupled simulations are not an essential part of this text. Can you describe qualitatively (perhaps using a flow chart) what a simulation of coupled physics looks like, using electromagnetic heating as an example? Such simulations can often require long computation times. What are some of the factors to consider in reducing computation times?

8.12.8 Example of coupling: radiofrequency heating to destroy a tumor

Describe the governing equation and boundary conditions to be used for the problem described in Section 1.12.12. What is the reason we do not need to use the full Maxwell's equations? Draw a schematic showing the coupling between the electrical heat generation and the heat equation when the conductivities are a function of temperature. How would you implement this in COMSOL?

8.12.9 Implementation of any rate process as a species equation

The mass species equation (Eqn. 8.2) can be used in a number of ways. Consider the extent of burn, Ω, described by the following equation:

$$\frac{d\Omega}{dt} = Ae^{-(E/RT)} \tag{8.41}$$

where t is time, E is some sort of activation energy, T is absolute temperature and A is a constant. Describe how you would implement Eqn. 8.41 to calculate the extent of burn, Ω, using a software that can solve Eqn. 8.2.

9 Material properties and other input parameters

Since our modeling is physics based, that is, using governing equations of energy, mass or momentum conservation and their associated boundary conditions, we need whatever material properties and process parameters that are present in these equations. For example, in solving a flow problem, the governing equations of momentum conservation contain viscosity and density as property values. In order to solve a flow problem numerically, we need data for viscosity and density at a minimum. This chapter (see Figure 9.1) provides a very short introduction to all of the relevant properties in solving fluid flow, heat transfer and mass transfer. Material properties associated with related physics, such as dielectric properties in electromagnetic heating, are discussed in Chapter 8.

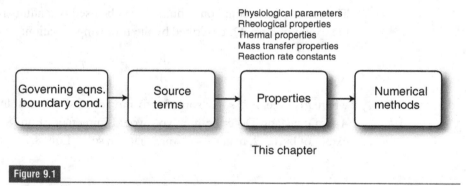

Figure 9.1

This chapter introduces material properties as part of the theoretical framework for modeling.

9.1 What material property data and input parameters do we need?

Since we are solving problems related to fluid flow, heat transfer and mass transfer, the relevant material properties are summarized in Table 9.1. Additionally, we need input parameters such as anatomical parameters (i.e., size, shape) and physiological parameters (i.e., reaction rates, tissue destruction temperatures).

Table 9.1 Summary of property data and parameters needed for simulations, in addition to anatomical and physiological parameters.

Type of physics	Properties that may be needed	Parameters that may be needed
Fluid flow	Viscosity (Newtonian fluid) m, n or τ_0 (non-Newtonian fluid)	
Heat transfer	Thermal conductivity, density specific heat, latent heat, apparent specific heat, reflectivity, emissivity	heat transfer coefficient
Mass transfer	Mass diffusivity, density, equilibrium vapor pressure, psychrometric partition coefficient, reaction rate constant	mass transfer coefficient

> **Box 9.1 Software implementation**
>
> Implementation of properties in COMSOL is discussed in Chapter 2 starting at Section 2.11. Table 1.4 lists the COMSOL implementations of properties in various case studies.

9.2 Where do we get data?

9.2.1 Measurement

Own measurement If possible, the best way is to collect your own data. This way you have full control over the quality. As easy as it sounds, this has two problems – expense and time. Many input parameters require careful experimentation with complex experimental setups that require significant time and resources. The advantages of modeling can quickly evaporate if the only way to incorporate it is to obtain input data through such detailed experimentation.

Outsourcing Specialized laboratory services exist that can make measurements for you, for a fee and provide the data. This can save time and be cost effective as the particular company may specialize in collecting such data and probably has a setup that is ready to go.

9.2.2 Literature

Databases Although material property databases exist for other material processing (Anonymous, 2002), for biomaterials, this is barely the beginning. See, for example, the dental properties database (O'Brien, 2008) which was previously available on the web.

Databases are still a long way from providing data specific to a particular tissue, its physiological condition, and so on.

Handbooks and research publications   Handbooks, such as that by Bronzino (1995), are not primarily data collections, although many chapters have significant relevant data. Handbooks from other fields may provide some of the data (e.g., Green and Perry, 2007; Rohsenow *et al.*, 1997). Due to the specific nature of many applications, the most effective source of data in past computational projects at Cornell University (Datta, 2009) have been scientific research papers and specialized books. Specific research papers can be searched in large bibliographic databases such as the *Web of Knowledge* (http://isiwebofknowledge.com/). Cited papers featuring in the *Web of Knowledge* allow one to track the publications that follow from a particular piece of work and this feature has been quite effective in finding past information as well as the most recent information on a specialized topic. Specialized books (e.g., Saltzman, 2001) are also quite effective in locating specific data needed for modeling.

9.3   How accurate should the data be?

Of course, the obvious answer to this is "as accurate as possible." However, as discussed under sensitivity analysis (see next section), a process may not be equally (and highly) sensitive to all input parameters. For input parameters that the process is less sensitive to, the need for accuracy is less critical. In fact, sensitivity analysis, discussed below, can point us to the parameters that the solution is most sensitive to and therefore the ones for which more accurate data is needed. An example of such sensitivity analysis which considers 10% variation of the parameters is discussed in Section 5.4.1.

9.4   What to do when accurate data is not available

Particularly when it comes to biomaterials and bioprocesses, detailed and accurate properties data are often unavailable. It is not a question of looking in the right place. A practical approach to allow us to make progress in modeling would be as follows:

Start with a reasonable guess   Based on the discussions in previous sections on available data, start with properties of a material similar in composition, at a comparable temperature etc. For example, meat properties are available in food literature and transport properties of beef can be used as a starting point for modeling transport in human tissue.

Perform sensitivity analysis Computations are repeated for a range of values around the property value guessed and the change in the solution variable is noted. For example, if thermal conductivity was guessed and temperature is the solution variable, computations are repeated for a range of thermal conductivity values around the guessed value and temperatures are noted in each case. This procedure is called sensitivity analysis. Sensitivity analysis can show how sensitive the process is to the particular property value. If the process changes a lot due to a small variation of the property, it implies that an accurate value of the particular property is needed. On the other hand, if the process does not change much as the property value is changed, it implies that more accurate data may not be necessary as its effect on the solution would be minimal. Such sensitivity analysis can be repeated with respect to any of the properties or parameters. An example of how a sensitivity analysis is performed is discussed in Section 5.4.

9.5 Anatomical and physiological parameters

9.5.1 Standard anatomical data

Standard anatomical and physiological data are available in the literature. Table 9.10 at the end of this chapter shows an example that is a compilation of various data for a Reference Man and Reference Woman that are both Caucasian (Heymsfield et al., 2005). The original source of the data being the International Commission of Radiological Protection (Snyder et al., 1975). Another example of standard anatomical data is the eye shown in Figure 9.2.

9.5.2 Body surface area

The surface area of a human body can be estimated from (DuBois and DuBois, 1916)

$$A = 0.202 m^{0.425} H^{0.725} \tag{9.1}$$

where m is the mass of the body in kg and H is the height of the body, in m. This value has been known to provide an underestimate of body surface area and does not hold for 'extremes' (Parsons, 2003).

418 Material properties and other input parameters

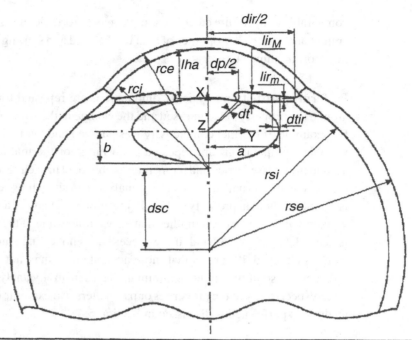

dp : pupilary diameter;	$dp = 4$ mm
a : long halfdiagonal of the lens;	$a = 4.5$ mm
b : short halfdiagonal of the lens;	$b = 2$ .mm
lir_M : maximum thickness of the iris at the pupilary border;	$lir_M = 0.6$ mm
lir_m : maximum thickness of the iris at the iris periphery;	$lir_m = 0.25$ mm
dir : length of the vertical axis of the iris;	$dir = 11$ mm
rce : radius of the cornea epithelium;	$rce = 8$ mm
rci : radius of the cornea endothelium;	$rci = 7.4$ mm
lha : aqueous humor thickness;	$ha = 3$ mm
rse : outer radius of the backside layer of the eye;	$rse = 12$ mm
rsi : inner radius of the backside layer of the eye;	$rsi = 11$ mm
dsc : distance between cornea and sclera centers;	$dsc = 5$ mm
$dtir$: laser spot diameter;	$dtir = 0.5$ mm
dr : catching-angle between iris and lens;	$dt = 3°$

Figure 9.2

Various dimensions of the human eye (Apiou-Sbirlea and L'Huiller. 1998).

9.5.3 Blood perfusion

Representative blood perfusion data are shown in Table 9.2. Blood perfusion data in the literature is often reported in various units, one of them is ml/min · g, which would be ml of blood flowing per minute per gram of tissue. To convert from these units, for example, to the units needed in the bioheat equation (m³ of blood per second per m³ of tissue), the density of tissue would be needed. Significant variations in the blood flow rate are possible.

Temperature-dependent blood perfusion has been used in the modeling literature. For example, the following has been assumed in modeling hyperthermia (Erdmann et al., 1998), based on literature values:

In muscle,

$$\dot{m}_b = \begin{cases} 0.45 + 3.55 \exp\left(-\frac{(T-45)^2}{12}\right) & T \leq 45\,°C \\ 4.00 & T > 45\,°C \end{cases} \quad (9.2)$$

where $\dot{m}_b = \rho_b \dot{V}_b^v$ is the mass flow rate of blood, ρ_b being the density of blood. The units of $\dot{m}_b$ in Equations 9.2–9.4 are kg/s · m³. In fat,

$$\dot{m}_b = \begin{cases} 0.36 + 0.36 \exp\left(-\frac{(T-45)^2}{12}\right) & T \leq 45\,°C \\ 0.72 & T > 45\,°C \end{cases} \quad (9.3)$$

Finally, in the tumor, they used the following term:

$$\dot{m}_b = \begin{cases} 0.833 & T < 37\,°C \\ 0.833 - (T-37)^{4.8}/5.438 \times 10^3 & 37 \leq T \leq 42\,°C \\ 0.416 & T > 42\,°C \end{cases} \quad (9.4)$$

Table 9.2 Blood perfusion data for specific tissues and organs for humans. See also Table 9.4.

Organ	Blood flow (ml/min/g)	Reference
Kidney (whole cortex)	4.0–5.0 (SE)	Thurau and Levine (1971)
Kidney (outer medulla)	1.2	Thurau and Levine (1971)
Kidney (inner medulla)	0.25	Thurau and Levine (1971)
Muscle (skeletal)	0.027 (average)	Keele and Neil (1971)
Muscle (resting thigh, 1.5 cm depth)	0.018 ± 0.011	Sekins et al., 1980
Muscle (resting thigh, 3.0 cm depth)	0.026 ± 0.013	Sekins et al., 1980
Muscle (resting thigh)	0.020–0.022	Lassen et al., 1964
Skin (forearm, in the cold)	≈0.02	Johnson et al., 1986
Skin (forearm, thermoneutral)	0.04–0.05	Johnson et al., 1986
Skin (forearm, hyperthermic)	near or > 0.20	Johnson et al., 1986

In another example, in studying breast tumor, the response of the blood vessels to the tumor temperature is assumed to be (He *et al.*, 2006)

$$A = A_0 e^{b(T-T_0)} \quad \text{where} \quad b = \begin{cases} 0.1 & T = 39 - 42\,°C \\ -0.1 & T > 42\,°C \end{cases} \quad (9.5)$$

where A_0 and T_0 are the cross-sectional area and blood temperature before heating (37 °C), respectively, and b is the parameter whose values are provided based on experimental results.

9.5.4 Metabolic heat generation

Metabolic heat generation is the difference between total energy produced due to metabolism, called the metabolic rate (M), and the total external work performed (W), i.e.,

$$Q_m = M - W \quad (9.6)$$

where Q_m is the metabolic heat generation. The external work performed is the part of the total energy produced by the body that is not given off as heat and can be considered in terms of the activity of the muscle cells in the body. The *maximum* value of W is about 20–25% of M and is difficult to measure (Parsons, 2002). Sometimes, for estimation of Q_m, W is ignored. Equations are available for estimating the metabolic rate (Parsons, 2003) and some representative data are shown in Tables 9.3 and 9.4.

Basal metabolic rate, the rate of energy consumption required to maintain life at complete rest, can be calculated from

$$BMR = 70 m^{0.75} \quad \text{kcal/kg/day} \quad (9.7)$$

where BMR is the basal metabolic rate and m is the mass in kg. Typical metabolic rates for various activities can be seen in Table 9.3 and those for various organs are listed in Table 9.4. As mentioned above, only a portion of the BMR calculated this way goes into heat production, i.e., BMR is not equal to the heat generation by the animal.

Metabolic heat generation is implemented as a source term, as discussed in Section 8.1. If the metabolic heat generation (and therefore the heat source term) is a function of other variables, this is also readily implemented in COMSOL by defining these functions (see Section 2.8.2).

9.5 Anatomical and physiological parameters

Table 9.3 Typical metabolic rate for various human activities.

Activity	Metabolic rate per unit body surface area W/m²
Sleeping	41
Reclining	33
Sitting quietly	58
Standing relaxed	70
Dressing and undressing	74
Walking on the level at 3–6 km/hr	116–221
Driving	
car	87
motorcycle	116
heavy vehicle	186
House cleaning	116–198
Shopping	81–105
Carpentry, metal working, industrial painting	150
Dancing	140–256
Swimming	314
Jogging (at about 9 km/hr)	357
Walking up stairs	690

Adapted from Shitzer, A. and R.C. Eberhart. 1985. Heat generation, storage, and transport processes. 1:137–151. In: *Heat Transfer in Medicine and Biology* edited by A. Shitzer and R.C. Eberhart. Plenum Press, New York. Reprinted with kind permission from Springer Science+Business Media.

Table 9.4 Blood perfusion (flow rate), oxygen consumption and metabolic rate in various organs in the human body.

Organ	Mass kg	Blood flow ml/min	Oxygen consumption ml/min	Metabolic rate W
Heart muscle	0.3	250–1800	30–90	10–31
Skeletal muscle	31.0	1200–24 000	50–1000	17–350
Skin	3.6	400–2800	12–85	4–30
Liver	2.6	1500	51	18
Kidney	0.3	1260	18	6
Brain	1.4	750	49	17

Adapted from Shitzer, A. and R.C. Eberhart. 1985. Heat generation, storage, and transport processes. 1:137–151. In: *Heat Transfer in Medicine and Biology* edited by A. Shitzer and R.C. Eberhart. Plenum Press, New York. Reprinted with kind permission from Springer Science+Business Media.

9.5.5 Growth rate and metabolic heat generation in a tumor

The diameters of some tumors have been modeled in relation to their doubling time as

$$D = 0.01 e^{0.002134(\tau - 50)} \text{m} \quad (9.8)$$

where D is the diameter and τ is the time in days required for the volume to double.

Metabolic heat generation in a tumor has been related by the equation

$$Q_m \tau = 3.27 \times 10^6 \text{W/m}^3 \cdot \text{days} \tag{9.9}$$

Here τ is the time required (in days) for the tumor to double its volume.

9.5.6 Tissue shrinkage or swelling

Shrinkage or swelling can be thermally driven (due to increased temperature) or driven by moisture loss. There is no universal relationship between temperature and moisture change and the amount of shrinkage (or swelling). This is partly due to variations in tissue properties but also due to the dependence of shrinkage on the matrix mechanical properties. The latter require solution of a solid mechanics problem which would provide the deformation.

Collagen is the most abundant protein in the body. Severe heating results in shrinkage due to a time-dependent irreversible transformation of the native triple-helix structure into a more random (coiled) structure. Experimental data for tendons show that shrinkage is initially slow, in a small shrinkage regime, followed by a rapid, large shrinkage regime and finally a slow, continuing shrinkage regime (Wright and Humphrey, 2002). The rate of shrinkage depends on many factors, including the hydration level. An example of shrinkage data that has only the temperature effect is shown in Figure 9.3. Empirical curve fit data, such as in Figure 9.3, can be used in combination with heat transfer calculations for simple estimations of shrinkage.

A linear dependence of shrinkage with moisture has been used outside of biomedical applications. For example, in the drying of beef tissue that included a large change in moisture content, a linear relationship between volume and moisture content was used (Trujillo et al., 2007), which can be developed as follows. Let v be the specific volume of a sample (m³/kg dry material), β the volume-shrinkage coefficient, ρ_{b0} the bulk density of the sample (kg/m³) at zero value of moisture content (kg dry material/m³) and X the moisture content on a dry basis (kg of water per kg of dry material). We can write (Zogzas et al., 1994)

$$\beta = \frac{1}{v} \frac{\partial v}{\partial X} \tag{9.10}$$

$$= \frac{1}{v_0} \frac{v - v_0}{X - X_0} \tag{9.11}$$

where the subscript 0 refers to the state of zero moisture content. The units for β are the inverse of those for X. Since $X_0 = 0$ and $v_0 = 1/\rho_{b0}$, we can simplify

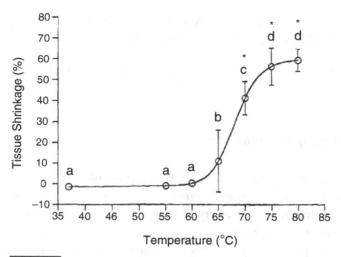

Figure 9.3

Shrinkage (%) in human glenohumeral joint capsule/ligament specimens heated to various temperatures. Maximum shrinkage took about 3 minutes to achieve, at 75 °C and 80 °C. The range is for mean±SD. Reprinted from Hayashi and Markel (1998) Copyright (1998), with permission from Elsevier.

Eqn. 9.11 as

$$v = \frac{1}{\rho_{b0}} (1 + \beta X) \qquad (9.12)$$

A value of $\beta = 0.936$ for beef tissue was reported by Trujillo *et al.* (2007). For small changes in moisture (for example, in possible drying-out of the corneal flap during a LASIK operation), a linear change of dimension with water content can be assumed.

In simulation, simple shrinkage or swelling using the above equations can be included as a deforming mesh. Implementation of a deforming mesh in COMSOL is discussed in Case study x.

9.5.7 Temperatures for tissue destruction

Tissue destruction, whether intended (for example, with radiofrequency ablation or in cryosurgery) or unintended (in a burn injury or frostbite), is really a function of time-temperature history, rather than a single temperature. However, for many practical uses, tissue at a location is considered destroyed when some extreme (high or low) temperature value is reached at that location. When temperature at a location increases to 43 °C or higher, the tissue at that location is considered destroyed (Sapareto and Dewey, 1984). When the temperature is decreasing, tissue at a location is considered destroyed when the temperature at that location reaches $-45\,°C$ or lower (see, for example, the study of Rabin, 2003). Thus,

$$T_{\substack{\text{temperature for}\\\text{tissue destruction}}} = \begin{cases} < & -45\,°C \\ > & 43\,°C \end{cases} \tag{9.13}$$

9.6 Rheological properties

To obtain velocities in a fluid, we need to solve the momentum equation or the Navier–Stokes equations (Eqns. 7.14–7.16). The momentum equation needs the flow properties of viscosity and density. Density is discussed later in this chapter. Viscosity is a special case of the more general class of properties known as rheological properties. Rheological properties are the properties that relate to flow and deformation. They can relate to solids and fluids. Here the discussion is limited to properties of fluids.

Fluid properties are typically divided into two classes, Newtonian and non-Newtonian, depending on how shear stress relates to the shear rate (velocity gradient). Fluids that follow the relation

$$\tau = -\mu \frac{\partial u}{\partial y} \tag{9.14}$$

are called Newtonian fluids. Here τ is the shear stress, μ is the viscosity, u is the velocity in the direction x and $\partial u/\partial y$ is the velocity gradient (or shear rate). Thus, for Newtonian fluids, shear stress is proportional to shear rate. Fluids that do not follow this relationship are approximated by more general equations and are called non-Newtonian fluids. Examples of such relation are:

$$\tau = m\left(-\frac{\partial u}{\partial y}\right)^n, \; n < 1 \text{ Pseudoplastic fluid} \tag{9.15}$$

$$\tau = m\left(-\frac{\partial u}{\partial y}\right)^n, \; n > 1 \text{ Dilatant fluid} \tag{9.16}$$

$$\tau = \tau_0 + m\left(-\frac{\partial u}{\partial y}\right)^n, \; \text{Bingham plastic fluid} \tag{9.17}$$

$$\sqrt{\tau} = \sqrt{\tau_0} + k\sqrt{-\frac{\partial u}{\partial y}} \;\; \text{Casson fluid} \tag{9.18}$$

Here m is called the consistency coefficient, n the flow behavior index, and τ_0 the yield stress. The relationships for Newtonian and pseudoplastic fluids are

9.6 Rheological properties

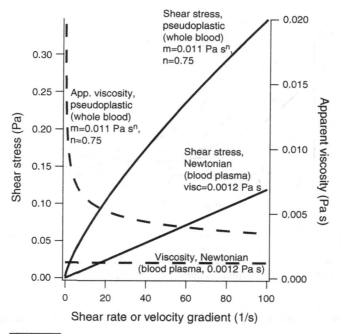

Figure 9.4

Illustration of a non-Newtonian (pseudoplastic) fluid behavior in contrast with a Newtonian fluid behavior. Values used for whole blood and blood plasma are approximate, at a body temperature of 37 °C

plotted in Figure 9.4, which shows that unlike Newtonian fluids, where the shear stress–shear rate relation is a straight line with its slope given by a constant value of viscosity,

$$\mu = \frac{\tau}{-\partial u/\partial y} \quad (9.19)$$

the shear stress–shear rate relation for other fluids is not a straight line and its slope changes with the value of shear rate. This slope is termed an apparent viscosity given by

$$\mu_{app} = \frac{\tau}{-\partial u/\partial y} \quad (9.20)$$

in analogy to viscosity, but its values depend on the shear rate which, in turn, depend on the flow rate, geometry and other factors.

Of the various body fluids, the flow properties of blood are perhaps the most important and are the most studied (see, for example, Neofytou, 2004). In general, blood is a non-Newtonian fluid, as shown in Figure 9.5, where the apparent viscosity, μ_{app}, depends on the shear rate. In many literature studies, blood is treated as a Casson fluid, discussed later. The shear rate in typical arteries is of the order of 100. Figure 9.5 shows that in this range of shear rate, the apparent

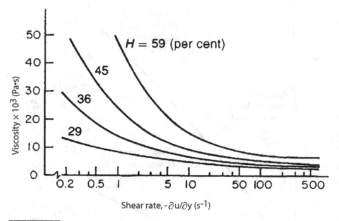

Figure 9.5

Apparent viscosity of blood as a function of shear rate for various hematocrit fractions, showing the decrease in apparent viscosity with shear rate, a non-Newtonian behavior. Figure 3.18 from "Introduction to Bioengineering", edited by Berger, Goldsmith & Lewis (1996). By permission of Oxford University Press.

viscosity is somewhat constant. Therefore, for flow in arteries, blood can be considered a Newtonian fluid. By some estimate (Mates, 1995), in vessels smaller than approximately 100 μm, blood exhibits significant non-Newtonian behavior.

Whole blood consists of formed elements and plasma. The formed elements include red blood cells or erythrocytes (99.9%), and white blood cells or leukocytes and platelets (0.1%). The blood plasma consists of mostly water (92%), plasma proteins (7%) and other solutes (1%). The term hematocrit refers to the percentage of whole blood occupied by the cellular elements. Of the suspended material, since the red blood cells are the most abundant, they contribute the most to rheology. As expected, blood apparent viscosity would depend on the hematocrit content of the blood, as can be seen in Figure 9.5. Blood viscosity also depends on plasma viscosity which in turn, depends on the protein concentration in the plasma. Higher temperature also reduces viscosity, however, within the circulation, temperature changes are perhaps insignificant.

One of the common expressions used to describe non-Newtonian behavior is a Casson-type equation given by (cited in Berger, 1996):

$$\sqrt{\tau} = 0.33 + 0.166\sqrt{-\frac{\partial u}{\partial y}} \qquad (9.21)$$

The yield stress represented by the square of the number 0.33 is thought to be due to formation of rouleaux (aggregations of two or more red blood cells due to possible electrostatic forces between the red blood cells). At higher shear rate,

Table 9.5 Normal average values of blood velocity and Reynolds number in the arteries of various sizes for a man (70 kg, 1.8 m^2). The numbers in parentheses for velocity are the systolic/diastolic and those in parentheses for Reynolds number are the peak values. 'Main vein' refers to one of the usually four terminal pulmonary veins. Data from Milnor (1989).

Vessel	Mean velocity [cm/s]	Mean Reynolds number
Systemic		
Ascending aorta	18 (112/0)	1500 (9400)
Abdominal aorta	14 (75/0)	640 (3600)
Renal artery	40 (73/26)	700 (1300)
Femoral artery	12 (52/2)	200 (860)
Femoral vein	4	104
Superior vena cava	9 (23/0)	550 (1400)
Inferior vena cava	21 (46/0)	1400 (3000)
Pulmonary		
Main artery	19 (96/0)	1600 (7800)
Main vein	19 (38/10)	800 (2200)

these aggregations break down, leading to more of a Newtonian behavior. Average values of blood velocities in vessels of various sizes are noted in Table 9.5.

9.7 Thermal conductivity

Thermal conductivity is one of the properties needed to obtain temperatures and heat fluxes, by solving Eqn. 7.22. The thermal conductivity of materials reduces as we move from solid to liquid to gas. Such general understanding is, however, insufficient to predict thermal properties of tissue, blood etc., whose composition varies significantly while, at the same time, the absolute magnitude of change is small. Almost all thermal properties data come from measurements. Typical values of thermal conductivity for human materials are shown in Table 9.6.

It is important to note that there can be many sources of variability in measured thermal properties data. Significant sample-to-sample and species-to-species biological variabilities exist. When a measurement probe is inserted into living tissue, a fluid pool may form around the probe because of the mechanical trauma and this can significantly affect the results. Secondly, tissue structure is quite heterogenous and therefore the probe can only measure a spatial average of the tissue properties

Table 9.6 Thermal properties of biomaterials of human origin. For the expressions having temperature dependence, T is in °C. The data for Valvano *et al.* (1985) are taken in vitro and therefore do not include perfusion effects. Their data not being in vivo, may also include some degeneration effects and the possibility of tissue not being in perfectly isotonic condition. Details of the rest of the data were not available.

Material	Thermal conductivity k (W/m·K)	Density ρ W/m³	Specific heat C_p (kJ/kg·K)	Thermal diffusivity $k/\rho C_p \times 10^6$ (m²/s)	Temp (°C)	Source
Kidney	0.545	1050	3.89		5	2,3
Heart	0.587	1060	3.72		5	2,3
Spleen	0.545	1050	3.72		5	2,3
Liver	0.566	1050	3.59		5	2,3
Liver	0.4692+0.001161T			0.1279+0.00036T	3–45	1
Lung	0.4071+0.001076T			0.1192+0.00031T	3–45	1
Spleen	0.4913+0.001300T			0.1270+0.00047T	3–45	1
Pancreas	0.4365+0.002844T			0.1391+0.00084T	3–45	1
Myocardium	0.4925+0.001195T			0.1289+0.00050T	3–45	1
Renal cortex	0.4989+0.001288T			0.1266+0.00055T	3–45	1
Renal medulla	0.4994+0.001102T			0.1278+0.00055T	3–45	1
Renal pelvis	0.4795+0.001923T			0.1329+0.00011T	3–45	1
Cerebral cortex	0.5043+0.000296T			0.1283+0.00050T	3–45	1
Brain, grey	0.566	1050	3.68		5	2,3
Brain, whole	0.528	1050	3.68		5	2,3
Brain, white	0.503	1040	3.59		5	2,3
Tooth, enamel	0.932	2800	0.711		5	2,3
Tooth, dentin	0.569	1960	1.59		5	2,3
Tooth, pulp	0.419	1050	3.77		5	2,3
Bone, cancellous	0.36	1080	2.238			
Bone, cortical	0.36	1850	1.3			
Aorta	0.476					4
Blood (whole)	0.492					4
Blood (plasma)	0.570					4
Fat	0.201–0.217					4
Tumor (colon cancer)	0.545					1
Skin (epidermis)	0.209					4
Skin (dermis)	0.293–0.322					4

[1]Valvano *et al.* (1985); [2]Cooper and Trezck (1970); [3]Cooper and Trezck (1971); [4]cited in Diller *et al.* (1999);

which can be quite nonuniform. Finally, blood flow, extracellular water and local metabolism are factors that strongly affect heat transfer in living tissue, but are difficult to determine or control experimentally – these can lead to variability in the measured thermal properties data.

An example of an empirical equation relating thermal conductivity of tissue to its composition is given by (Cooper and Trezek, 1971)

$$k[\text{W/m} \cdot \text{K}] = \rho \left(0.628 m_{\text{water}} + 0.117 m_{\text{protein}} + 0.231 m_{\text{fat}}\right) \quad (9.22)$$

where m_{water}, m_{protein} and m_{fat} are mass fractions of water, protein and fat, respectively. One possible source of these mass fractions can be the standard composition data for tissues (see Table 9.10).

Thermal conductivity values are also a function of temperature, as expected. An example of temperature dependence is shown in Figure 9.6(a) which is an average of a number of human, dog, pig and rabbit tissues, in the temperature range 3–45 °C. Examples of data at higher temperatures can be seen in Figure 9.6(b) for a temperature range of 35–90 °C. Another example of temperature dependency (Valvano et al., 1985) is given by Eqn. 9.23 obtained through linear regression of data for a number of tissues (human, dog, pig and rabbit):

$$k = 0.4882 + 0.001265T \quad r = 0.642 \quad (9.23)$$

where k is in W/m·K and T is in °C. Being in equation form, it can be particularly useful in simulations.

Heating also causes irreversible changes. Thus, Figure 9.7 shows that while heating to 80 °C did not cause irreversible change (heating and cooling provides same thermal conductivity), heating to 90 °C did change the tissue irreversibly (heating and cooling provides different thermal conductivity values).

Thermal conductivity of frozen tissues can be obtained in a manner analogous to Eqn. 9.22, considering the mass fractions of ice and water and using their respective thermal conductivities.

9.8 Specific heat

Specific heat is the quantity of heat required to raise the temperature of a unit mass of a substance by one degree. One possible unit of measurement of specific heat is J/kg · K. Some typical values of specific heat are shown in Table 9.6. Like thermal conductivity, specific heat is also a function of composition. An example of an empirical relationship to calculate specific heat from composition is given by (Cooper and Trezek, 1971)

$$C_p[\text{J/kg} \cdot \text{K}] = 4200 m_{\text{water}} + 1090 m_{\text{protein}} + 2300 m_{\text{fat}} \quad (9.24)$$

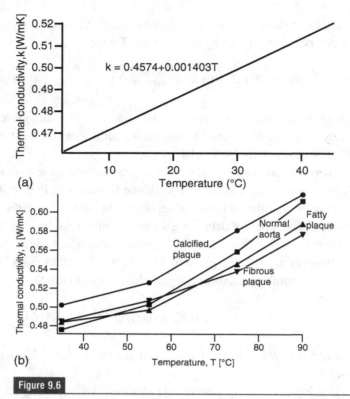

Figure 9.6

(a) Variation of tissue thermal conductivity with temperature; (b) variation of arterial tissue thermal conductivity at higher temperatures. Data from Diller *et al.* (1999).

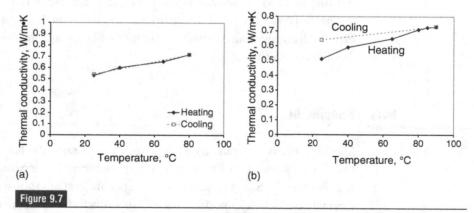

Figure 9.7

Variation of (cow liver) tissue thermal conductivity with temperature showing thermal conductivities can change from heating to cooling due to irreversible changes in tissue above a certain temperature (as in (b)). From Bhattacharya and Mahajan (2003). Reproduced with permission.

9.8.1 Specific heat during phase change of biomaterials: apparent specific heat

Specific heat refers to sensible (as opposed to latent) heat. When phase change is involved, as in freezing, latent heat needs to be considered in heat transfer calculations. For a partially frozen biomaterial, where some (but not all) of the water is frozen, an apparent specific heat can be defined as

$$C_{pa} = dH/dT \qquad (9.25)$$

where H is enthalpy at temperature T. This apparent specific heat includes the latent heat effects. An example of such data for animal tissue is shown in Figure 9.8. Apparent specific heat can also be estimated from its definition, by treating the tissue as a mixture of ice, unfrozen water and the rest of the solid matrix, as (see Section 7.4.1 for derivation)

$$C_{pa} = w\left((1-f)C_{pu} + fC_{pi} + \lambda_f \frac{\partial f}{\partial T}\right) + (1-w)C_{ps} \qquad (9.26)$$

where C_{pu}, C_{pi} and C_{ps} are the specific heats, respectively, of unfrozen water, ice and the remaining solid part (includes dissolved solutes). The quantity f is the mass fraction of water that is ice at any temperature, T; w is the water content of the tissue before freezing and λ_f is the latent heat of pure water.

9.8.2 Specific heat during phase change of pure materials: adaptation of apparent specific heat

In some applications, as in icing the body, one could be interested in the melting of pure water. For pure water, the use of apparent specific heat, as just discussed, has problems as $C_{pa} = dH/dT$ would lead to an infinite value of C_{pa} because all the enthalpy change (say from ice to water) takes place without any change of temperature (at 0 °C). In other words, $\Delta H = 339$ but $\Delta T = 0$, leading to $C_{pa} = \Delta H/\Delta T \rightarrow \infty$. This is illustrated in Figure 9.9(a). Obviously, the numerical computations cannot work with an infinite value of specific heat. In practice, an approximation is made, as shown in Figure 9.9(b), where the phase change is assumed to occur over a small temperature range, like the 2 °C difference assumed in this figure. The area under the curve, ΔH, which is equal to the enthalpy change due to freezing, is kept equal to the true latent heat, λ, but spreading the phase change over a temperature range causes the peak to be a finite value, as illustrated

432 Material properties and other input parameters

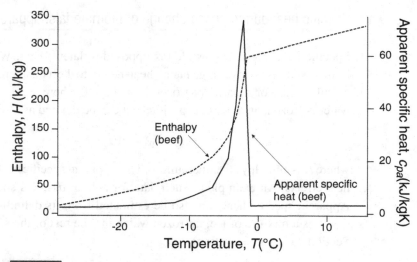

Figure 9.8

Enthalpy and apparent specific heat data for beef tissue during phase change. Apparent specific heat (solid line) is the slope of the enthalpy–temperature curve (dashed line) and therefore includes specific heat as well as latent heat due to phase change.

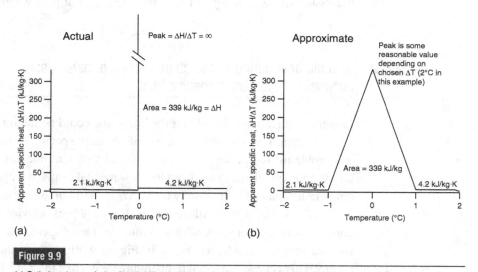

Figure 9.9

(a) Enthalpy change during freezing/thawing of pure water and (b) its approximation for computational purposes.

Figure 9.9(b). Approximation of the phase change, as shown in Figure 9.9, has been extended to biomaterials as well, for simplicity. When using this approximation for biomaterials, the latent heat would need to be multiplied by the weight fraction of water, i.e.,

$$\Delta H = w\lambda_{\text{water}} \tag{9.27}$$

where w is the mass fraction of water in the biomaterial. See also discussion of latent heat in tissue in Section 9.12.

9.9 Density

An empirical correlation for density as a function of composition is given by (Cooper and Trezek, 1971)

$$\rho\,[\text{kg/m}^3] = 1000/\left(m_{\text{water}} + 0.649 m_{\text{protein}} + 1.227 m_{\text{fat}}\right) \tag{9.28}$$

where m is the mass fraction of the particular component noted in its subscript.

Density of frozen materials can also be estimated in a manner analogous to Eqn. 9.28, by considering ice and water as separate components and using their respective densities. This, however, can have considerable error. Density values for various biomaterials of human origin are shown in Table 9.6.

9.10 Thermal diffusivity

Thermal diffusivity is related to thermal conductivity, k, density, ρ and specific heat, C_p, as

$$\alpha = \frac{k}{\rho C_p} \tag{9.29}$$

where α is the thermal diffusivity. Thermal diffusivity data corresponding to the same tissues as in Eqn. 9.23 is given by

$$\alpha = (0.1304 + 0.000519T) \times 10^{-6} \quad r = 0.510 \tag{9.30}$$

where α is in m^2/s and T is in °C. Thermal diffusivity values for various biomaterials of human origin are shown in Table 9.6.

Table 9.7 Thermal properties of some cryoprotectants.

	Toxic level (%)	Thermal conductivity (W/m·K)	Specific heat (J/kg·K)	Density (kg/m^3)
Glycerol	14.28	0.4494	3459.19	1097.42
DMSO	6.06	0.4593	3532.65	1072.76
Propylene glycol	9.30	0.4674	3275.37	1067.20
Ethylene glycol	7.86	0.4591	3560.32	1071.55

9.11 Thermal properties of related materials

Thermal Properties of Cryoprotectants Cryoprotectants are used in cryopreservation of cells or tissues. Thermal properties of some common cryoprotectants are given in Table 9.7.

9.12 Latent heat of fusion and evaporation

In freezing/thawing or evaporation/condensation, it is the water in the biomaterial that is undergoing phase change. Therefore, as a first order of approximation, latent heats of fusion and evaporation for a tissue are related to the water content of the tissue:

$$\lambda_{\text{tissue}} = w\lambda_{\text{water}} \tag{9.31}$$

where λ_{tissue} and λ_{water} are latent heats of tissue and pure water, respectively, and w is the water content. For freezing/thawing, the latent heat of fusion for pure water is 333 kJ/kg while for evaporation/condensation, the latent heat of evaporation for pure water is 2257 kJ/kg.

9.13 Radiative properties

Surface radiative properties are needed when considering radiative heat exchange. There is not a significant amount of data available in this area. An example of data for human skin is shown in Figure 9.10. The variation with wavelength (spectral variation) is generally ignored and a total emittance value between 0.98–0.99 is used.

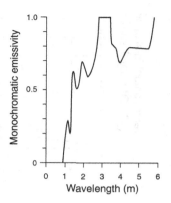

Figure 9.10

The spectral emissivity of human skin at room temperature. The corresponding total emittance is 0.993. Most authors use values between 0.98 and 0.99. Adapted from Elan et al. (1963). Reprinted with permission from Macmillan Publishers Ltd.

9.14 Equilibrium vapor pressure

When modeling diffusion mass transfer in a gas phase, such as moisture loss from a surface, we need the vapor pressure of the adsorbed liquid on the surface to set the boundary condition at an air–tissue interface (see also Section 7.13.4). For example, moisture loss from a surface is given by

$$\text{Moisture loss} = h_m A (c_{v,surface} - c_{v,\infty}) \quad (9.32)$$

The concentration of vapor at the surface in the gas phase is related to the vapor pressure at the surface, and is given by

$$c_{v,surface} = \frac{p_v}{RT} \quad (9.33)$$

Using SI units for p_v, R and T, units for $c_{v,surface}$ are kg/m^3. At the boundary between tissue and air (see Figure 9.11(a), the vapor in the air will be in equilibrium with moisture in the tissue, which will provide the vapor pressure, p_v (see also Section 7.13.4). This information is obtained experimentally. A curve fit to the data available for human skin in contact with air provides the following equation (Kasting and Barai, 2003)

$$\log w = \log(0.0386) + \log\left(\frac{10.4(a_w/1.010)}{(1 - a_w/1.010)(1 + (10.4 - 1)a_w/1.010)}\right) \quad (9.34)$$

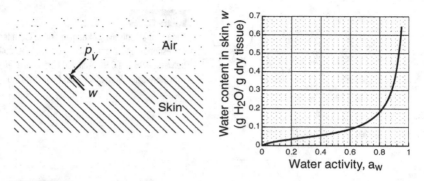

Figure 9.11

Equilibrium data for water in skin in contact with water vapor in the air.

Here w is the water content in the skin tissue (g of H_2O per g of dry tissue). The vapor pressure in the air at the skin–air interface is given in terms of *water activity*, a_w, defined as

$$a_w = \frac{p_v}{p_{v,\text{sat}}} \tag{9.35}$$

where $p_{v,\text{sat}}$ is the vapor pressure of pure water at a particular temperature, as provided in Table 9.12. Being a ratio, a_w is dimensionless. Equation 9.34 is plotted in Figure 9.11(b) for convenience.

9.15 Properties of an air–water vapor mixture

The properties of an air–water vapor mixture are needed, for example, in setting boundary conditions when air surrounds a tissue. Thus, in convective mass transfer from a surface (Eqn. 9.32), the term $c_{v,\infty}$ would require this information. Such air–water vapor mixture properties are provided in a psychrometric chart, as shown in Figure 9.13.

9.16 Mass diffusivity

Diffusivity, which appears in the diffusion mass transfer equation, typically incorporates many different mechanisms of transport that are analogous to molecular diffusion, but can be quite complex for a particular material (such as a tissue) and process combination. Since it can incorporate many different mechanisms, diffusivity is often referred to as *effective diffusivity*. From fundamental considerations, diffusivity in gases, liquids and solids can be discussed separately.

Table 9.8 Diffusivities of gases in air at 1 atm (1.013×10^5 Pa).

Diffusing species	Diffusivity $D_{A,\text{air}} \times 10^4 \text{m}^2/\text{s}$	Temp °C	Source
Water vapor	0.2538	27	1
Oxygen	0.188	27	1
Carbon dioxide	0.157	27	1
Carbon monoxide	0.202	27	1
Sulfur dioxide	0.126	27	1
Hydrogen	0.78	25	2
Ammonia	0.22	25	2
Helium	0.70	25	2
Ethanol	0.14	25	2
Benzene	0.09	25	2
Acetic acid	0.12	25	2
Mercury	0.13	25	2
Carbon tetrachloride	0.083	25	2

[1] Mills (1995); [2] Basmadjian (2004).

9.16.1 Diffusivity in gases

Data for diffusivity in gases is widely available in textbooks and handbooks. As an example, Table 9.8 shows the diffusivity of various gases in air at one atmosphere.

Temperature and other effects on diffusivity can be estimated using the equation

$$D_{AB} = \frac{0.001858 T^{3/2} (1/M_A + 1/M_B)^{1/2}}{p \sigma_{AB}^2 \Omega_{D,AB}} \quad (9.36)$$

where D_{AB} is the diffusivity of A through B, T is the absolute temperature, M_A and M_B are molecular weights, p is the absolute pressure in atm, σ_{AB} is the collision diameter in Å, and $\Omega_{D,AB}$ is a dimensionless function of the temperature and the intermolecular potential.

9.16.2 Diffusivity in liquid

The diffusivity of proteins and other molecules in water can be estimated based on the Stokes–Einstein equation:

$$D_{AB} = \frac{k_B T}{6\pi \mu R} \quad (9.37)$$

where A is the diffusing species (such as a protein), B is the fluid through which diffusion is taking place, k_B is the Boltzmann constant, 1.38054×10^{-23} J/K, T is the absolute temperature, μ is the viscosity and R is the radius of the molecule. While this equation is strictly applicable only in cases where the diffusing molecule is large compared to the surrounding solvent molecules, it has proven useful for solute–solvent pairs in which the radius, R, is only two to three times the solvent radius (Saltzman, 2001). For small solute molecules that are identical to solvent, Eqn. 9.37 is modified as:

$$D_{AB} = \frac{k_B T}{4\pi \mu R} \tag{9.38}$$

Molecular weight and diffusivity values available in the literature are plotted in Figure 9.12. Empirical equations, in lieu of the ones mentioned above, are also used (Saltzman, 2001). For example, for proteins and larger molecules in water, the following equation is used:

$$D_{A,\text{water}} = \frac{9.40 \times 10^{-15} T}{\mu M_w^{1/3}} \tag{9.39}$$

while for the situation when the solute and solvent are similar, the following equation is used.

$$D_{A,\text{water}} = 7.4 \times 10^{-17} \frac{T \sqrt{\psi_{\text{solvent}} M_{w,\text{solvent}}}}{\mu \tilde{V}_A} \tag{9.40}$$

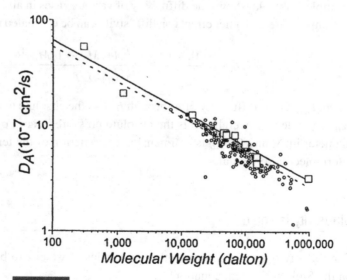

Figure 9.12

Variation of diffusivity with molecular weight for globular proteins in water. The solid line indicates the best fit while the dashed line is Eqn. 9.39. From Salzman (2001). By permission of Oxford University Press.

In both equations, $D_{A,\text{water}}$ is in cm^2/s, T is in K, μ is in Pa·s and M_w is in Daltons (1 Dalton = 1.65×10^{-27} kg). In Equation 9.40, which is for a dilute solution, $\tilde{V}$ is in m^3/mol and ψ_{solvent} is 2.6 for water. Besides the factors noted in the above equations, the pH of the liquid can also change the diffusivity significantly. Diffusivity values for various materials through water and biological fluids are shown in Tables 9.14 and 9.15, respectively.

9.16.3 Diffusion through solids (gels and tissues)

Diffusion through gels and tissues can be quite complex. The mechanism of liquid diffusion through solid is capillarity, while gas diffusion through solid can be molecular diffusion through the gas filled pores of the solid. For the purpose of this text, it suffices to say that such data has to come from measurements and not many reliable prediction equations are commonly available. Sample diffusivity values for such systems are shown in Table 9.9.

For gas diffusion through the pores of a solid, the effective gas diffusivity, D_{eff}, is related to the molecular diffusivity of the gas by the expression

$$D_{\text{eff}} = \frac{D_{AB}}{\tau} \tag{9.41}$$

where D_{AB} is the molecular diffusivity of gas A through gas B (which occupies the pores in the solid) and τ is the tortuosity of the pores in the solid. For liquid diffusion in solid (capillary diffusion), diffusivity can vary strongly with moisture content.

9.17 Partition coefficient

As discussed in Section 7.13, when two different phases are in contact, concentrations of a component in the two phases are not the same and their ratio is termed the partition or distribution coefficient. The best source for data on partition coefficient appears to be the appropriate journal papers.

9.18 Diffusive permeability and transmissibility

Diffusive permeability, often referred to as just permeability, is a transport property related to mass diffusivity. If Fick's law is written for flux over a thickness, as shown in Figure 7.12, as

Table 9.9 Some representative values of diffusivity in solids.

Diffusing through	Diffusing species	Diffusivity $D \times 10^9$ m^2/s	Temp °C	Source
Human dermis	Glucose	0.264	-	1
Human viable epidermis	Glucose	0.1	-	1
Stratum Corneum	Aldosterone	0.00018	-	2
Stratum Corneum	Corticosterone	0.00025	-	2
Stratum Corneum	Decanol	0.047	-	2
Stratum Corneum	Estradiol	0.0035	-	2
Stratum Corneum	Lidocaine	0.005	-	2
Stratum Corneum	Napthol	0.005	-	2
Stratum Corneum	Octanol	0.04	-	2
Stratum Corneum	Progesterone	0.006	-	2
Stratum Corneum	Testosterone	0.002	-	2
Fresh porcine sclera	Ethacrynic acid (ECA)	0.485	4	3
Rabbit cornea	Human serum albumin	0.022	35	4
Human cornea	Glucose	0.25	35	4
Fresh porcine sclera	Sodium Fluorescein (NaF)	0.0523	4	3
Human sclera	Sulforhodamine	0.128	37	3
Myelin–parallel to nerve	Water	0.37	-	5
Myelin–perpendicular to nerve	Water	0.13	-	5
Normal liver	Water	1.83–1.51	-	6
Cirrhotic liver	Water	1.37–1.09	-	6
Metastases	Water	0.94–0.85	-	6
Hepatocellular carcinoma	Water	1.33–1.39	-	6
Hemangiomas	Water	2.95–2.84	-	6
Cysts	Water	3.63–2.89	-	6
Benign hepatocellular lesions	Water	1.75–1.64	-	6
Benign lesions	Water	2.45–2.14	-	6
Malignant lesions	Water	1.08–1.06	-	6
HT29 colon cancer multicellular layers	Tirapazamine (TPZ)	0.04	-	7
Muscle cell in rabbit aorta	Nitric oxide (NO)	3.3	37	8
Endothelial cell in rabbit aorta	Nitric oxide (NO)	0.3	37	8
Collagen	Albumin	0.016	35	4
Collagen	Glucose	0.27	35	4
Glyceral monooleate gel (pH 5)	Bupivacaine	0.0527	-	9
Glyceral monooleate gel (pH 9)	Bupivacaine	0.0179	-	9

[1]Khalil et al. (2006); [2]Mitragotri (2000); [3]Lin et al. (2007); [4]Liu et al. (2006); [5]Andrews et al. (2006); [6]Taouli et al. (2003); [7]Hicks et al. (2006); [8]Malinski et al. (1993); [9]Shah et al. (2001)

$$\text{Flux} = D \frac{c_1^* - c_2^*}{t} \tag{9.42}$$

$$= DK \frac{c_1 - c_2}{t} \tag{9.43}$$

$$= \frac{DK}{t}(c_1 - c_2) \tag{9.44}$$

Table 9.10 Human body composition.

	Weight (kg)	Water (kg)	Ash (kg)	Fat (kg)	Protein (kg)	Density $\times 10^{-3}$ (kg/m^3)
Total body	70	42	3.7	13.3	10.6	1.7
Total soft tissue	60	38.7	0.4	11.4	8.7	
Adipose tissue	15	2.3	0.3	12	0.75	0.92
subcutaneous	7.5	1.1	0.02	6	0.38	0.92
other separable	5	0.75	0.01	4	0.25	0.92
interstitial	1	0.15	0.002	0.80	0.05	0.92
Blood (whole)	5.5	4.4	0.06	0.04	0.99	1.06
plasma	3.1	2.9	0.03	0.02	0.21	1.03
Connective tissue	3.4	2.1	0.14	0.04	1.2	1.2
CNS	1.43	1.1	0.02	0.16	0.11	
brain	1.4	1.1	0.02	0.15	0.11	1.03
GI tract	1.2	0.95	0.01	0.7	0.16	1.04
Heart	0.33	0.24	0.0036	0.03	0.6	1.03
Kidneys	0.31	0.24	0.0034	0.02	0.05	1.05
Liver	1.8	1.3	0.02	0.12	0.32	
Lung	1	0.78	0.01	0.01	0.18	1.05
Muscle (skeletal)	28	22	0.34	0.62	4.8	1.04
Pancreas	0.10	0.07	0.0012	0.01	0.01	1.05
Skeleton	10	3.3	2.8	1.9	1.9	1.4
bone	5.0	0.85	2.7	0.05	1.3	2.2
cortical	4	0.60	2.2	0.04	1	1.85
trabecular	1	0.23	0.50	0.01	0.24	1.08
red marrow	1.5	0.60	0.01	0.60	0.30	1.03
yellow marrow	1.5	0.23	0.003	1.2	0.06	0.98
cartilage	1.1	0.86	0.05	0.01	0.18	1.1
periarticular tissue	0.90	0.57	0.04	0.01	0.14	1.1
Skin	2.6	1.6	0.02	0.26	0.75	1.10
Spleen	0.18	0.14	0.0025	0.0029	0.04	1.06

Data from Heymsfield *et al.* 2005.

where D is diffusivity, K is the partition coefficient ($= c^*/c$, see Eqn. 7.82), c^* and c are concentrations in the solid and the surrounding fluid, respectively and t is the thickness of the material. The term DK/t is termed permeability (Truskey *et al.*, 2004). Since K is dimensionless, the units of permeability are typically m/s or equivalent. As an example, a large collection of permeability data for cornea, sclera and conjunctiva can be seen in Prausnitz and Noonan (1998).

In some literature, such as that on contact lens, DK/t is referred to as *transmissibility*, while permeability is defined as DK, that is, without the thickness

Table 9.11 Data for reference man and woman. The current reference man is between 20 and 30 years of age, weighs 70 kg, is 170 cm in height and lives in a climate with an average temperature of 10°C to 20°C. Reference man is a Caucasian and Western European or North American in habitat and custom. Original data from Elia, 1992; Snyder et al., 1975; cited in Heymsfield et al., 2005.

	Unit	Reference man	Reference woman
Body mass (BM)	kg	70	58
Body height	cm	170	160
Body surface area	cm^2	18,000	16,000
Total body specific gravity	g/cm^3	1.07	1.04
Water content	ml/kg BM	600	500
extracellular water	ml/kg BM	260	200
intracellular water	ml/kg BM	340	300
Total blood	ml	5,200	3,900
	g	5,500	4,100
Red blood cells	ml	2,200	1,350
	g	2,400	1,500
Plasma	ml	3,000	2,500
	g	3,100	2,600
Total body fat	kg	13.5	16.0
Total body adipose tissue	kg	15.0	19.0
subcutaneous AT	kg	7.5	13.0
separable AT	kg	5.0	4.0
yellow marrow	kg	1.5	1.3
interstitial AT	kg	1.0	0.7
Total body connective tissue	kg	5.05	4.10
cartilage	kg	2.50	2.0
tendons and fascia	kg	0.85	0.70
other connective tissue	kg	1.70	1.40
Total body skin	kg	2.60	1.79
Total skin thickness	μm	1,300	1,300
Wet skeletal weight	kg	10.0	6.8
Fat-free wet skeletal weight	kg	8.0	5.8
Dry skeletal weight	kg	5.0	3.4
Percent ash of wet skeletal weight		28	28
Total bone marrow	kg	3.00	2.60
red bone marrow	kg	1.50	1.30
yellow bone marrow	kg	1.50	1.30

(Brennan, 2001). When using literature data it is therefore critical to consider the definition used in a particular context. A number of units are used for permeability and transmissibility, depending on the application. For example, in oxygen flux through contact lenses, the unit of flux used is $\mu L/cm^2 \cdot h$.

Table 9.12 Thermophysical properties of saturated water.

Temp K	Pressure $P \times 10^{-5}$ Pa	Specific heat C_p kJ/kg·K	Viscosity N·s/m² $\mu \times 10^6$	Thermal conduc. k W/m·K	Prandtl number Pr	Expansion coefficient $\beta \times 10^6$ K^{-1}
273.15	0.00611	4.217	1750	0.569	12.99	−68.05
275	0.00697	4.211	1652	0.574	12.22	−32.74
280	0.00990	4.198	1422	0.582	10.26	46.04
285	0.01387	4.189	1225	0.590	8.81	114.1
290	0.01917	4.184	1080	0.598	7.56	174.0
295	0.02617	4.181	959	0.606	6.62	227.5
300	0.03531	4.179	855	0.613	5.83	276.1
305	0.04712	4.178	769	0.620	5.20	320.6
310	0.06221	4.178	695	0.628	4.62	361.9
315	0.08132	4.179	631	0.634	4.16	400.4
320	0.1053	4.180	577	0.640	3.77	436.7
325	0.1351	4.182	528	0.645	3.42	471.2
330	0.1719	4.184	489	0.650	3.15	504.0
335	0.2167	4.186	453	0.656	2.88	535.5
340	0.2713	4.188	420	0.660	2.66	566.0
345	0.3372	4.191	389	0.668	2.45	595.4
350	0.4163	4.195	365	0.668	2.29	624.2
355	0.5100	4.199	343	0.671	2.14	652.3
360	0.6209	4.203	324	0.674	2.02	697.9
365	0.7514	4.209	306	0.677	1.91	707.1
370	0.9040	4.214	289	0.679	1.80	728.7
373.15	1.0133	4.217	279	0.680	1.76	750.1
375	1.0815	4.220	274	0.681	1.70	761
380	1.2869	4.226	260	0.683	1.61	788

Adapted from Mills, A. F. (1995), *Basic Heat and Mass Transfer*, Irwin, Chicago.

9.19 Reaction rate constants

Diffusion of a species is often accompanied by its utilization, for example, oxygen diffusion through a tissue is accompanied by its metabolic consumption. Depending on the application, this utilization is also termed clearance. Reactions such as consumption or generation of a species are implemented in the computation as source terms and are discussed in Section 8.10. Rate constants are specific to a reaction and need to be either measured or obtained from published literature.

Table 9.13 Thermal properties of air at atmospheric pressure.

Temp K	Density ρ kg/m^3	Specific heat C_p kJ/kg·K	Viscosity $\mu \times 10^5$ kg/m·s	Thermal conductivity k W/m·K	Thermal diffusivity $\alpha \times 10^5$ m^2/s	Prandtl number Pr
200	1.7690	1.0064	1.3286	0.01809	1.0161	0.739
250	1.4133	1.0054	1.5992	0.02227	1.5673	0.722
260	1.3587	1.0054	1.6504	0.02308	1.6896	0.719
270	1.3082	1.0055	1.7005	0.02388	1.8154	0.716
280	1.2614	1.0057	1.7504	0.02467	1.9447	0.714
290	1.2177	1.0060	1.7985	0.02547	2.0792	0.710
300	1.1769	1.0063	1.8465	0.02624	2.2156	0.708
310	1.1389	1.0068	1.8929	0.02701	2.3556	0.705
320	1.1032	1.0073	1.9392	0.02779	2.5008	0.703
330	1.0697	1.0079	1.9855	0.02853	2.6462	0.701
340	1.0382	1.0085	2.0302	0.02928	2.7965	0.699
350	1.0086	1.0092	2.0748	0.03003	2.9503	0.697
360	0.9805	1.0100	2.1177	0.03078	3.1081	0.695
370	0.9539	1.0109	2.1606	0.03150	3.2666	0.693
380	0.9288	1.0120	2.2018	0.03223	3.4289	0.691
390	0.9050	1.0130	2.2447	0.03295	3.5942	0.690
400	0.8822	1.0142	2.2859	0.03365	3.7609	0.689
450	0.7842	1.0212	2.4849	0.03710	4.6327	0.684
500	0.7057	1.0300	2.6703	0.04041	5.5594	0.681

Adapted from *Tables of Thermal Properties of Gases*, National Bureau of Standards Circular 564, Washington, D.C. (1955).

9.20 Other parameters

In addition to equilibrium properties such as vapor pressure, and transport properties such as thermal conductivity and mass diffusivity, a model typically requires other parameters, some of which are listed below.

9.20.1 Convective heat and mass transfer coefficients

The surface convective heat transfer coefficient, h, and surface convective mass transfer coefficient, h_m, are needed for implementing boundary conditions. The quantities h and h_m are not material properties but parameters that describe the heat and mass transfer processes, respectively. In most cases, they are obtained from the well-known correlations available in most heat and mass transfer textbooks.

9.20 Other parameters

Table 9.14 Diffusivities in water.

Diffusing species	Diffusivity $D \times 10^9 \, m^2/s$	Temp °C	Source
Oxygen	2.42	25	2
Oxygen	2.7	37	1
Hydrogen	4.8	25	2
Hydrogen	7.8	37	1
Helium	7.3	25	2
Helium	8.4	37	1
Methane	1.8	25	2
Methane	2	37	1
Ammonia	2	25	2
Ammonia	2.7	37	1
Acetylene	2.4	37	1
Carbon monoxide	2.17	25	2
Carbon monoxide	2.7	37	1
Nitrogen	2	25	2
Nitrogen	2.5	37	1
Ethylene	2.5	37	1
Nitric oxide	3.5	37	1
Ethane	1.6	37	1
Methanol	1.28	25	2
Methanol	1.1	37	1
Hydrogen sulfide	1.9	37	1
Hydrogen chloride	3.1	25	2
Hydrogen chloride	4	37	1
Argon	2.7	37	1
Sodium hydroxide	2.3	37	1
Carbon dioxide	2	25	2
Carbon dioxide	2.7	37	1
Propane	1.3	37	1
Formic acid	1.41	25	5
Formic acid	2	37	1
Ethanol	1.24	25	2
Ethanol	1.1	37	1
Ethyl alcohol	1.5	37	1
Acetone	1.28	25	2
Acetone	1.5	37	1
Sodium chloride	2.1	37	1
Acetic acid	1.26	25	2
Acetic acid	1.6	37	1
Urea	1.4	25	2
Urea	1.8	37	1
1-Propanol	1.2	37	1

Table 9.14 (continued).

Diffusing species	Diffusivity $D \times 10^9 m^2/s$	Temp °C	Source
2-Propanol	1.2	37	1
Butyric acid	0.087	25	5
Nitric acid	3.5	37	1
Succinic acid	0.094	25	5
Sulfur dioxide	1.7	25	2
Chlorine	1.45	25	2
Chlorine	1.9	37	1
Creatinine	1.22	37	1
Propionic acid	1.06	25	5
Propionic acid	1.4	37	1
n-Butanol	1.0	37	1
Glycine	1.4	37	1
Benzene	1.02	25	2
Benzene	1.4	37	1
Glycerol	1.1	37	1
Phenol	1.3	37	1
Sulfuric acid	2.6	37	1
Benzyl alcohol	1.1	37	1
Chlorobenzene	0.91	25	2
Valine	0.83	25	5
Valine	1.1	37	1
Benzoic acid	1	25	5
Benzoic acid	1.3	37	1
Bromine	1.6	37	1
Glucose	1.3	25	2
Glucose	0.9	37	1
1,2,4-Trichlorobenzene	0.76	25	2
Mercury	2.9	25	2
2,4,2',4-Tetrachlorobiphenyl	0.55	25	2
^{51}Cr-ethylenediamine tetraacetate (EDTA)	0.74	37	1
Sodium fluorescein (NaF)	0.289	4	3
Sodium fluorescein (NaF)	0.728	37	3
Sucrose	0.7	37	1
Fluorescein	0.57	25	4
Fluorescein	0.7	37	1
DDT	0.49	25	2
Polypeptide	0.21	25	4

Table 9.14 (continued).

Diffusing species	Diffusivity $D \times 10^9 \mathrm{m}^2/\mathrm{s}$	Temp °C	Source
Lysozyme (egg white)	1.0	25	2
Lactalbumin	0.13	25	4
Ovalbumin	0.087	25	4
BSA	0.071	-	6
BSA	0.083	25	1
Hemoglobin	0.69	25	2
IgA (human)	0.052	25	4
IgG (human)	0.044	25	4
IgM (human)	0.032	25	4
S-IgA (human)	0.017	25	4
Fibrinogen	0.2	25	2
Tobacco mosaic virus	0.044	25	2
Air	2.7	37	1
Glycine	1.1	25	5
Glutamine	0.76	25	5
Alanine	0.91	25	5
Leucine	0.73	25	5
Serine	0.88	25	5
Asparagine	0.83	25	5
Threonine	0.8	25	5
Vinblastine	0.33	37	7
Cl^-	2.03	25	5
Cu^{2+}	0.71	25	2
Fe^{3+}	0.6	25	2
$H_2PO_4^{2-}$	0.88	25	5
H^+	9.3	25	2
Li^+	1.03	25	5
K^+	2	25	2
Mg^{2+}	0.71	25	2
Na^+	1.3	25	2
NH_4^+	1.97	25	5
NO_3^-	1.9	25	2
OH^-	5.3	25	2
HPO_4^-	0.76	25	5
SO_4^{2-}	1.1	25	2
HCO_3^-	1.18	25	5

[1] Saltzman (2001); [2] Basmadjian (2004); [3] Lin *et al.* (2007); [4] Saltzman *et al.* (1994); [5] Stewart (2003); [6] Gutenwik *et al.* (2004); [7] Modok *et al.* (2006)

Table 9.15 Diffusivities in biological fluids.

Solvent	Diffusing species	Diffusivity $D \times 10^9 \text{m}^2/\text{s}$	Temp °C	Source
Free solution	DMNB-HPTS[†] (Caged fluorescent dye)	0.33	37	1
Free solution	DMNB-caged fluorescein dextran[‡] (Caged fluorescent dye)	0.098	37	1
Free solution	Sodium fluorescein	0.7	37	1
Lateral intercellular spaces of canine kidney cells	DMNB-HPTS	0.28	37	1
Lateral intercellular spaces of canine kidney cells	DMNB-caged fluorescein dextran	0.06	37	1
Plasma	Dextran (16000 MW)	0.241	37	2
Multicell layer	Sucrose	0.0042	37	3
Multicell layer	Vinblastine	0.0019	37	3
Plasma	Inulin	0.217	37	2
Plasma	Sucrose	0.59	37	2
Plasma	Uric acid	0.745	37	2
Plasma	Urea	1.46	37	2
Plasma	Creatinine	0.871	37	2

[1]Xia (1998); [2]Saltzman (2001); [3]Modok *et al.* (2006)
[†] 8-((4,5-dimethoxy-2-nitrobenzyl)oxy)pyrene-1,3,6-trisulfonic acid
[‡] (4,5-dimethoxy-2-nitrobenzyl) fluorescein dextran (10 000 MW)

9.20.2 Radiative heat transfer coefficient

For special cases of radiative heat transfer, when the radiative surface temperatures are close to each other, a radiative heat transfer coefficient can be defined that is a simplified form of the complete radiative balance equations:

$$h_r = 4\sigma T^3 \tag{9.45}$$

Here h_r is the radiative heat transfer coefficient and T is the surface temperature.

9.20.3 Combined heat transfer coefficient

The convective and radiative heat transfer coefficients described above can be combined into a single surface heat transfer coefficient as $h + h_r$. Using this coefficient, heat flux from a surface can be written as

$$q'' = (h + h_r)(T_s - T_a) \tag{9.46}$$

9.20 Other parameters

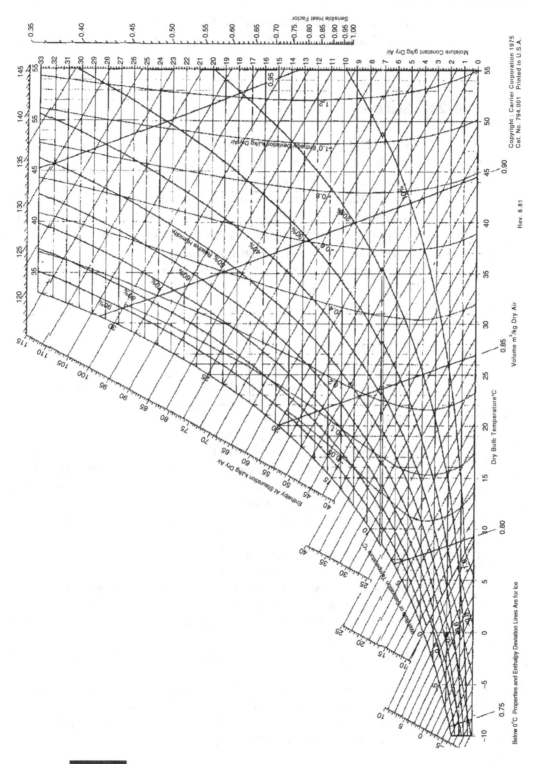

Figure 9.13

Psychrometric chart at normal temperatures (Courtesy of Carrier Corporation).

where q'' is the heat flux from a surface, T_s is the surface temperature and T_a represents both the ambient fluid temperature and surrounding surface temperatures.

9.21 Summary

Some of the properties and parameters needed for simulation of transport processes are shown in Table 9.1. These are in addition to anatomical and physiological parameters that may be necessary. Properties data generally come from measurements. Detailed properties data are hard to find for biomedical systems. Individual research papers are perhaps the best source for data. Unavailability of appropriate properties and parameter data and their expected variabilities makes sensitivity analysis with respect to data an important part of modeling. Such analysis provides the insight into how the properties and parameters affect a particular process.

References

Andrews, T. J., Osborne, M. T., and Does, M. D. (2006). Diffusion of myelin water. *Magnetic Resonance in Medicine*, **56**(2):381–385.

Anonymous. (2002). Some material property databases. On the web at http://www.gi.rwth-aachen.de/mebsp-b/datab-list.html

Apiou-Sbirlea, G. and J-P. L'Huiller. (1998). Simulating and optimizing of argon laser iridectomy. Influence of irradiation duration on the corneal and lens thermal injury. Part of the EUROPTO Conference on Lasers in Opthalmology, Stockholm, Sweden, Sept. 1998. SPIE Vol. 3564.

Basmadjian, D. (2004). *Mass Transfer: Principles and Applications*. Boca Raton, FL., CRC Press.

Berger, S. A., W. Goldsmith and E. R. Lewis (1996). *Introduction to Bioengineering*. Oxford, Oxford University Press.

Bhattacharya, A. and R. L. Mahajan. (2003). Temperature dependence of thermal conductivity of biological tissues. *Physiological Measurement*, **24**:769–783.

Brennan, N. A. (2001). A model of oxygen flux through contact lenses. *Cornea*, **201**(1):104–108.

Bronzino, J. D. (1995). *The Biomedical Engineering Handbook*. Boca Raton, Florida, CRC Press.

Cooper, T. E., and Trezck, G. J. (1970). A probe technique for determining the thermal conductivity of tissue. ASME Paper No. 70-WA/HT-18.

Cooper, T. E., and Trezck, G. J. (1971). Correlation of thermal properties of some human tissues with water content. *Aerospace. Med.*, **42**, 24–27.

Datta, A. K. (2009). Computer-Aided Engineering: Application to Biomedical Processes. Course home page. Dept. of Biological and Environmental Engineering, Cornell University. 1 Jan. 2009 <http://courses.cit.cornell.edu/bee4530/>

Diller, K. R., J. W. Valvano and J. A. Pearce (1999). Bioheat transfer. In *CRC Handbook of Thermal Engineering*, edited by Frank Kreith. Pages 4–114 to 4–187.

DuBois, D. and DuBois, E. F. (1916). A formula to estimate surface area if height and weight are known. *Archives of Internal Medicine*, **17**:863.

Elam, R., D. W. Goodwin, and K. L. Williams (1963). Optical properties of human epidermis. *Nature*, **198**(4884):1001–1002.

Elia M. (1992). Organ and tissue contribution to metabolic rate: In: Kinney, J. M., Tucker H. N. (eds.) Energy metabolism: Tissue determinants and cellular corollaries. Raven Press, New York, 19–60.

Erdmann, B., J. Lang, M. Seebass (1998). Optimization of temperature distributions for regional hyperthermia based on a nonlinear heat transfer model. Annals of the New York Academy of Sciences, **858**:36–46.

Green, D. W. and Perry, R. H. (2007). *Perry's Chemical Engineers Handbook*. New York, McGraw-Hill.

Gutenwik, J., Nilsson, B., and Axelsson, A. (2004). Coupled diffusion and adsorption effects for multiple proteins in agarose gel. *AIChE Journal*, **50**(12):3006–3018.

Hayashi, K. and M. D. Markel (1998). Thermal modification of joint capsule and ligamentous tissues. *Operative Techniques in Sports Medicine*, **6**(3):120–125.

He, Y; Shirazaki, M; Liu, H, Himeno, R., Sun, Z. (2006). A numerical coupling model to analyze the blood flow, temperature, and oxygen transport in human breast tumor under laser irradiation *Computers in Biology and Medicine*: **36**(12):1336–1350.

Heymsfield, S. B., T. G. Lohman, Z. Wang and S. B. Going., Eds. (2005). *Human Body Composition. Human Kinetics*, Illinois, Champaign.

Hicks, K. I., Pruijn, F. B., Secomb, T. W., Hay, M. R., Hsu, R., Brown, J. M., Denny, W. A., Dewhirst, M. W., and Wilson, W. R. (2006). Use of three-dimensional tissue cultures to model extravascular transport and predict in vivo activity of hypoxia-targeted anticancer drugs. *Journal of the National Cancer Institute*, **98**(16): 1118–1128.

Johnson, J. M., Brengelmann, G. L., Hales, J. R. S., Vanhoutte, P. M., and Wenger, C. B. (1986). Regulation of the cutaneous circulation, *Fed. Proc.*, **45**, 2841–2850.

Kasting, G. B. and N. D. Barai (2003). Equilibrium water sorption in human stratum corneum. *Journal of Pharmaceutical Sciences* **92**(8):1624–1631.

Keele, C. A. and Neil, E., Eds. (1971) *Samson Wright's Applied Physiology*, 12th ed. London, Oxford Press.

Khalil, E., Kretsos, K., and Kasting, G. B. (2006). Glucose partition coefficient and diffusivity in the lower skin layers. *Pharm Res*, **23**(6):1227–1234.

Lassen, N. A., Lindbjerg, J., and Munck, O. (1964) Measurement of bloodflow through skeletal muscle by intramuscular injection of xenon-133, *Lancet*, **1**, 686–689.

Lin, C. W., Wang, Y., Challa, P., Epstein, D. L., and Yuan, F. (2007). Transscleral diffusion of ethacrynic acid and sodium fluorescein. *Molecular Vision*, **13**(27–28): 243–251.

Liu, Y., Griffith, M., Watsky, M. A., Forrester, J. V., Kuffova, L., Grant, D., Merrett, K., and Carlsson, D. J. (2006). Properties of porcine and recombinant human collagen matrices for optically clear tissue engineering applications. *Biomacromolecules*, **7**(6):1819–1828.

Malinski, T., Taha, Z., Grunfeld, S., Patton, S., Kapturczak, M., and Tomboulian, P. (1993). Diffusion of Nitric-Oxide in the Aorta Wall Monitored in-Situ by Porphyrinic Microsensors. *Biochemical and Biophysical Research Communications*, **193**(3):1076–1082.

Mates, R. E. (1995). Arterial macrocirculatory hemodynamics. In *The Biomedical Engineering Handbook*, Ed. J. D. Bronzino. Boca Raton, CRC Press.

Mills, A. F. (1995). *Basic Heat and Mass Transfer*. Chicago, Irwin.

Milnor, W. R. (1989). *Hemodynamics*. Baltimore, Williams and Wilkins.

Mitragotri, S. (2000). In situ determination of partition and diffusion coefficients in the lipid bilayers of stratum corneum. *Pharmaceutical Research*, **17**(8):1026–1029.

Modok, S., Hyde, P., Mellor, H. R., Roose, T., and Callaghan, R. (2006). Diffusivity and distribution of vinblastine in three-dimensional tumour tissue: Experimental and mathematical modelling. *European Journal of Cancer*, **42**(14):2404–2413.

Neofytou, P. (2004). Comparison of blood rheological models for physiological flow simulation. *Biorheology*, **41**:693–714.

O'Brien, W. J. (2008). *Dental Materials and their Selection*. Hanover Park, IL, Quintessence Pub. Co.

Parsons, K. (2003). *Human Thermal Environment*. New York, Taylor & Francis.

Prausnitz, M. R. and J. S. Noonan (1998). Permeability of cornea, sclera, and conjunctiva: A literature analysis for drug delivery to the eye. *Journal of Pharmaceutical Sciences* **87**(12):1479–1488.

Rabin, Y. (2003). A general model for the propagation of uncertainty in measurements into heat transfer simulations and its application to cryosurgery. *Cryobiology*, **46**:109–120.

Rohsenow, W. M., J. P. Hartnett and Y. I. Cho (1997). *Handbook of Heat Transfer*. New York, McGraw-Hill.

Saltzman, M. (2001). *Drug Delivery*. New York, Oxford University Press.

Saltzman, W. M., Radomsky, M. L., Whaley, K. J., and Cone, R. A. (1994). Antibody diffusion in human cervical mucus. *Biophysical Journal*, **66**(2 Pt 1):508–515.

Sapareto, S. A. and W. C. Dewey (1984). Thermal dose determination in cancer therapy. *International Journal of Radiation Oncology Biology Physics*, **10**:787–800.

Sekins, K. M., Dundore, D., Emery, A. F., McGrath, P. W. and Nelp, W. B. (1980). Muscle blood flow changes in response to 915 MhZ diathermy with surface cooling as measured by Xe^{133} clearance. *Arch Phys. Med. Rehabil.*, **61**, 105–113.

Shah, J. C., Sadhale, Y., and Chilukuri, D. M. (2001). Cubic phase gels as drug delivery systems. *Advanced Drug Delivery Reviews*, **47**(2–3):229–250.

Shitzer, A. and R. C. Eberhart (1985). Heat generation, storage, and transport processes. 1:137–151. In: *Heat Transfer in Medicine and Biology* edited by A. Shitzer and R. C. Eberhart. Plenum Press, New York.

Snyder, W. S., M. J. Cook, E. S. Nasset, L. R. Karhausen, G. P. Howells and I. H. Tipton (1975). Report on the task group of reference Man. Oxford, UK, Pergamon Press.

Stewart, P. S. (2003). Diffusion in biofilms. *Journal of Bacteriology*, **185**(5):1485–1491.

Taouli, B., Vilgrain, V., Dumont, E., Daire, J. L., Fan, B., and Menu, Y. (2003). Evaluation of liver diffusion isotropy and characterization of focal hepatic lesions with two single-shot echo-planar MR imaging sequences: Prospective study in 66 patients. *Radiology*, **226**(1):71–78.

Thurau, K. and Levine, D. Z. (1971). The renal circulation, in *The Kidney: Morphology, Biochemistry, Physiology,* Rouiller, C. and Muller, A. F., Eds., New York, Academic Press, 1–70.

Trujillo, F. J., C. Wiangkaew, *et al.* (2007). Drying modeling and water diffusivity in beef meat. *Journal of Food Engineering* **78**(1):74–85.

Truskey, G. A., F. Yuan and D. F. Katz (2004). *Transport Phenomena in Biological Systems*. New Jersey, Pearson Prentice Hall.

Valvano, J. W., J. R. Cochran and K. R. Diller (1985). Thermal conductivity and diffusivity of biomaterials. *International Journal of Thermophysics*, **6**(3):301–311.

Wright, N. T. and Humphrey, J. D. (2002). Denaturation of collagen via heating: An irreversible rate process. *Annual Review of Biomedical Engineering* **4**, 109–128.

Xia, P., Bungay, P. M., Gibson, C. C., Kovbasnjuk, O. N., and Spring, K. R. (1998). Diffusion coefficients in the lateral intercellular spaces of Madin-Darby canine kidney cell epithelium determined with caged compounds. *Biophysical Journal*, **74**(6):3302–3312.

Zogzas, N. P., Z. B. Maroulis, and D. Marinos-Kouris (1994). Densities, shrinkage and porosity of some vegetables during air drying. *Drying Technology*, **12**(7):1653–1666.

9.22 Problems (short questions)

(1) Which properties are needed for each one of the problem formulations in Chapter 1, starting at Section 1.12.2?

(2) For the problem formulations in Chapter 1, mention when constant properties data would not be the best solution and, instead, properties varying with temperature and moisture would be preferred.

(3) Can there be situations where unavailability of properties data would lead one to abandon a physics-based model altogether (and go for an observation-based or empirical model)?

(4) While performing simulations, what is the most common approach to determine if accurate data is needed for a particular process?

(5) Would you treat blood as a (choose one and explain): (1) Newtonian; (2) non-Newtonian?

(6) Is the heat transfer coefficient a material property? Explain.

(7) Use the composition-based equation (Eqn. 9.22) to predict the thermal conductivity of liver tissue and compare with a prediction for the same tissue using correlation of the measured data provided in Table 9.6.

(8) Outside the range of ice formation (completely frozen or completely unfrozen), how does apparent specific heat relate to specific heat?

(9) To model freezing or thawing of a biomaterial, would the thermal conductivity and density be functions of temperature? How would you formulate the equations that provide the variations in these two properties with temperature during a freezing or thawing process?

(10) What are some of the reasons that composition-based correlations may not predict properties very well?

(11) Does the latent heat of fusion of a tissue have the same value as that of water?

(12) For diffusivities in water, what equation would you use to correct for temperature?

(13) Diffusivity in a solid can be quite small. Do you expect this to lead to computational difficulties? If there is such difficulty, are there steps that can be taken to reduce such difficulty?

(14) Large values for heat and mass transfer coefficients can lead to numerical difficulties. How can you avoid using a very large value of heat transfer coefficient on a surface?

10 Solving the equations: numerical methods

In the spirit of using the software less as a blackbox, it is essential to have at least a rudimentary knowledge of how the equations developed in Chapters 7–9 are solved. However, solution methods can be complex and entire books have been written and dedicated courses exist in most universities on these methods. Instead of digging into details that are clearly outside the scope of this book, in this chapter we introduce the basics of two of the most common numerical methods in their simplest forms. The reader is referred to dedicated books on numerical methods for further details (see, for example, Jaluria and Torrance, 1986; Johnson, 1987; Owen, 2009).

In Chapters 7–9, a biomedical process was replaced by its mathematical model, consisting of the governing differential equations and additional equations called the boundary conditions. The methods to solve these equations can be divided into two broad classes – analytical and numerical. Analytical solution refers to solutions that can be obtained using algebra and calculus. These are the majority of the solutions we learn in a first course in fluid mechanics or heat transfer. Analytical solutions are extremely powerful for understanding some of the fundamentals, but they typically require drastic simplification of the problem which restricts their applicability. Since numerical methods are significantly more flexible, almost all engineering analysis softwares are developed using such methods. Perhaps the two most popular numerical methods are the finite difference and the finite element methods. A gentle introduction to these two methods is presented in this chapter (see Figure 10.1).

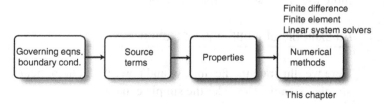

Figure 10.1

This chapter introduces techniques (numerical methods) for solving model equations.

10.1 Flexibility of numerical methods

Analytical solutions often place a severe restriction on the geometry, material properties, boundary conditions etc., of a model. For example, heat conduction in a slab is described by the governing equation

$$\frac{\partial T}{\partial t} = \alpha \frac{\partial^2 T}{\partial x^2} \qquad (10.1)$$

Consider the boundary conditions given by

$$\left.\frac{\partial T}{\partial x}\right|_{x=0,t} = 0 \quad \text{(from symmetry)} \qquad (10.2)$$

$$T(L, t > 0) = T_s \quad \text{(surface temperature is specified)} \qquad (10.3)$$

$$(10.4)$$

and the initial condition

$$T(x, t = 0) = T_i \qquad (10.5)$$

where T_i is the constant initial temperature and T_s is the constant temperature at the two surfaces of the slab at time $t > 0$. The analytical solution to Eqn. 10.1, for these boundary and initial conditions, is given by

$$\frac{T - T_s}{T_i - T_s} = \sum_{n=0}^{\infty} \frac{4(-1)^n}{(2n+1)\pi} \cos \frac{(2n+1)\pi x}{2L} e^{-\alpha\left(\frac{(2n+1)\pi}{2L}\right)^2 t} \qquad (10.6)$$

This solution makes the following assumptions: the slab is infinitely long in the other two dimensions; the thickness $2L$ does not change; all thermal properties (α) are constant throughout the material and remain constant as the temperature changes non-uniformly throughout the material; the initial condition of the slab is a constant temperature etc. Such restrictions make it quite difficult to apply analytical methods to obtain accurate answers for a real process. It will be shown that numerical solutions can routinely overcome such restrictions and are therefore more useful for solving real-life problems.

10.1.1 An example of a numerical solution

Here we illustrate the flexibility (and power) of a numerical solution using a very simple example. Consider the simple equation

$$\frac{df}{dx} = e^{x^2} \qquad (10.7)$$

10.1 Flexibility of numerical methods

which cannot be integrated directly, i.e., we cannot solve for $f(x)$ analytically. However, from the definition of a derivative,

$$\frac{df}{dx} \approx \frac{f(x+\Delta x) - f(x)}{\Delta x} \quad (10.8)$$

we can write (approximately)

$$f(x+\Delta x) = f(x) + \Delta x \frac{df}{dx} \quad (10.9)$$

This is an algebraic equation where, if we know the function f at x, we can find its value at $x + \Delta x$. Suppose we start from $x = 0$ where the function is known to be $f(0) = 1$. From Eqn. 10.9, assuming $\Delta x = 1$, we can calculate $f(x)$ at $x = 1$ as

$$f(1) = f(0) + (1)(e^{0^2})$$
$$= 1 + 1$$
$$= 2$$

We can repeat the process to calculate $f(2)$ as

$$f(2) = f(1) + (1)(e^{1^2})$$
$$= 4.718$$

We can continue this process to calculate $f(3)$ and so on. This is plotted in Figure 10.2.

If we now interpolate between the calculated points, we have f for all values of x, i.e., there was no need to know how to integrate Eqn. 10.7. We can make several

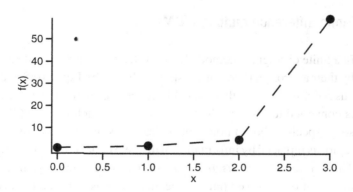

Figure 10.2

A numerical solution of Eqn. 10.7 (solid circles). The dashed lines show interpolation between the solid circles.

observations from this simple example, that can be generalized for numerical solutions:

- The process did not require any knowledge of analytical solution techniques, i.e., integration of $\int e^{x^2} dx$ in this case. Thus, numerical methods can work for arbitrary equations.
- We can improve the accuracy of the solution by simply taking smaller steps Δx. The smaller the Δx, the closer are the computed values to the true solution.
- As we take smaller Δx to improve the accuracy, computation time will increase for the same range of x values for which we are computing.

Biological processes often involve multiple materials or material properties that vary significantly with position, complex shapes, significant variation in material properties with time during the process and phenomena that couple more than one type of physics. Numerical methods routinely handle such complexities and are therefore often the preferred way to solve problems involving biological processes.

In the following sections, we will develop the two most commonly used numerical methods, which are the finite difference method and the finite element method, in their simplest versions, applying the methods to the one-dimensional governing heat equation and boundary conditions developed in Chapter 7. Both the finite difference method and the finite element method eventually convert the heat equation (a partial differential equation) to a set of algebraic equations. A short discussion on solving the set of algebraic equations is also included below. Our goal here is to see the complete solution process, starting from equations in Chapters 7 and 8, in its most elementary form.

10.2 Finite difference method (FDM)

In a finite difference method, the derivatives in a differential equation are replaced by their approximations, as illustrated already by Eqns. 10.8 and 10.9 in the previous section. Through this finite difference approximation, the differential equation is converted to a set of algebraic equations which is solved for the unknowns at some specific chosen points or nodes. Thus, information in a continuous spatial region is replaced by information at discrete nodal points covering the same region. This is referred to as *discretization* of the space. Similarly, time is discretized, i.e., the equation is solved only at specific times as opposed to solving for all times, as in an analytical solution. Information at other locations or times is obtained through interpolation of the nodal values with respect to space or time. The finite

10.2 Finite difference method (FDM)

difference method is perhaps the simplest of the numerical methods to solve differential equations and has been quite popular for simpler geometries. Numerous textbooks are available on finite difference methods (e.g., Jaluria and Torrance, 1986). The reader is referred to such excellent books for details and only a very limited introduction is provided here using a simple example.

10.2.1 FDM: A few common formulations

As illustrated in Figure 10.3 for a function $f(x)$, replacing the derivative of $f(x)$ at any point x, $\partial f / \partial x$, by its finite difference, can be achieved in a number of ways, three of which are:

forward difference

$$\left.\frac{\partial f}{\partial x}\right|_i \cong \frac{f_{i+1} - f_i}{\Delta x} \tag{10.10}$$

backward difference

$$\left.\frac{\partial f}{\partial x}\right|_i \cong \frac{f_i - f_{i-1}}{\Delta x} \tag{10.11}$$

central difference

$$\left.\frac{\partial f}{\partial x}\right|_i \cong \frac{f_{i+\frac{1}{2}} - f_{i-\frac{1}{2}}}{\Delta x} \tag{10.12}$$

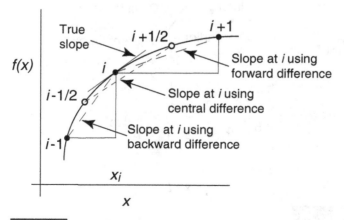

Figure 10.3

Graphical representation of various approximations of the derivative of a function at x_i. The forward, backward and central difference approximations shown in the figure are given by Eqns. 10.10, 10.11 and 10.12, respectively.

All three finite difference methods just described will give different numerical results for the derivative $\partial f/\partial x$, as illustrated in Figure 10.3, but as Δx decreases, the results become closer and closer.

10.3 FDM: converting the 1D heat equation to algebraic equations

As an illustration of using the finite difference method to solve a partial differential equation in heat and mass transfer applications, consider transient heating of a 1D slab-shaped material with known surface temperature, as shown in Figure 10.4. The governing equation and boundary conditions are therefore given by:

Governing equation

$$\frac{\partial T}{\partial t} = \alpha \frac{\partial^2 T}{\partial x^2} \tag{10.13}$$

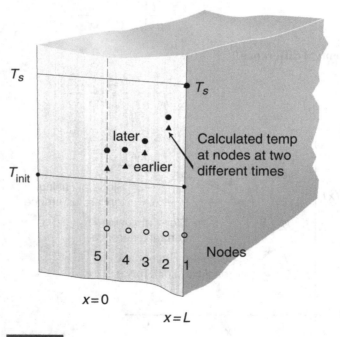

Figure 10.4

Illustration of discretization of a 1D slab geometry using a finite difference method into nodes (hollow circles) and computed temperatures at the nodes (solid circles) at two different times.

10.3 FDM: converting the 1D heat equation to algebraic equations

Boundary conditions

$$T(x = L) = T_s. \tag{10.14}$$

$$\frac{\partial T}{\partial x} = 0 \quad \text{at } x = 0 \tag{10.15}$$

Initial condition

$$T(t = 0) = T_{init} \tag{10.16}$$

The symmetry condition, given by Eqn. 10.15 is a statement that the slab is being heated or cooled symmetrically and therefore the temperature profile will be symmetric about the centerline. Since in numerical solution we are always trying to reduce the amount of computation, it would suffice to compute temperatures for one half of the slab thickness (the other half will have the mirror image of the same temperatures).

10.3.1 Discretization of the spatial term

The spatial derivative (right-hand side of Eqn. 10.13) is written using 10.12 as

$$\left.\frac{\partial^2 T}{\partial x^2}\right|_x = \frac{\partial}{\partial x}\left(\frac{\partial T}{\partial x}\right)$$

$$\cong \frac{\left.\frac{\partial T}{\partial x}\right|_{x+\Delta x/2} - \left.\frac{\partial T}{\partial x}\right|_{x-\Delta x/2}}{\Delta x} \tag{10.17}$$

The two terms in the numerator are further converted using Eqn. 10.12 as

$$\left.\frac{\partial^2 T}{\partial x^2}\right|_x \cong \frac{\frac{T_{x+\Delta x} - T_x}{\Delta x} - \frac{T_x - T_{x-\Delta x}}{\Delta x}}{\Delta x} \tag{10.18}$$

$$= \frac{T_{x+\Delta x} - 2T_x + T_{x-\Delta x}}{(\Delta x)^2} \tag{10.19}$$

10.3.2 Discretization of the transient term

Notice that the temperatures on the right-hand side of Eqn. 10.19, $T_x, T_{x+\Delta x}, T_{x-\Delta x}$, are also functions of time. Equation 10.13 can be discretized at location x using temperatures at discrete time level t

$$\left.\frac{\partial T}{\partial t}\right|_{x,t} = \alpha \left.\frac{\partial^2 T}{\partial x^2}\right|_{x,t} \tag{10.20}$$

or it can be discretized at the discrete time level $t + \Delta t$, as

$$\left.\frac{\partial T}{\partial t}\right|_{x,t+\Delta t} = \alpha \left.\frac{\partial^2 T}{\partial x^2}\right|_{x,t+\Delta t} \qquad (10.21)$$

Depending on which one of this we choose, the final equations will be different and we will have two alternative ways of computing.

If we choose to use temperatures from time level t, i.e., Eqn. 10.20, we can write the time derivative (left-hand side of Eqn. 10.13) at the discrete location x using the forward difference method (Eqn. 10.10) as

$$\left.\frac{\partial T}{\partial t}\right|_{x,t} \cong \frac{T_x^{t+\Delta t} - T_x^t}{\Delta t} \qquad (10.22)$$

10.3.3 Complete finite difference formulation of the 1D heat equation

The spatial term (right-hand side of Eqn. 10.13) at the discrete time level t is rewritten from Eqn. 10.19 as

$$\left.\frac{\partial^2 T}{\partial x^2}\right|_{x,t} = \frac{T_{x+\Delta x}^t - 2T_x^t + T_{x-\Delta x}^t}{(\Delta x)^2} \qquad (10.23)$$

Plugging Eqns. 10.22 and 10.23 into Eqn. 10.20, we get

$$\frac{T_x^{t+\Delta t} - T_x^t}{\Delta t} = \alpha \left[\frac{T_{x+\Delta x}^t - 2T_x^t + T_{x-\Delta x}^t}{(\Delta x)^2}\right] \qquad (10.24)$$

Since the values at time level t are known, the unknown in the above equation is $T_x^{t+\Delta t}$. Rewriting the equation in terms of the unknown,

$$T_x^{t+\Delta t} = T_x^t + \frac{\alpha \Delta t}{(\Delta x)^2} \left[T_{x+\Delta x}^t - 2T_x^t + T_{x-\Delta x}^t\right] \qquad (10.25)$$

We now change notation to denote the position x as i and $x+\Delta x$ and $x-\Delta x$ as $i+1$ and $i-1$, respectively. Likewise, for time, the notation is changed to superscript j, for time t, and $j+1$ for time $t + \Delta t$,

$$T_i^{j+1} = T_i^j + \frac{\alpha \Delta t}{(\Delta x)^2} \left[T_{i+1}^j - 2T_i^j + T_{i-1}^j\right] \qquad (10.26)$$

Referring to Figure 10.4, the boundary condition (Eqn. 10.14) is now used to set

$$T_1 = T_s \qquad (10.27)$$

10.3 FDM: converting the 1D heat equation to algebraic equations

and the initial condition (Eqn. 10.16) is used to set values at $t = 0$ which is denoted as time level 0 for all i.

$$T^0 = T_{\text{init}} \tag{10.28}$$

To find the temperature at the node next to the surface at the next time, T_2^1, Eqn. 10.26 is written for $i = 2$ and $j = 0$ as

$$T_2^1 = T_2^0 + \frac{\alpha \Delta t}{(\Delta x)^2} \left[T_3^0 - 2T_2^0 + T_1^0 \right] \tag{10.29}$$

Likewise, for node 3:

$$T_3^1 = T_3^0 + \frac{\alpha \Delta t}{(\Delta x)^2} \left[T_4^0 - 2T_3^0 + T_2^0 \right] \tag{10.30}$$

For node 4:

$$T_4^1 = T_4^0 + \frac{\alpha \Delta t}{(\Delta x)^2} \left[T_5^0 - 2T_4^0 + T_3^0 \right] \tag{10.31}$$

For Node 5 we use the other boundary condition (Eqn. 10.15)

$$\frac{T_5^1 - T_4^1}{\Delta x} = 0 \tag{10.32}$$

which leads to

$$T_5^1 = T_4^1 \tag{10.33}$$

Using the notation, the calculated quantities for two time steps are illustrated in Table 10.1.

Table 10.1 Illustration of temperatures at the first two time steps of the finite difference method.

(known) Initial $j=0$	(calculate) Δt $j=1$	(calculate) $2\Delta t$ $j=2$
T_1^0	T_1^1	T_1^2
T_2^0	T_2^1	T_2^2
T_3^0	T_3^1	T_3^2
T_4^0	T_4^1	T_4^2
T_5^0	T_5^1	T_5^2

Table 10.2 Illustration of computed numerical values of temperatures at the first two time steps of the finite difference method

(known) Initial $j=0$	(calculate) $t=0.5$ s $j=1$	(calculate) $t=1.0$ s $j=2$
0	0	0
37	36.982	36.963
37	37	36.999
37	37	37
37	37	37

For a sample calculation, consider a half-thickness of slab as 0.04 m which leads to $\Delta x = 0.01$ m and a time step of $\Delta t = 0.5$ s, initial temperature of 37 °C, boundary temperature of 0 °C, and thermal diffusivity of 10^{-7} m²/s. Note that the Δx and Δt were chosen arbitrarily (discussed in the next section). For these parameters, the computations lead to the numerical values shown in Table 10.2.

10.4 FDM: stability (limitations in choosing step sizes)

One characteristic of Eqn. 10.26 is that the computations will not be stable, i.e., they will provide garbage answers for some choices of Δx and Δt. This is referred to as *instability* of the computations for the particular formulation. Analysis shows that Δx and Δt need to be chosen such that the following condition is satisfied:

$$\frac{\alpha \Delta t}{(\Delta x)^2} \leq \frac{1}{2} \tag{10.34}$$

This condition implies that Δt has an upper limit. Choosing a smaller Δt also increases the *accuracy* of the solution. However, a smaller Δt requires more computation to reach the same final time t, thus increasing computation time. In practice, once the time step is small enough so that the computation is stable, choice of Δt is often a balance between accuracy and computing time.

10.5 FDM: summary

The finite difference method is a numerical approach to solve differential equations where the equations are solved at discrete positions and times, with interpolations at other locations (and times). It was shown that the governing partial differential equation for heat transfer in a 1D slab, i.e.,

$$\frac{\partial T}{\partial t} = \alpha \frac{\partial^2 T}{\partial x^2}$$

can be written, by replacing the derivatives with their finite difference approximation as (here i denotes location and j denotes time)

$$T_i^{j+1} = T_i^j + \frac{\alpha \Delta t}{(\Delta x)^2}\left[T_{i+1}^j - 2T_i^j + T_{i-1}^j\right]$$

where we solve for discrete values of temperature, T_i^j. To keep computations stable, the Δt and Δx in the equation need to be chosen such that

$$\frac{\alpha \Delta t}{\Delta x^2} \leq \frac{1}{2}$$

Like many other numerical methods, the finite difference method can adjust to most governing equations, geometry, boundary conditions and properties. In general, the smaller the Δt and Δx, the more accurate is the solution but the computing time is increased, often disproportionately. These comments are qualitatively similar for other numerical methods, such as the one described next.

10.6 Finite element method (FEM)

Like the finite difference method just described, another and certainly the most popular numerical method to solve partial differential equations originating in engineering is the finite element method. The finite element method has many advantages over the finite difference method – for example, it generally provides a greater flexibility for modeling complex geometries and when the geometry is changing (as in a moving boundary). It has been widely used in solving structural, mechanical, heat transfer and fluid dynamics problems.

10.7 FEM: converting the 1D heat equation to algebraic equations

Here we develop a simple finite element method for the same 1D heat transfer in a slab described in the previous section for the finite difference method. This is illustrated in Figure 10.5. The governing equations and boundary conditions are repeated here for convenience.

466 Solving the equations: numerical methods

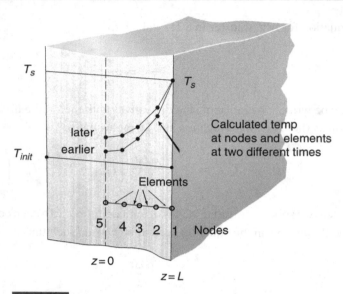

Figure 10.5

Illustration of discretization using a finite element method of a 1D slab geometry into nodes and elements. Temperatures computed by the method include not only the nodal values (solid circles), but also the interpolated values in between (solid lines joining the circles).

Governing equation

$$\frac{\partial T}{\partial t} = \alpha \frac{\partial^2 T}{\partial z^2} \tag{10.35}$$

Boundary conditions

$$T(z = L) = T_s \tag{10.36}$$

$$\frac{\partial T}{\partial z} = 0 \quad \text{at } z = 0 \tag{10.37}$$

Initial condition

$$T(t = 0) = T_{\text{init}} \tag{10.38}$$

10.7.1 Discretization of the spatial term

Due to the symmetry in the slab, we only need to compute for one half of the slab. As shown in Figure 10.5, the half thickness is divided into a number of nodes (illustration shows five). Let $\overline{T}_i (i = 1, 2, \ldots, n)$ be the approximate solution (as opposed to the true solution T that is assumed to be unavailable). Thus the T_is are the nodal unknowns (temperatures) that we will solve for using this numerical method.

10.7 FEM: converting the 1D heat equation to algebraic equations

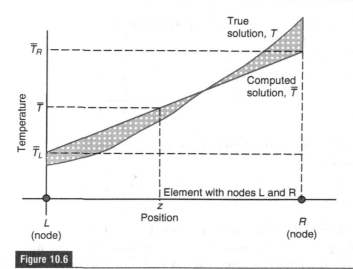

Figure 10.6

Illustration of one two-noded finite element from Figure 10.5 with computed solution and true solution, showing the error (shaded region) as the difference between the two solutions.

Temperature values in between the nodes will be approximated using a straight line, but unlike the finite difference method where this interpolation is accomplished after the solution, here the interpolation scheme is built into the solution process. Since $\overline{T}$ is an approximate solution, it will, in general, not satisfy Eqn. 10.35 exactly. This is illustrated in Figure 10.6, where one of the elements from Figure 10.5 is shown enlarged, with nodes at $z = L$ (left node) and $z = R$ (right node). Note the difference between the true solution, T, and the computed approximate solution, $\overline{T}$, denoted as the shaded region, which will be the error in the computed solution.

Using linear interpolation between the computed nodal temperatures, $\overline{T}_L$ and $\overline{T}_R$, any other temperature $\overline{T}$ within the element can be expressed. We can write

$$\frac{\overline{T} - \overline{T}_L}{z - L} = \frac{\overline{T}_R - \overline{T}_L}{R - L} \tag{10.39}$$

which is solved for $\overline{T}$ as

$$\begin{aligned}
\overline{T} &= \overline{T}_L + \frac{z - L}{R - L}\left(\overline{T}_R - \overline{T}_L\right) \\
&= \left(1 - \frac{z - L}{R - L}\right)\overline{T}_L + \frac{z - L}{R - L}\overline{T}_R \\
&= \frac{R - z}{R - L}\overline{T}_L + \frac{z - L}{R - L}\overline{T}_R \\
&= p_L \overline{T}_L + p_R \overline{T}_R
\end{aligned} \tag{10.40}$$

where p_L and p_R are the two polynomials given by

$$p_L = \frac{R-z}{R-L} \quad (10.41)$$

$$p_R = \frac{z-L}{R-L} \quad (10.42)$$

that are used to interpolate between $\overline{T}_L$ and $\overline{T}_R$. Equation 10.40 can also be written in a general form as

$$\overline{T}(z,t) = \sum_{i=1,n} p_i(z)\overline{T}_i(t) \quad (10.43)$$

where i refers to the nodal points so that the $\overline{T}_i$s are the temperatures at node i and the p_is are the polynomial terms at any node i. The polynomials, p_is are functions of z while the temperature at any node, $\overline{T}_i$, can change with time in this transient problem.

Let $\tilde{\epsilon} = T - \overline{T}$ denote the error (also called the residual) at any location and time, as illustrated in Figure 10.6. Thus,

$$\tilde{\epsilon} = \frac{\partial \overline{T}}{\partial t} - \alpha \frac{\partial^2 \overline{T}}{\partial z^2} \quad (10.44)$$

so that $\tilde{\epsilon}$ will be zero if $\overline{T} = T$, the true solution (since T satisfies Eqn. 10.35). One finite element method, known as the Galerkin method, sets the weighted error to zero, i.e.,

$$\int w_i \tilde{\epsilon} \, dz = 0 \quad (10.45)$$

Here w_i is the weighting factor at a node i. Thus, the error $\tilde{\epsilon}$ is made zero on average, but not necessarily at every location. A common practice is to set the weights w_i to be equal to the p_is, the polynomials (for example, p_L and p_R, given by Eqns. 10.41 and 10.42, respectively, for the element shown in Figure 10.6). Substituting for w_i and $\tilde{\epsilon}$, Eqn. 10.45 becomes

$$\int_{\text{length}} p_i \left[\frac{\partial \overline{T}}{\partial t} - \alpha \frac{\partial^2 \overline{T}}{\partial z^2} \right] dz = 0 \quad (10.46)$$

For an element e

$$\int_e p_i \left[\frac{\partial \overline{T}^e}{\partial t} - \alpha \frac{\partial^2 \overline{T}^e}{\partial z^2} \right] dz = 0 \quad i = 1, 2, \ldots, n \quad (10.47)$$

10.7 FEM: converting the 1D heat equation to algebraic equations

For the two nodes L and R, Eqn. 10.47 becomes

$$\int_L^R p_L \left[\frac{\partial \overline{T}}{\partial t} - \alpha \frac{\partial^2 \overline{T}}{\partial z^2} \right] dz = 0 \qquad (10.48)$$

$$\int_L^R p_R \left[\frac{\partial \overline{T}}{\partial t} - \alpha \frac{\partial^2 \overline{T}}{\partial z^2} \right] dz = 0 \qquad (10.49)$$

Since we have shown for the two-noded element,

$$\overline{T} = p_L \overline{T}_L + p_R \overline{T}_R \qquad (10.50)$$

we can take the derivative with respect to time as

$$\frac{\partial \overline{T}}{\partial t} = p_L \frac{\partial \overline{T}_L}{\partial t} + p_R \frac{\partial \overline{T}_R}{\partial t} \qquad (10.51)$$

Substituting Eqn. 10.51 into Eqn. 10.48 (remembering that $\overline{T}_L$ and $\overline{T}_R$ are not functions of z)

$$\frac{\partial \overline{T}_L}{\partial t} \left[\int p_L^2 dz \right] + \frac{\partial \overline{T}_R}{\partial t} \left[\int p_L p_R dz \right] - \alpha \int p_L \frac{\partial^2 \overline{T}}{\partial z^2} dz = 0 \qquad (10.52)$$

To evaluate the third term, we note that

$$\frac{\partial}{\partial z} \left(p_L \frac{\partial \overline{T}}{\partial z} \right) = \frac{\partial p_L}{\partial z} \frac{\partial \overline{T}}{\partial z} + p_L \frac{\partial^2 \overline{T}}{\partial z^2}$$

$$p_L \frac{\partial^2 \overline{T}}{\partial z^2} = \frac{\partial}{\partial z} \left(p_L \frac{\partial \overline{T}}{\partial z} \right) - \frac{\partial p_L}{\partial z} \frac{\partial \overline{T}}{\partial z}$$

Therefore, integrating both sides,

$$\int p_L \frac{\partial^2 \overline{T}}{\partial z^2} dz = \int_L^R \frac{\partial}{\partial z} \left(p_L \frac{\partial \overline{T}}{\partial z} \right) dz - \int_L^R \frac{\partial p_L}{\partial z} \frac{\partial \overline{T}}{\partial z} dz$$

$$= p_L \frac{\partial \overline{T}}{\partial z} \bigg|_R - p_L \frac{\partial \overline{T}}{\partial z} \bigg|_L - \int_L^R \frac{\partial p_L}{\partial z} \left[\overline{T}_L \frac{\partial p_L}{\partial z} + \overline{T}_R \frac{\partial p_R}{\partial z} \right] dz$$

$$= p_L \frac{\partial \overline{T}}{\partial z} \bigg|_R - p_L \frac{\partial \overline{T}}{\partial z} \bigg|_L - \overline{T}_L \int_L^R \left(\frac{\partial p_L}{\partial z} \right)^2 dz$$

$$- \overline{T}_R \int_L^R \left(\frac{\partial p_L}{\partial z} \right) \left(\frac{\partial p_R}{\partial z} \right) dz$$

$$= -p_L \frac{\partial \overline{T}}{\partial z} \bigg|_L - \overline{T}_L \int_L^R \left(\frac{\partial p_L}{\partial z} \right)^2 dz - \overline{T}_R \int_L^R \left(\frac{\partial p_L}{\partial z} \right) \left(\frac{\partial p_R}{\partial z} \right) dz$$

$$(10.53)$$

Note that $p_L(R) = 0$ has been used in the right-hand side of the last equation above. Substituting Eqn. 10.53 into Eqn. 10.52, we get

$$\frac{\partial \overline{T}_L}{\partial t}\left[\int p_L^2 dz\right] + \frac{\partial \overline{T}_R}{\partial t}\left[\int p_L p_R dz\right] + \alpha p_L \left.\frac{\partial \overline{T}}{\partial z}\right|_L + \overline{T}_L \alpha \int_L^R \left(\frac{\partial p_L}{\partial z}\right)^2 dz$$
$$+ \overline{T}_R \alpha \int_L^R \left(\frac{\partial p_L}{\partial z}\right)\left(\frac{\partial p_R}{\partial z}\right) dz = 0 \qquad (10.54)$$

Equation 10.54 is the transformed version of Eqn. 10.48. Likewise, Eqn. 10.49 can be transformed by substituting for $\overline{T}$ from Eqn. 10.50 and carrying out the integration to obtain

$$\frac{\partial \overline{T}_L}{\partial t}\left[\int p_L p_R dz\right] + \frac{\partial \overline{T}_R}{\partial t}\left[\int p_R^2 dz\right]$$
$$- \alpha p_R \left.\frac{\partial \overline{T}}{\partial z}\right|_R + \overline{T}_L \alpha \int_L^R \left(\frac{\partial p_L}{\partial z}\right)\left(\frac{\partial p_R}{\partial z}\right) dz + \overline{T}_R \alpha \int_L^R \left(\frac{\partial p_R}{\partial z}\right)^2 dz = 0$$
$$(10.55)$$

In matrix form, Eqns. 10.54 and 10.55 are collected as

$$\begin{bmatrix} \int p_L^2 dz & \int p_L p_R dz \\ \int p_L p_R dz & \int p_R^2 dz \end{bmatrix} \begin{bmatrix} \frac{\partial \overline{T}_L}{\partial t} \\ \frac{\partial \overline{T}_R}{\partial t} \end{bmatrix}$$
$$+ \begin{bmatrix} \alpha \int_L^R \left(\frac{\partial p_L}{\partial z}\right)^2 dz & \alpha \int_L^R \left(\frac{\partial p_L}{\partial z}\right)\left(\frac{\partial p_R}{\partial z}\right) dz \\ \alpha \int_L^R \left(\frac{\partial p_L}{\partial z}\right)\left(\frac{\partial p_R}{\partial z}\right) dz & \alpha \int_L^R \left(\frac{\partial p_R}{\partial z}\right)^2 dz \end{bmatrix} \begin{bmatrix} \overline{T}_L \\ \overline{T}_R \end{bmatrix}$$
$$= \begin{bmatrix} -\alpha p_L \left.\frac{\partial \overline{T}}{\partial z}\right|_L \\ \alpha p_R \left.\frac{\partial \overline{T}}{\partial z}\right|_R \end{bmatrix} \qquad (10.56)$$

In a compact form, we can write the above matrix having the equations for one element as

$$[c]^e \frac{\partial [\overline{T}]^e}{\partial t} + [k]^e [\overline{T}]^e = [b]^e \qquad (10.57)$$

10.7 FEM: converting the 1D heat equation to algebraic equations

where the superscript e stands for element, $[\overline{T}] = \begin{bmatrix} \overline{T}_L \\ \overline{T}_R \end{bmatrix}$ and c and b are the matrices that represent the corresponding items in Eqn. 10.56. We recall that p_L and p_R are given by:

$$p_L = \frac{z-R}{L-R}, \quad p_R = \frac{z-L}{R-L}$$

The integrations in Eqn. 10.56, such as $\int p_L^2 dz$, are simple. These are performed as below:

$$\int_L^R p_L^2 dz = \int_L^R \frac{(z-R)^2}{(L-R)^2} dz$$

$$= \frac{1}{(L-R)^2} \left[\frac{z^3}{3} - z^2 R + R^2 z \right]_L^R$$

$$= \frac{1}{(L-R)^2} \left[\frac{R^3}{3} - R^3 + R^3 - \frac{L^3}{3} - L^2 R + R^2 L \right]$$

$$= \frac{1}{(L-R)^2} \left[\frac{R^3 - 3L^2R + 3R^2L - L^3}{3} \right]$$

$$= \frac{1}{(L-R)^2} \left[\frac{-(L-R)^3}{3} \right]$$

$$= \frac{R-L}{3}$$

$$\int_L^R p_R^2 dz = \int_L^R \frac{(z-L)^2}{(R-L)^2} dz$$

$$= \frac{1}{(R-L)^2} \left[\frac{z^3}{3} - z^2 L + L^2 z \right]_L^R$$

$$= \frac{1}{(R-L)^2} \left[\frac{R^3}{3} - R^2 L + L^2 R - \frac{L^3}{3} - L^3 + L^3 \right]$$

$$= \frac{1}{(R-L)^2} \left[\frac{R^3 - 3R^2L + 3L^2R - L^3}{3} \right]$$

$$= \frac{1}{(R-L)^2} \left[\frac{(R-L)^3}{3} \right]$$

$$= \frac{R-L}{3}$$

Solving the equations: numerical methods

$$\int_L^R p_L p_R \, dz = \int_L^R \left(\frac{z-R}{L-R}\right)\left(\frac{z-L}{R-L}\right) dz$$

$$= -\frac{1}{(R-L)^2}\left[\frac{z^3}{3} - (L+R)\frac{z^2}{2} + LRz\right]_L^R$$

$$= -\frac{1}{(R-L)^2}\left[\frac{R^3}{3} - (L+R)\frac{R^2}{2} + LR^2 - \frac{L^3}{3} + (L+R)\frac{L^2}{2} - L^2R\right]$$

$$= -\frac{1}{(R-L)^2}\left[\frac{2R^3 - 3LR^2 - 3R^3 + 6LR^2 - 2L^3 + 3L^3 + 3L^2R - 6L^2R}{6}\right]$$

$$= \frac{1}{(R-L)^2}\left[\frac{R^3 - 3LR^2 + 3L^2R - L^3}{6}\right]$$

$$= \frac{1}{(R-L)^2}\frac{(R-L)^3}{6}$$

$$= \frac{R-L}{6}$$

$$\int_L^R \left(\frac{\partial p_L}{\partial z}\right)^2 dz = \int_L^R \left(\frac{1}{L-R}\right)^2 dz$$

$$= \frac{1}{(L-R)^2}(R-L)$$

$$= \frac{1}{R-L}$$

$$\int_L^R \left(\frac{\partial p_R}{\partial z}\right)^2 dz = \int_L^R \left(\frac{1}{R-L}\right)^2 dz$$

$$= \frac{1}{(R-L)^2}(R-L)$$

$$= \frac{1}{R-L}$$

$$\int_L^R \left(\frac{\partial p_L}{\partial z}\right)\left(\frac{\partial p_R}{\partial z}\right) dz = \int_L^R \frac{1}{(L-R)(R-L)} dz$$

$$= -\frac{R-L}{(R-L)^2}$$

$$= -\frac{1}{R-L}$$

We then substitute the values of the integrals just obtained into Equation 10.56 to obtain the two equations for one element. Writing this for element 1,

10.7 FEM: converting the 1D heat equation to algebraic equations

Element 1

$$\begin{bmatrix} \frac{h_1}{3} & \frac{h_1}{6} \\ \frac{h_1}{6} & \frac{h_1}{3} \end{bmatrix} \begin{bmatrix} \frac{d\overline{T}_1}{dt} \\ \frac{d\overline{T}_2}{dt} \end{bmatrix} + \begin{bmatrix} \frac{\alpha}{h_1} & -\frac{\alpha}{h_1} \\ -\frac{\alpha}{h_1} & \frac{\alpha}{h_1} \end{bmatrix} \begin{bmatrix} \overline{T}_1 \\ \overline{T}_2 \end{bmatrix} = \begin{bmatrix} -\alpha \frac{\partial \overline{T}}{\partial z}\big|_1 \\ \alpha \frac{\partial \overline{T}}{\partial z}\big|_2 \end{bmatrix}$$

where $h_1 = R_1 - L_1$, $\overline{T}_1 = \overline{T}_L$ and $\overline{T}_2 = \overline{T}_R$. Similarly, we can write the two equations for element 2 as

Element 2

$$\begin{bmatrix} \frac{h_2}{3} & \frac{h_2}{6} \\ \frac{h_2}{6} & \frac{h_2}{3} \end{bmatrix} \begin{bmatrix} \frac{d\overline{T}_2}{dt} \\ \frac{d\overline{T}_3}{dt} \end{bmatrix} + \begin{bmatrix} \frac{\alpha}{h_2} & -\frac{\alpha}{h_2} \\ -\frac{\alpha}{h_2} & \frac{\alpha}{h_2} \end{bmatrix} \begin{bmatrix} \overline{T}_2 \\ \overline{T}_3 \end{bmatrix} = \begin{bmatrix} -\alpha \frac{\partial \overline{T}}{\partial z}\big|_2 \\ \alpha \frac{\partial \overline{T}}{\partial z}\big|_3 \end{bmatrix}$$

where $h_2 = R_2 - L_2$. If we skip the matrix notation, the equations for the elements are written as

Element 1

$$\frac{h_1}{3} \frac{d\overline{T}_1}{dt} + \frac{h_1}{6} \frac{d\overline{T}_2}{dt} + \frac{\alpha}{h_1} \overline{T}_1 - \frac{\alpha}{h_1} \overline{T}_2 = -\alpha \frac{\partial \overline{T}}{\partial z}\bigg|_1 \tag{10.58}$$

$$\frac{h_1}{6} \frac{d\overline{T}_1}{dt} + \frac{h_1}{3} \frac{d\overline{T}_2}{dt} - \frac{\alpha}{h_1} \overline{T}_1 + \frac{\alpha}{h_1} \overline{T}_2 = \alpha \frac{\partial \overline{T}}{\partial z}\bigg|_2 \tag{10.59}$$

Element 2

$$\frac{h_2}{3} \frac{d\overline{T}_2}{dt} + \frac{h_2}{6} \frac{d\overline{T}_3}{dt} + \frac{\alpha}{h_2} \overline{T}_2 - \frac{\alpha}{h_2} \overline{T}_3 = -\alpha \frac{\partial \overline{T}}{\partial z}\bigg|_2 \tag{10.60}$$

$$\frac{h_2}{6} \frac{d\overline{T}_2}{dt} + \frac{h_2}{3} \frac{d\overline{T}_3}{dt} - \frac{\alpha}{h_2} \overline{T}_2 + \frac{\alpha}{h_2} \overline{T}_3 = \alpha \frac{\partial \overline{T}}{\partial z}\bigg|_3 \tag{10.61}$$

Notice that terms such as $-\alpha \frac{\partial \overline{T}}{\partial z}\big|_1$ are unknowns, but we can eliminate them (except the first and the last one) by combining equations. Thus Eqns. 10.59 and 10.60 can be combined to eliminate $\alpha \frac{\partial \overline{T}}{\partial z}\big|_2$, and so on. Let us try to do this in an organized way (which is what is needed if you really want to write the computer code for the solution). Equations 10.59 and 10.60 from the equations for the first two elements are added to obtain

$$\frac{h_1}{6} \frac{d\overline{T}_1}{dt} + \left(\frac{h_1}{3} + \frac{h_2}{3}\right) \frac{d\overline{T}_2}{dt} + \frac{h_2}{6} \frac{d\overline{T}_3}{dt} - \frac{\alpha}{h_1} \overline{T}_1 + \left(\frac{\alpha}{h_1} + \frac{\alpha}{h_2}\right) \overline{T}_2 - \frac{\alpha}{h_2} \overline{T}_3 = 0 \tag{10.62}$$

This leads to three equations for the first two elements which are now written in a matrix form as

$$\begin{bmatrix} \frac{h_1}{3} & \frac{h_1}{6} & 0 \\ \frac{h_1}{6} & \frac{h_1}{3}+\frac{h_2}{3} & \frac{h_2}{6} \\ 0 & \frac{h_2}{6} & \frac{h_2}{3} \end{bmatrix} \begin{bmatrix} \frac{d\overline{T}_1}{dt} \\ \frac{d\overline{T}_2}{dt} \\ \frac{d\overline{T}_3}{dt} \end{bmatrix} + \begin{bmatrix} \frac{\alpha}{h_1} & -\frac{\alpha}{h_1} & 0 \\ -\frac{\alpha}{h_1} & \frac{\alpha}{h_1}+\frac{\alpha}{h_2} & -\frac{\alpha}{h_2} \\ 0 & -\frac{\alpha}{h_2} & \frac{\alpha}{h_2} \end{bmatrix} \begin{bmatrix} \overline{T}_1 \\ \overline{T}_2 \\ \overline{T}_3 \end{bmatrix} = \begin{bmatrix} -\alpha\frac{\partial \overline{T}}{\partial z}\big|_1 \\ 0 \\ \alpha\frac{\partial \overline{T}}{\partial z}\big|_3 \end{bmatrix}$$

(10.63)

To repeat, these equations are for the first two elements (three nodes). If we combine all the equations for the $n-1$ elements (n nodes) into one global matrix, we get

$$\underset{n \times n}{[C]} \underset{n \times 1}{\left[\frac{d\overline{T}}{dt}\right]} + \underset{n \times n}{[K]} \underset{n \times 1}{[\overline{T}]} = \underset{n \times 1}{[B]} \qquad (10.64)$$

The dimensions of the matrices are shown underneath each term. Notice that Eqn. 10.64 is an ordinary differential equation in time. The derivatives in spatial dimension, z, have been eliminated. Also, there are n equations in it, with n unknowns, $\overline{T}_1, \overline{T}_2, \ldots, \overline{T}_n$. Our task now is to convert the time derivative also into an algebraic expression. This is easily done using the finite difference method described in the earlier section of this chapter.

10.7.2 Discretizing the transient term

The transient term, $d\overline{T}/dt$ in Eqn. 10.64, is often discretized using some form of finite difference method, discussed earlier in Section 10.2. Two of the most common forms of finite difference method are the explicit and the implicit methods, depending on how we treat the terms containing $\overline{T}$, other than the derivative in Eqn. 10.64.

Explicit (also called forward) method The explicit method is the one discussed in Section 10.2. Here temperature in terms other than the one containing the derivative $d\overline{T}/dt$ are taken at the current time level, t, i.e., in terms of the known temperatures, so that we can write

$$[C]\frac{[\overline{T}^{t+\Delta t}] - [\overline{T}^t]}{\Delta t} + [K][\overline{T}^t] = [B] \qquad (10.65)$$

10.7 FEM: converting the 1D heat equation to algebraic equations

which is simplified as

$$[C][\overline{T}^{t+\Delta t}] = [[C] - \Delta t[K]][\overline{T}^{t}] + \Delta t[B] \qquad (10.66)$$

This method offers significant computational savings since temperatures are computed directly (contrasted with the implicit method described below). However, this method is conditionally stable, i.e., the time step has to be smaller than a critical value, as discussed in Section 10.4. Use of a time step larger than the critical value can lead to unstable computational results that are garbage.

Implicit (also called backward) method Here temperature in terms other than the one containing the derivative $d\overline{T}/dt$ are taken at the new time level, $t + \Delta t$, i.e., in terms of the unknown temperatures, so that we write

$$[C]\frac{[\overline{T}^{t+\Delta t}] - [\overline{T}^{t}]}{\Delta t} + [K][\overline{T}^{t+\Delta t}] = [B] \qquad (10.67)$$

which is simplified as

$$[[C] + \Delta t[K]][\overline{T}^{t+\Delta t}] = [C][\overline{T}^{t}] + \Delta t[B] \qquad (10.68)$$

This method requires more computations but is unconditionally stable, i.e., time steps can be as large as possible. However, too large a time step introduces spurious oscillations. Thus, for both the implicit and the explicit methods, time step selection comes from trial and error for a particular problem.

Since the implicit method is used more commonly, further development of the finite element method in this chapter will use the implicit method. Note that at this stage, the governing differential equation for heat transfer has been completely converted into a set of algebraic equations, to be solved for the unknowns $\overline{T}_1, \overline{T}_2, \ldots, \overline{T}_n$.

10.7.3 Inclusion of boundary conditions

Note that the set of linear algebraic equations given by Eqn. 10.63 came from the governing equation. We have not yet used the two boundary conditions. Note also that the quantities $-\alpha \partial T/\partial z|_1$ and $-\alpha \partial T/\partial z|_n$ in the first and last equation are still unknowns and need to be known before we can solve the set of equations given by matrix 10.63. *The boundary conditions will provide these two additional unknowns.* Using the implicit formulation (Eqn. 10.68), if we rewrite the matrix 10.63, we get

$$\begin{bmatrix} \frac{h_1}{3} + \Delta t \frac{\alpha}{h_1} & \frac{h_1}{6} - \Delta t \frac{\alpha}{h_1} & 0 \\ \frac{h_1}{6} - \Delta t \frac{\alpha}{h_1} & \frac{h_1}{3} + \frac{h_2}{3} + \Delta t \left(\frac{\alpha}{h_1} + \frac{\alpha}{h_2} \right) & \frac{h_2}{6} - \Delta t \frac{\alpha}{h_2} \\ 0 & \frac{h_2}{6} - \Delta t \frac{\alpha}{h_2} & \frac{h_2}{3} + \Delta t \frac{\alpha}{h_2} \end{bmatrix} \begin{bmatrix} \overline{T}_1^{t+\Delta t} \\ \overline{T}_2^{t+\Delta t} \\ \overline{T}_3^{t+\Delta t} \end{bmatrix}$$

$$= \begin{bmatrix} \frac{h_1}{3} & \frac{h_1}{6} & 0 \\ \frac{h_1}{6} & \frac{h_1}{3} + \frac{h_2}{3} & \frac{h_2}{6} \\ 0 & \frac{h_2}{6} & \frac{h_2}{3} \end{bmatrix} \begin{bmatrix} \overline{T}_1^t \\ \overline{T}_2^t \\ \overline{T}_3^t \end{bmatrix} + \begin{bmatrix} -\Delta t\, \alpha \frac{\partial \overline{T}}{\partial z}\Big|_1 \\ 0 \\ \Delta t \alpha \frac{\partial \overline{T}}{\partial z}\Big|_3 \end{bmatrix} \qquad (10.69)$$

To introduce the boundary conditions, we consider the three types of boundary conditions separately

Temperature specified at the boundary This temperature at the boundary can be used to reduce the number of equations to be solved. This, however, turns out to be not as convenient for practical implementation, as it involves major restructuring of computer storage. Instead, the following is done. Let us say the boundary condition is given by

$$\overline{T}_1^{t+\Delta t} = \overline{T}_1^t = \overline{T}_s$$

where $\overline{T}_s$ is the surface temperature. Now since we know the value of $\overline{T}_1$, we do not need to solve Equation 10.58. We therefore make appropriate changes in the first row of matrices to incorporate the above equation and the system of equations changes to:

$$\begin{bmatrix} 1 & 0 & 0 \\ \frac{h_1}{6} - \Delta t \frac{\alpha}{h_1} & \frac{h_1}{3} + \frac{h_2}{3} + \Delta t \left(\frac{\alpha}{h_1} + \frac{\alpha}{h_2} \right) & \frac{h_2}{6} - \Delta t \frac{\alpha}{h_2} \\ 0 & \frac{h_2}{6} - \Delta t \frac{\alpha}{h_2} & \frac{h_2}{3} + \Delta t \frac{\alpha}{h_2} \end{bmatrix} \begin{bmatrix} \overline{T}_1^{t+\Delta t} \\ \overline{T}_2^{t+\Delta t} \\ \overline{T}_3^{t+\Delta t} \end{bmatrix}$$

$$= \begin{bmatrix} 1 & 0 & 0 \\ \frac{h_1}{6} & \frac{h_1}{3} + \frac{h_2}{3} & \frac{h_2}{6} \\ 0 & \frac{h_2}{6} & \frac{h_2}{3} \end{bmatrix} \begin{bmatrix} \overline{T}_s \\ \overline{T}_2^t \\ \overline{T}_3^t \end{bmatrix} + \begin{bmatrix} 0 \\ 0 \\ \Delta t \alpha \frac{\partial \overline{T}}{\partial z}|_3 \end{bmatrix} \qquad (10.70)$$

Now we rewrite the matrices by bringing the terms related to $T_1^{t+\Delta t}$ in all other equations to the right side and add those to the respective elements of $[B]$. Thus the new matrix system becomes:

10.7 FEM: converting the 1D heat equation to algebraic equations

$$\begin{bmatrix} 1 & 0 & 0 \\ 0 & \frac{h_1}{3}+\frac{h_2}{3}+\Delta t\left(\frac{\alpha}{h_1}+\frac{\alpha}{h_2}\right) & \frac{h_2}{6}-\Delta t\frac{\alpha}{h_2} \\ 0 & \frac{h_2}{6}-\Delta t\frac{\alpha}{h_2} & \frac{h_2}{3}+\Delta t\frac{\alpha}{h_2} \end{bmatrix} \begin{bmatrix} \overline{T}_1^{t+\Delta t} \\ \overline{T}_2^{t+\Delta t} \\ \overline{T}_3^{t+\Delta t} \end{bmatrix}$$

$$= \begin{bmatrix} 1 & 0 & 0 \\ 0 & \frac{h_1}{3}+\frac{h_2}{2} & \frac{h_2}{6} \\ 0 & \frac{h_2}{6} & \frac{h_2}{3} \end{bmatrix} \begin{bmatrix} \overline{T}_s \\ \overline{T}_2^t \\ \overline{T}_3^t \end{bmatrix} + \begin{bmatrix} 0 \\ \Delta t\frac{\alpha}{h_1}\overline{T}_s \\ \Delta t\alpha\frac{\partial \overline{T}}{\partial z}|_3 \end{bmatrix} \quad (10.71)$$

Heat flux specified at the boundary This boundary condition is given by

$$-k\frac{\partial \overline{T}}{\partial z}\bigg|_1 = q_s''$$

and is handled in the most straightforward way since the term in [B] is written, using the boundary condition, as

$$-\alpha\frac{\partial \overline{T}}{\partial z}\bigg|_1 = \frac{1}{\rho c_p}\left(-k\frac{\partial \overline{T}}{\partial z}\bigg|_1\right)$$

$$= \frac{1}{\rho c_p}q_s''$$

where q_s'' is the specified heat flux at the boundary.

Convective boundary condition The convective boundary condition is given by

$$-k\frac{\partial \overline{T}}{\partial z}\bigg|_1 = h(\overline{T}_1 - \overline{T}_\infty)$$

For a transient problem, this term is handled slightly differently, depending on whether the time derivative is treated in an explicit or implicit way (see Section 10.7.2). For an explicit formulation

$$-k\frac{\partial \overline{T}}{\partial z}\bigg|_1^t = h(\overline{T}_1^t - \overline{T}_\infty)$$

Thus, $-\alpha \partial T/\partial z|_1$ in the $[B]$ matrix is replaced by $-\frac{h}{\rho c_p} T_\infty$ and the coefficient of T_1^t in the $[K]$ matrix is replaced as shown below:

$$\begin{bmatrix} \frac{h_1}{3} + \Delta t \frac{\alpha}{h_1} & \frac{h_1}{6} - \Delta t \frac{\alpha}{h_1} & 0 \\ \frac{h_1}{6} - \Delta t \frac{\alpha}{h_1} & \frac{h_1}{3} + \frac{h_2}{3} + \Delta t \left(\frac{\alpha}{h_1} + \frac{\alpha}{h_2}\right) & \frac{h_2}{6} - \Delta t \frac{\alpha}{h_2} \\ 0 & \frac{h_2}{6} - \Delta t \frac{\alpha}{h_2} & \frac{h_2}{3} + \Delta t \frac{\alpha}{h_2} \end{bmatrix} \begin{bmatrix} \overline{T}_1^{t+\Delta t} \\ \overline{T}_2^{t+\Delta t} \\ \overline{T}_3^{t+\Delta T} \end{bmatrix}$$

$$= \begin{bmatrix} \frac{h_1}{3} + \frac{h}{\rho c_p} \Delta t & \frac{h_1}{6} & 0 \\ \frac{h_1}{6} & \frac{h_1}{3} + \frac{h_2}{3} & \frac{h_2}{6} \\ 0 & \frac{h_2}{6} & \frac{h_2}{3} \end{bmatrix} \begin{bmatrix} \overline{T}_1^t \\ \overline{T}_2^t \\ \overline{T}_3^t \end{bmatrix} + \begin{bmatrix} -\Delta t \frac{h}{\rho c_p} \overline{T}_\infty \\ 0 \\ \Delta t \alpha \frac{\partial \overline{T}}{\partial z}|_3 \end{bmatrix} \quad (10.72)$$

For an implicit formulation

$$-k \frac{\partial \overline{T}}{\partial z}\bigg|_1^{t+\Delta t} = h\left(\overline{T}_1^{t+\Delta t} - \overline{T}_\infty\right)$$

The coefficient of $T_1^{t+\Delta t}$ is changed instead and the resulting matrix is

$$\begin{bmatrix} \frac{h_1}{3} + \Delta t \frac{\alpha}{h_1} - \frac{h}{\rho c_p} \Delta t & \frac{h_1}{6} - \Delta t \frac{\alpha}{h_1} & 0 \\ \frac{h_1}{6} - \Delta t \frac{\alpha}{h_1} & \frac{h_1}{3} + \frac{h_2}{3} + \Delta t \left(\frac{\alpha}{h_1} + \frac{\alpha}{h_2}\right) & \frac{h_2}{6} - \Delta t \frac{\alpha}{h_2} \\ 0 & \frac{h_2}{6} - \Delta t \frac{\alpha}{h_2} & \frac{h_2}{3} + \Delta t \frac{\alpha}{h_2} \end{bmatrix} \begin{bmatrix} \overline{T}_1^{t+\Delta t} \\ \overline{T}_2^{t+\Delta t} \\ \overline{T}_3^{t+\Delta t} \end{bmatrix}$$

$$= \begin{bmatrix} \frac{h_1}{3} & \frac{h_1}{6} & 0 \\ \frac{h_1}{6} & \frac{h_1}{3} + \frac{h_2}{3} & \frac{h_2}{6} \\ 0 & \frac{h_2}{6} & \frac{h_2}{3} \end{bmatrix} \begin{bmatrix} \overline{T}_1^t \\ \overline{T}_2^t \\ \overline{T}_3^t \end{bmatrix} + \begin{bmatrix} -\Delta t \frac{h}{\rho c_p} \overline{T}_\infty \\ 0 \\ \Delta t \alpha \frac{\partial \overline{T}}{\partial z}|_3 \end{bmatrix} \quad (10.73)$$

This way, we have a set of n linear algebraic equations to be solved for the n unknowns $\overline{T}_1^{t+\Delta t}, \overline{T}_2^{t+\Delta t}, \ldots, \overline{T}_n^{t+\Delta t}$. We will now discuss the various methods of solving the algebraic equations.

10.8 FEM: solving the linear system of algebraic equations

In Eqns. 10.70, 10.72 or 10.73, we see that the finite element method leads to a linear algebraic system of equations. Let us observe this by plugging in the values of various parameters in this set of equations. The initial temperature is 37 °C. The fluid temperature is 70 °C, the heat transfer coefficient at the surface is 20 W/m²°C, thermal diffusivity, α, is 10^{-7} m²/s, $\rho c_p = 10^6$ J/m³·°C and a uniform element size of 0.01 m and time step of 1 s is considered. Substituting these values in Eqn. 10.73, we obtain Eqn. 10.74.

$$\begin{bmatrix} 0.0033 & 0.0017 & 0 \\ 0.0017 & 0.0067 & 0.0017 \\ 0 & 0.0017 & 0.0033 \end{bmatrix} \begin{bmatrix} T_1^1 \\ T_2^1 \\ T_3^1 \end{bmatrix} = \begin{bmatrix} 0.1843 \\ 0.3700 \\ 0.1850 \end{bmatrix} \quad (10.74)$$

In this set of equations, the unknowns are T_1^1, T_2^1 and T_3^1, the temperatures at 1 s (the superscript refers to this 1 s). As said earlier, the temperatures at 0 s (initial values) have been plugged in to obtain the above equation, so the equation allows us to get the temperatures at 1 s using temperatures at 0 s.

Of course, you already know how to solve this set of equations (written in the matrix form). There are many different ways to solve these equations. For large sets of equations that would be more common than the three-equation system above, computer solutions have many alternatives that are faster or slower, depending on the situation. The study of solutions of linear algebraic equations is a branch of research by itself and we cannot go into the details. We will, instead, do a very quick overview. In general, we can think of two classes of methods – direct and iterative. These are now discussed.

10.8.1 Direct methods

Direct methods for solving the set of equations of the form shown in Eqn. 10.74 (i.e., $Ax = b$) are based on Gaussian elimination. As you may recall from your numerical analysis course, Gaussian elimination involves two steps. In the first step, elementary row operations are performed on the augmented matrix (an augmented matrix is obtained by combining the matrices A and b) to reduce it to a triangular form. Elementary row operations in matrices involve row switching, row multiplication and row addition. Back-substitution is then performed to find the final solution. Use of Gaussian elimination to obtain a solution to Eqn. 10.74 is shown below.

$$A = \begin{bmatrix} 0.0033 & 0.0017 & 0 \\ 0.0017 & 0.0067 & 0.0017 \\ 0 & 0.0017 & 0.0033 \end{bmatrix} \quad x = \begin{bmatrix} T_1^1 \\ T_2^1 \\ T_3^1 \end{bmatrix} \quad b = \begin{bmatrix} 0.1843 \\ 0.3700 \\ 0.1850 \end{bmatrix}$$

The augmented matrix is:

$$\begin{bmatrix} 0.0033 & 0.0017 & 0 & 0.1843 \\ 0.0017 & 0.0067 & 0.0017 & 0.3700 \\ 0 & 0.0017 & 0.0033 & 0.1850 \end{bmatrix}$$

Now, we perform the elementary row operations. The first operation is multiplying the first row by (0.0017/0.0033) and subtracting the product from the second row, i.e., $R2 - R1 * 0.0017/0.0033 \rightarrow R2$. The augmented matrix becomes:

$$\begin{bmatrix} 0.0033 & 0.0017 & 0 & 0.1843 \\ 0 & 0.0058 & 0.0017 & 0.2751 \\ 0 & 0.0017 & 0.0033 & 0.1850 \end{bmatrix}$$

Then multiplying the second row by (0.0017/0.0058) and subtracting the product from the third row, i.e., $R3 - R2 * 0.0017/0.0058 \rightarrow R3$, gives:

$$\begin{bmatrix} 0.0033 & 0.0017 & 0 & 0.1843 \\ 0 & 0.0058 & 0.0017 & 0.2751 \\ 0 & 0 & 0.0028 & 0.1047 \end{bmatrix}$$

Now, writing the matrices in the original form:

$$\begin{bmatrix} 0.0033 & 0.0017 & 0 \\ 0 & 0.0058 & 0.0017 \\ 0 & 0 & 0.0028 \end{bmatrix} \begin{bmatrix} T_1^1 \\ T_2^1 \\ T_3^1 \end{bmatrix} = \begin{bmatrix} 0.1843 \\ 0.2751 \\ 0.1047 \end{bmatrix}$$

The unknowns, T_1^1, T_2^1 and T_3^1, the temperatures at the three nodes after 1 s, can now be obtained by back substitution giving $T_3^1 = 37.3476$, $T_2^1 = 36.3252$ and $T_1^1 = 37.1355$.

LU decomposition Another method of performing Gaussian elimination is through LU decomposition (or LU factorization). In this case, the matrix A is transformed into a product of a lower triangular matrix L and an upper triangular matrix U.

$$A = LU \tag{10.75}$$

Matrices L and U are of the form:

$$L = \begin{bmatrix} l_{11} & 0 & 0 & 0 & 0 \\ l_{21} & l_{22} & 0 & 0 & 0 \\ \cdot & \cdot & \cdot & 0 & 0 \\ \cdot & \cdot & \cdot & \cdot & 0 \\ l_{m1} & \cdot & \cdot & \cdot & l_{mm} \end{bmatrix} \quad U = \begin{bmatrix} u_{11} & u_{12} & \cdot & \cdot & u_{1m} \\ 0 & u_{22} & \cdot & \cdot & \cdot \\ 0 & 0 & \cdot & \cdot & \cdot \\ 0 & 0 & 0 & \cdot & \cdot \\ 0 & 0 & 0 & 0 & u_{mm} \end{bmatrix}$$

$$\tag{10.76}$$

The equation system $Ax = b$ is then solved by using forward substitution followed by backward substitution:

$$Ly = b \\ Ux = y \qquad (10.77)$$

Different direct methods (in COMSOL or any other finite element software) differ in the way LU factorization is performed. As mentioned earlier, discussion of the different algorithms is beyond the scope of this book.

10.8.2 Iterative methods

For a very large system of equations, the use of direct methods for the solution discussed above would take too much time and computer memory and therefore, may not be possible in certain cases. Iterative methods of solution are useful in problems with large numbers of variables. In the iterative methods, we start with an initial approximation of the exact solution and change this approximation in successive steps to bring it closer to the exact solution (Johnson, 1987).

$$x^{i+1} = x^i + \alpha_i d^i, \quad i = 0, 1, \ldots, n \qquad (10.78)$$

Here, x is the exact solution, x^i are the successive approximations, d^i is the search direction and $\alpha_i (> 0)$ is the step length. The different iterative methods are characterized by the choice of different search direction and step length. The rate of convergence or the number of steps/iterations needed to arrive close to the actual solution (or to reduce the initial error $x - x^0$, where x^0 is the initial guess) depends on the nature of the problem and the choice of the iterative method. In iterative methods, the error of the solution is calculated at each step and once the error estimate is sufficiently small, as determined by the convergence criteria, the final solution is returned. It must also be noted here that iterative solvers do not always converge or arrive at a solution, unlike direct solvers.

Preconditioners are used with iterative solvers to improve the rate of convergence for a large system of equations. The preconditioner P of a matrix A (for the system $Ax = b$) is a matrix defined as:

For left preconditioning, $P^{-1}Ax = P^{-1}b$

For right preconditioning, $AP^{-1}Px = b$

Therefore, instead of solving the original set of equations given by $Ax = b$, the iterative solver solves for the preconditioned system as obtained above, either by left preconditioning, or right preconditioning.

We now demonstrate the use of a common iterative scheme, the Jacobi method, to solve Eqn. 10.74. For this set of equations (i.e., of the form, $Ax = b$), the following iteration scheme is used in the Jacobi method:

$$x_j^{(i+1)} = \frac{1}{a_{jj}} \left(b_j - \sum_{k \neq j} a_{jk} x_k^{(i)} \right) \tag{10.79}$$

Here, x_j are the elements of the matrix x (nodal temperatures T_1^1, T_2^1 and T_3^1 in this case), a_{jk} are the elements of the matrix A and i is the iteration number of the Jacobi method. Writing Eqn. 10.74 in the equation form:

$$\begin{aligned} 0.0033T_1^1 &+ 0.0017T_2^1 &+ 0 &= 0.1843 \\ 0.0017T_1^1 &+ 0.0067T_2^1 &+ 0.0017T_3^1 &= 0.3700 \\ 0 &+ 0.0017T_2^1 &+ 0.0033T_3^1 &= 0.1850 \end{aligned}$$

Now, dropping the superscripts of T (which denote a time of 1 s) and introducing superscripts, (i), $(i + 1)$, for the iteration number of the Jacobi method, we write the iteration scheme based on Eqn. 10.79:

$$\begin{aligned} T_1^{(i+1)} &= \tfrac{1}{0.0033} \left(0.1843 - 0.0017 T_2^{(i)} \right) \\ T_2^{(i+1)} &= \tfrac{1}{0.0067} \left(0.3700 - 0.0017 T_1^{(i)} - 0.0017 T_3^{(i)} \right) \\ T_3^{(i+1)} &= \tfrac{1}{0.0033} \left(0.1850 - 0.0017 T_2^{(i)} \right) \end{aligned}$$

We start with an initial guess of 37 °C for the three nodal temperatures, i.e., $T_1^{(1)} = T_2^{(1)} = T_3^{(1)} = 37$. Then we use the iteration scheme shown above to solve for $T_1^{(i+1)}$, $T_2^{(i+1)}$, and $T_3^{(i+1)}$ using values from from the previous step. The values obtained after successive iteration are shown below:

	i=1	i=2	i=3	i=4	i=5	i=6	i=7	i=8
T_1	37	36.7879	37.0724	37.0446	37.1190	37.1118	37.1312	37.1293
T_2	37	36.4478	36.5016	36.3572	36.3713	36.3335	36.3372	36.3274
T_3	37	37.0000	37.2845	37.2568	37.3311	37.3239	37.3433	37.3414

	i=9	i=10	i=11	i=12	i=13	i=14	i=15
T_1	37.1344	37.1339	37.1352	37.1351	37.1354	37.1354	37.1355
T_2	36.3283	36.3257	36.3260	36.3253	36.3254	36.3252	36.3252
T_3	37.3465	37.3460	37.3473	37.3472	37.3476	37.3475	37.3476

As you can observe the values of T_1^1, T_2^1 and T_3^1, the temperatures at the three nodes after 1 s, obtained by the Jacobi method after 15 iterations are the same as those obtained by Gaussian elimination (direct method, Section 10.8.1).

10.9 FEM: choice between linear solvers

Unfortunately, too many factors influence whether a particular method will work faster or slower for a given problem. Thus, recommendations for the choice of method are difficult. The coefficient matrix is often sparse. A sparse matrix is a matrix in which the elements are primarily zeros. Iterative methods are usually preferred for sparse matrices unless they have a tridiagonal structure (Gerald and Wheatley, 2004). The reason is that elimination does not preserve the sparseness, unless the matrix is tridiagonal, so that one cannot work with only non-zero terms.

Often computer memory is the bottleneck in simulations. Since different solvers utilize the memory differently, the choice of solver becomes critical. In COMSOL, both direct and iterative methods are implemented, as discussed in Section 3.5. For problems of the type discussed in the book, i.e., 1D/2D problems or small 3D problems with not many degrees of freedom (about 100 000), the direct solver UMFPACK should be used. If the UMFPACK solver runs out of memory, the PARDISO direct solver should be used.

For larger problems with many degrees of freedom (more than 100 000), the direct solvers run out of memory since they scale badly with problem size. Iterative solvers are memory efficient and can therefore be used for large problems. However, these solvers are less stable than direct solvers and do not always guarantee a solution. In COMSOL, the GMRES iterative solver should be selected if the direct solvers do not work (run out of memory), and the appropriate preconditioner should be used. As discussed earlier, preconditioners improve the convergence of iterative solvers. The preconditioners that should be tried with the GMRES iterative solver are AMG (algebraic multigrid), GMG (geometric multigrid) or ILU (incomplete LU).

10.10 FEM: linearization of non-linear equations

Nonlinear equations are equations that involve higher-order polynomial functions or transcendental functions (functions that include logarithmic, trigonometric and exponential functions). Solution of these types of equations involves the use of iterative procedures such as the Newton–Raphson method (also known as Newton's

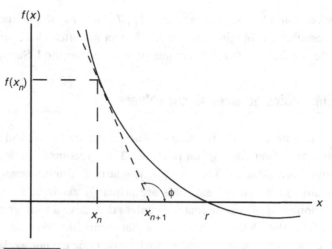

Figure 10.7

An illustration of the Newton–Raphson method to solve a nonlinear equation.

method). The method is based on linear approximation of the nonlinear function determined using the tangent to the curve of the function. Using successive approximations we try to obtain the solution to the equation. Suppose, we want to solve the non-linear equation $f(x) = 0$, shown in Figure 10.7: r is the root of the equation such that $f(r) = 0$; x_n is the current approximation and x_{n+1} is the next best approximation. We can relate x_n and x_{n+1} using the tangent to the curve at x_n, $f'(x_n)$, as follows:

$$f'(x_n) = \tan \phi = -\frac{f(x_n)}{x_{n+1} - x_n}$$

Rearranging we can write,

$$x_{n+1} = x_n - \frac{f(x_n)}{f'(x_n)} \tag{10.80}$$

The discussion in Sections 10.8.1–10.9 above relates to the solution of the system of linear equations obtained after finite element discretization of differential equations. However, in certain problems, such as in case of heat transfer, when the thermal properties are not constant but change with temperature, or when the heat source term depends on temperature, the algebraic set of equations that is obtained after discretization is nonlinear. The same is true for the mass transfer equation when the diffusivity or the reaction term is not constant. For example, in Case study II on cryosurgery, both thermal conductivity and specific heat are functions of temperature. Also, in Case study VIII on burn injury, the source terms in the heat and burn injury equations both depend on temperature. Discretization of such problems using the finite element method leads to a set of nonlinear equations. Another class of problems that always leads to nonlinearity are those involving the Navier–Stokes equation (Case study VI) for momentum balance. For such problems, the Newton–Raphson method described above, or its modified form, is used to linearize the set of nonlinear equations. The set of linear equations obtained thereafter are solved using either a direct or iterative solver (as described in Section 10.8) for each Newton–Raphson (or modified Newton–Raphson) iteration. The process is continued until the solution converges (refer to the next section for more details about convergence).

10.11 FEM: error in the finite element method and its reduction

In Section 5.1.2, we discussed the different types of errors; this section is concerned only with discretization errors (Slater, 2003). Since error refers to a comparison with a true solution, the only way to obtain an absolute magnitude of error in a finite element calculation is to compare with experiment. Sometimes it is also possible to compare with an analytical solution, but this is typically for a simpler problem (than the one at hand), since the finite element solution is sought for situations when analytical solution is not available in the first place. However, experimental data or analytical solutions are not always available. Thus, to have some idea of the error, relative error calculations are made with the primary purpose of improving the mesh or the time step used. We can think of discretization errors coming from three sources: (1) discretization of space; (2) discretization of time; and (3) linearization of non-linear equations.

One procedure to estimate the discretization error in space can be as follows. Figure 10.8(a) shows the nodal temperatures calculated for a certain 1D finite element computation of the heat equation using line elements. Since the temperature varies linearly (the slope is constant) within an element, the heat flux will be constant within the element. This is illustrated in Figure 10.8(b). Thus, the heat flux, computed from the temperatures, will vary discontinuously from one element to the next. Since there is no physical reason for such discontinuity in flux in this problem, this discontinuity is attributed to the error due to spatial discretization. The degree of this discontinuity at a particular location will depend on the element size and how rapidly the temperature is changing at that location. Since the real flux would vary continuously, an estimate of a continuous flux (somehow obtained) that closely follows the discontinuous flux, would be an improvement (over the discontinuous flux). The difference between the continuous flux (estimated) and discontinuous flux (from computed temperatures) would provide an estimate of the spatial discretization error in each element. This estimate can be used to refine

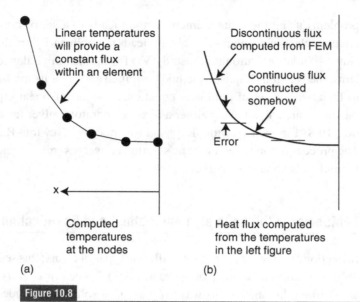

Figure 10.8

An illustration of a basis for error calculation in finite element method.

the mesh in regions where the error is large. In adaptive mesh refinement, this error is used to guide where the mesh needs to be made finer, i.e., more elements are needed. Another method, which is the default in COMSOL, is used to compare solutions computed on one mesh, and an identical (geometry-wise) mesh with one order higher shape functions. The one with higher-order shape function is assumed to be more accurate. It is important to note that such error estimates are generally qualitative, represent error due to spatial discretization only and should not be confused with errors in relation to the physical problem.

Error due to the approximation of a non-linear system as linear also need to be considered. The heat equation, for example, is non-linear when the thermal conductivity changes with temperature. Details of how this is implemented vary with the software. In COMSOL, the non-linear solver uses a damped Newton method. In this method, for a system of equations given by $f(T) = 0$, first a Newton step $\delta T = T^* - T_0$ is computed from

$$T^* - T_0 = \frac{-f(T_0)}{f'(T_0)} \tag{10.81}$$

about the linearization point T_0 (which is an initial guess of the solution). From this, T_1, a new estimate for T is guessed as $T_1 = T_0 + \lambda \delta T$ where $0 \geq \lambda \leq 1$. For this guessed value, we can write

$$T_1 - T_0 = \frac{f(T_1) - f(T_0)}{f'(T_0)} \qquad (10.82)$$

By subtracting Eqn. 10.82 from Eqn. 10.81, we can write

$$T^* - T_1 = \frac{-f(T_1)}{f'(T_0)} \qquad (10.83)$$

Since T^* was the intended Newton estimate, error, $E = T^* - T_1$, in the modified Newton, is estimated from Eqn. 10.83, i.e., from $f'(T_0)E = -f(T_1)$. From this error, E, a relative error is computed and compared with the relative error from the previous iteration. If the relative error is more than that in the previous iteration, the value of λ is reduced and this process is repeated until the relative error is less than that in the previous iteration. When this is true, the latest value of T_1 is taken as the solution.

Error due to discretization in time can be computed and controlled in a number of ways. For example, in COMSOL, this discretization error at each time step is controlled by using relative tolerance and absolute tolerance. If T is the solution vector corresponding to the solution of $f(T) = 0$ at a certain time step, and E is the solver's estimate of the (local) error in T, committed during this time step, as discussed in the previous paragraph, the time step is accepted if

$$\left(\frac{1}{N}\sum_i \left(\frac{|E_i|}{A_i + R|T_i|}\right)^2\right)^{1/2} < 1 \qquad (10.84)$$

where A_i is the absolute tolerance for DOF i, R is the relative tolerance and N is the number of degrees of freedom. Further details on this can be seen in the COMSOL User's Guide.

An example of quantitative calculation of some of the errors in COMSOL can be seen in Gobbert (2007).

10.12 FEM: convergence of the numerical solution as the mesh is refined

Strictly speaking, interpolation functions have to be chosen carefully to ensure convergence of the solution. Mathematically, convergence from mesh refinement can be proven only for the situations that satisfy the following conditions: (1) the elements are made smaller in such a way that every point of the solution domain can always be within an element, regardless of how small the element may be; (2) all previous meshes must be contained in the refined meshes; and (3) the form of interpolation function remains unchanged during the process of mesh refinement.

Causes of a solution not converging are many. Some of these can be: (1) physics inconsistent; (2) boundary conditions incorrect; (3) initial conditions incorrect; (4) incorrect material properties; (5) solution parameters needing adjustment; (6) solution strategy is impractical.

Improving convergence In practice, most commonly the mesh is refined and the computations are repeated until two successive computations produce nearly identical results.

10.13   FEM: stability of the numerical solution

Stability refers to transient problems and, since the time derivative in the finite element method is discretized using finite difference, the concept of stability essentially parallels that already discussed under the finite difference method (see Section 10.4). As we make $\Delta t \to 0$ in Eqn. 10.66, the solution approaches the exact solution for Eqn. 10.64. However, for the formulation given by Eqn. 10.66, as Δt is increased, it may exhibit entirely unrealistic behavior, including nonphysical oscillations, when the transient response is said to be unstable. Discussion of stability is definitely beyond the scope of this book. As a starter, the reader is referred to Huebner *et al.* (1994).

A more formal statement of the stability requirement of a solution process is as follows:

(1) The total error should remain bounded with time integration for fixed mesh intervals. Roundoff errors should not grow in a big way.
(2) The total error should approach zero as the mesh is refined at a fixed time level.

While the implicit formulation given by Eqn. 10.68 is unconditionally stable and stability criterion can be found for Eqn. 10.66 for simpler governing equations, in practice one simply tries a time step and element size combination to observe if the computations are indeed stable (i.e., results are not garbage). If the computations are not stable, they are repeated with a smaller time step until they become stable. One still needs to check for mesh convergence once the computations are stable.

10.14   FEM: generalization of methodology to more complex situations

The simple finite element methodology discussed in this section is for a 1D heat equation and for a constant thermal property, which are indeed very restrictive.

10.14 FEM: generalization of methodology to more complex situations

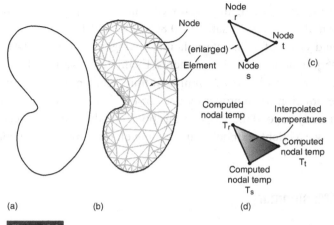

Figure 10.9

Illustration of (a) a 2D region; (b) how the 2D region can be divided using triangular elements; (c) enlarged view of one element with three nodes at the three vertices of the triangle where temperatures are calculated; (d) computed temperatures (levels of grey) within the element, obtained through interpolation of computed temperatures at the nodes.

For example, a heat transfer in a 2D region cannot be solved using 1D elements. However, the procedure developed here can be fairly easily generalized to more complex situations involving 2D or 3D, varying thermal property, and so on. For example, a 2D region can be divided using triangular elements, as shown in Figure 10.9, where the temperatures are computed at the three nodes, and, inside the triangle, temperatures are obtained through interpolation of the nodal values.

The interpolation equation can be written, analogous to Eqn. 10.40 as

$$\overline{T}(x,y) = p_r \overline{T}_r + p_s \overline{T}_s + p_t \overline{T}_t$$

$$= \begin{bmatrix} p_r & p_s & p_t \end{bmatrix} \begin{bmatrix} \overline{T}_r \\ \overline{T}_s \\ \overline{T}_t \end{bmatrix}$$

$$= [p][\overline{T}] \qquad (10.85)$$

where

$$p_r = \frac{x_s y_t - x_t y_s + (y_s - y_t)x + (x_t - x_s)y}{2(Area)} \qquad (10.86)$$

$$p_s = \frac{x_t y_r - x_r y_t + (y_t - y_r)x + (x_r - x_t)y}{2(Area)} \qquad (10.87)$$

$$p_t = \frac{x_r y_s - x_s y_r + (y_r - y_s)x + (x_s - x_r)y}{2(Area)} \qquad (10.88)$$

with x_r and y_r being the x and y-coordinates, respectively, at the node r. Similarly for nodes s and t. Equations 10.86, 10.87 and 10.88 are the three polynomials in x and y, used for 2D interpolation, analogous to Eqns. 10.41 and 10.42, used for interpolation for 1D line elements. Examples of various 1D, 2D and 3D elements are shown in Figure 3.2. Similarly, finite element formulation can be extended to the situation when properties are varying, i.e., α is not a constant. Such details are certainly beyond the scope of this book and the interested reader is referred to the many excellent textbooks on the finite element method (Huebner *et al.*, 1994).

10.15 FEM: summary

The finite element method is a numerical approach to solve differential equations where the equations are solved at discrete points (nodes) that define a region (element), with built-in interpolation over the element. In a transient problem, the method solves for temperatures at discrete points in time, typically using a finite difference approximation of the time derivative. It was shown that the governing partial differential equation for heat transfer in a 1D slab, i.e., Eqn. 10.35, can be transformed into a set of algebraic equations. These algebraic equations can be solved using many available techniques. In general, the accuracy of the solution is increased by reducing the element size and the time step, but this increases the computing time. From a practical standpoint, there is no way to decide on the element size and the time step before starting the computation. Typically, one chooses an element and time step size, performs the computation, reduces the size and recomputes until two successive computations do not show any appreciable difference in the computed values. True estimation of the error in the computation can only be done by comparing with experimental data, for most practical situations. Like many other numerical methods, the finite element method can adjust to most governing equations, geometry, boundary conditions and properties, although it is particularly suited to irregular shapes and complex boundary conditions.

References

Gerald, C. F. and P. O. Wheatley (2004). *Applied Numerical Analysis*. Boston, Pearson/Addison-Wesley.

Gobbert, M. K. (2007). A technique for the quantitative assessment of the solution quality on particular finite elements in COMSOL multiphysics. Proceedings of the COMSOL conference 2007, Boston.

Huebner, K. H., E. A. Thornton and T. G. Byrom (1994). *The Finite Element Method for Engineers*. New York, John Wiley & Sons.

Jaluria, Y. and K. E. Torrance (1986). *Computational Heat Transfer*. New York, Hemisphere Publishing Corporation.

Johnson, C. (1987). *Numerical Solution of Partial Differential Equations by the Finite Element Method*. Cambridge, UK, Cambridge University Press.

Owen, S. (2009). An introduction to unstructured mesh generation. Document on the web at http://www.andrew.cmu.edu/user/sowen/tutorial/tet_meshing_files/frame.htm and http://www.andrew.cmu.edu/user/sowen/survey/index.html

Slater, J. W. (2003). Uncertainty and error in CFD simulations. Document on the Web at http://www.grc.nasa.gov/WWW/wind/valid/tutorial/errors.html

10.16 Problems

10.16.1 Short questions

(1) How does a typical software handle non-linearity in the equations, such as that arising from property variations with temperature or concentration?

(2) What are the advantages and disadvantages of the Newton–Raphson method?

(3) How can one improve the accuracy in transient calculations?

(4) How does one decide the criteria for convergence in an iterative calculation?

(5) How does one typically decide the time step or the element size?

(6) What criterion did we use to develop the two algebraic equations for an element that would provide the two unknowns (nodal temperatures) for that element? Explain briefly.

10.16.2 Finite difference formulation of the variable property heat transfer equation

In practice, often the thermal properties change during heating or cooling, or the computational region may have two or more materials with different thermal conductivity values. Under these conditions, the heat conduction equation can be written in a more general form as

$$\rho C_p \frac{\partial T}{\partial t} = \frac{\partial}{\partial x}\left(k \frac{\partial T}{\partial x}\right) \quad (10.89)$$

Develop a finite difference approximation to this equation, analogous to Eqn. 10.26. Make appropriate assumptions on how to obtain the properties at mid-points of the nodes.

10.16.3 Finite difference formulation of the bioheat equation

The finite difference formulation developed in this chapter was for the heat equation without any source term. Develop a finite difference formulation for the bioheat equation (Eqn. 7.36) that has a temperature-dependent source term, $\rho_b c_b \dot{V}_b^v (T_a - T)$, arising from blood flow, and also a constant volumetric heat source, Q, arising from metabolism.

10.16.4 Finite difference formulation of the heat equation in 2D

Show that for a rectangular region, the 2D version of the finite difference formulation (Eqn. 10.26) of the heat equation (without the convection or the source term) is given by:

$$T_{i,j}^{k+1} = T_{i,j}^k + \frac{\alpha \Delta t}{(\Delta x)^2}\left[T_{i+1,j}^k - 2T_{i,j}^k + T_{i-1,j}^k\right] + \frac{\alpha \Delta t}{(\Delta y)^2}\left[T_{i,j+1}^k - 2T_{i,j}^k + T_{i,j-1}^k\right]$$
(10.90)

where the indices i, j stand for the x and y-directions, respectively, and k stands for time. Suppose there is a convective boundary condition in the two boundaries in the y-direction. How would you incorporate these two boundary conditions?

10.16.5 Checking the accuracy of a finite difference formulation by comparing with an analytical solution

The heat equation for an infinite slab with constant initial and boundary temperatures, for which the finite difference solution was developed in this chapter, is also one of the rare situations where we have an analytical solution (Eqn. 10.6). Thus, for this situation, the accuracy of the solution using the finite difference method can easily be quantified by comparing the solution with the corresponding analytical solution. Repeat the computations noted in Table 10.2, using $\Delta t = 360$ s, to obtain temperatures for a final time of $t = 60$ minutes, and find the error in this solution by comparing it with temperatures computed using the analytical solution for the same time (for this final time, to obtain the analytical solution, we need to keep only the first term in Eqn. 10.6). Is the error the same at all the nodes? Now repeat the finite difference computation with the time step halved ($\Delta t = 180$ s), and compute the error in temperatures at $t = 60$ minutes. How much did the error change? What factors decide whether it is worth going for a still smaller time step? Without halving the time step, i.e., using the time step from Table 10.2, repeat the computation with about twice as many elements, and compute the error in temperatures at $t = 60$ minutes. How much did the error change? What factors decide whether it is worth going for still more elements?

10.16.6 Programming the finite difference equation in a spreadsheet program

Implement the computations given by Eqn. 10.26 in EXCEL or any other spreadsheet program. Use the following data. Consider the half-thickness of the slab as 0.05 m, over which we have six equally spaced nodes. Initial temperature is 37 °C, boundary temperature is 10 °C, and thermal diffusivity is 10^{-7} m^2/s. Final time is 1800 s. What is the maximum Δt you can choose? Repeat computations with Δt higher than this value and comment on the results.

10.16.7 Finite element formulation of the bioheat equation

Rewrite Eqns. 10.48 and 10.49 for the bioheat equation by considering the two terms due to blood flow and metabolic heat generation as one combined source term which is temperature-dependent, i.e., for the equation

$$\frac{\partial T}{\partial t} = \alpha \frac{\partial^2 T}{\partial x^2} + Q' \qquad (10.91)$$

where Q' includes both terms and is temperature-dependent. Solve for the two additional integrals, i.e., terms arising due to Q', that would be part of Eqn. 10.56.

10.16.8 Finite element method for boundary temperature specified

Consider a slab heated symmetrically, as shown in the figure. In a finite element formulation, the 1D space is discretized using two 2-noded elements as shown in the figure. The temperature at the surface is specified to be 10 °C, so that the situation corresponds to Eqn. 10.70. Initial temperature is 37 °C. We want to calculate the temperature after the first time step of 1 s. (1) Write the set of equations (in a matrix form) for the three unknown temperatures at time 1 s, plugging in all the parameter values. The thermal diffusivity, α, is 10^{-7}m^2/s. Consider a uniform element size of 0.01 m. (2) Solve for the two unknowns. For the element given by the nodes 1 and 2, write the equation that provides temperatures at locations other than the nodes. (3) Would the temperatures obtained above at various locations be the same as that obtained using the analytical solution (the series solution, you may recall from your heat transfer course)? Explain. (4) How would you get the temperatures computed using the finite element solution closer to those

494 Solving the equations: numerical methods

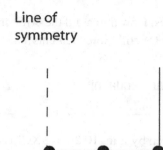

computed using the analytical solution? (5) Implement this problem in COMSOL and repeat the computations (use 3 nodes and $k = 0.1 \text{W/m} \cdot \text{K}$, $\rho = 1000 \text{kg/m}^3$ and $C_p = 1000 \text{J/kg} \cdot \text{K}$). 6) Verify if the coordinates of the nodes being used in COMSOL are indeed the same as what you used in your hand calculations (you can do this by `File>Export>Mesh to File` and creating a text file).

10.16.9 Finite element formulation for unequal element size

Rewrite Eqn. 10.69 for a 3-line element mesh, substituting the appropriate numerical values using the data provided. Also, write the interpolation equations, corresponding to Eqn. 10.43, for the three elements. The half-thickness of the slab is 1 cm. The element sizes decrease from the interior to the boundary, with the ratio of element size between two consecutive elements being 2. Time step is 10 s and the thermal diffusivity, α, is $10^{-7} \text{m}^2/\text{s}$. The slab is being cooled symmetrically, starting from an initial temperature of 37 °C. The boundary temperature is specified to be 5 °C.

10.16.10 Error in finite element computation and its reduction

Consider a transient 1D heat transfer in a region 5 cm thick, *during its early stages*, over the first 5 minutes. The region is being cooled, starting from an initial temperature of 37 °C. One of the surface temperatures is specified to be

5 °C, the other surface can be considered insulated. The thermal diffusivity, α, of the material is $10^{-7} m^2/s$. Solve this problem using COMSOL (use 21 nodes and $k = 0.1 W/m \cdot K$, $\rho = 1000 kg/m^3$ and $C_p = 1000 J/kg \cdot K$). Compare the temperature at each node with those computed from the analytical solution given by

$$\frac{T - T_i}{T_s - T_i} = 1 - \text{erf}\left[\frac{x}{2\sqrt{\alpha t}}\right] \quad (10.92)$$

where T_i is the initial temperature, T_s is the surface temperature, x is the distance from the surface, and erf is the error function. (1) Is there a region where the computed values are unphysical? (2) Do these unphysical values occur at an early time or later times, within the duration of 5 minutes? (3) Why do the unphysical values seem to disappear at later times? In comparison with the analytical solution, which region has the highest absolute error in the computed values? Can you reduce the error without increasing the number of elements?

10.16.11 Limitations of analytical solution

The analytical solution to the heat equation for an infinite slab for the specific boundary and initial conditions is given by Eqn. 10.6. If you are using this analytical solution to describe a real physical process, mention five restrictions that are automatically placed on the real process. For example, (1) the real process may involve a slab that has a hole in it, which would not be allowed if we are trying to use the series solution.

10.16.12 Convergence of finite element computations

Using a finite element software such as COMSOL, compute temperatures in a 1D slab for various numbers of elements. The slab half-thickness is 0.01 m. The initial temperature is 37 °C, fluid temperature is 70 °C, heat transfer coefficient at the surface is $20 W/m^2 \cdot °C$, $k = 0.1 W/m \cdot K$, $\rho = 1000 kg/m^3$ and $C_p = 1000 J/kg \cdot K$. Use a time step of 1 s and compute temperatures after 1 minute. Perform the computations for 2, 11 and 101 nodes and report (1) changes in the solution; (2) changes in computing time. (3) Now repeat the computation with a total of 11 nodes but spread non-uniformly, with many more nodes near the boundary and compare the results with results obtained previously for 2, 11 and 101 nodes.

Index

adaptive meshing, *see* error (discretization)
    example, 299
air
    thermophysical properties, 444
analytical solution
    limitations, 455
anatomical data, *see* property data
    body surface area, 417
    human body composition, 441
    reference man, 442
    reference woman, 442
    tissue shrinkage, 422
    tissue swelling, 422
apparent specific heat, 348
applications, *see* problem formulation
Arrhenius relationship, 405
arteries and veins
    heat transfer in, 349

basal metabolic rate, 420
bioheat equation, 349
    software implementation, 68
biomedical transport processes
    drug delivery and, 4
    goals, 4
    measures of success, 4
    thermal therapy and, 4
blood
    as Casson fluid, 425
    flow rates in organs, 421
    heat transfer due to, 349
    Reynolds number, 427
    rheological properties, 425
    thermal properties, 428
    velocity, 427
blood perfusion
    data, 419
    in fat, 419
    in muscle, 419
    temperature dependence, 419
blood plasma
    rheological properties, 425
body force, 339
boundary
    gas and liquid, 367
    liquid and solid, 367
    solid and gas, 367

Boundary conditions, *see also* boundary conditions (fluid flow), *see also* boundary conditions (heat transfer), *see also* boundary conditions (mass transfer)
    changing with time, 29
    software implementation, 73
    flux versus volumetric heat, 31
    how many, 27
    infinite region, 30
    interface between two materials, 30
    introduction, 27
    list, 28
    software implementation, 73
    what kind, 28
boundary conditions (fluid)
    inlet, 364
    inlet velocity profile, software implementation, 75
    pressure specified, 364
    software implementation, *see* boundary conditions
    symmetry, 363
    velocity continuity, 363
    velocity specified, 363
boundary conditions (heat)
    combined, 367
    software implementation, *see* boundary conditions
    surface convection, 366
    surface evaporation, 366
    surface heat flux, 365
    surface insulated (zero heat flux), 365
    surface radiation, 366
    surface temperature, 364
    symmetry, 365
boundary conditions (mass)
    equilibrium at interface, 367
    software implementation, *see* boundary conditions
    surface concentration, 369
    surface convection, 371
    surface impermeable (zero mass flux), 370
    surface mass flux, 370
    symmetry (zero mass flux), 370
brain
    thermal properties, 428
breast tumor detection, 173

burn injury
  boundary conditions, 280
  governing equations, 280
  input parameters, 282
  problem formulation, 280
  software implementation, 283
CAE, 51
cancer
  thermal dose for therapy, 408
cardiac ablation
  adaptive mesh refinement, 299
  boundary conditions, 290
  governing equations, 288
  input parameters, 290
  problem formulation, 288
  software implementation, 290
carotid artery (flow in)
  boundary conditions, 260
  governing equations, 259
  input parameters, 261
  problem formulation, 258
  software implementation, 262
case studies
  blood, oxygen flow examples, 185
  burn injury, see burn injury
  cardiac ablation, see cardiac ablation
  carotid artery, see carotid artery
  cryosurgery, see cryosurgery
  drug delivery (birth control patch), see drug delivery (birth control patch)
  drug delivery (therapeutic contact lens), see drug delivery (therapeutic contact lens)
  drug delivery examples, 184
  introduction, 179
  laser irradiation, see laser irradiation
  list with detailed description, 180
  metastatic melanoma, see radioimmunotherapy
  nitrogen elimination, see nitrogen elimination
  organization, 179
  radioimmunotherapy, see radioimmunotherapy
  thermal ablation, see thermal ablation
  thermal comfort examples, 184
  thermal therapy examples, 183
  web resources, 183
CEM43°C, 408
CFD, 51
circulatory system, 349
clinical applicator
  infrared heating, 393
  laser heating, 395
  microwave heating, 384
  radiofrequency heating, 389
  ultrasonic heating, 401
collagen
  diffusivity in, 440
computational electromagnetics, 51
computational fluid dynamics (CFD), 51
computational mechanics, 51
computer-aided engineering (CAE)
  defined, 50
  organization, 51

COMSOL, see also software implementation
  introduction, 55
  main window, 57
  model navigator, 56
  script, 84
  script for Monte Carlo simulation, 234
  variables, 110
conduction
  derivation of G.E., 342
  lumped parameter, 346
conservation
  energy, 342
  mass species, 353
  momentum, 338
  multiple species, 356
  total mass, 336
continuity equation, 336
convection
  derivation of G.E., 342
  software implementation, 65
convergence, see numerical solution (finite element)
cornea
  diffusivity in, 440
cryoprotectants, 434
cryosurgery, 170
  boundary conditions, 202
  governing equations, 201
  input parameters, 202
  optimization, 218
  problem formulation, 201
  software implementation, 202
cysts
  diffusivity in, 440
debugging, 157
  adding complexities in stages, 159
  checking for problem description, 158
  improving the solution process, 159
  simplifying the problem, 159
density, see property data
  air, 444
design
  biomedical, xxi
  modeling for, xxi
dielectric properties, see microwave heating
diffusivity, see property data, see property data
  software implementation, 77
  zero, software implementation, 81
discretization, see error (discretization)
distribution coefficient, 368
dropping terms, see governing equations
drug delivery
  examples, 184
  goals, 4
drug delivery (birth control patch)
  boundary conditions, 221
  governing equations, 221
  input parameters, 222
  Monte Carlo simulation, 234
  problem formulation, 221
  software implementation, 223

drug delivery (therapeutic contact lens)
  boundary conditions, 244
  governing equations, 244
  input parameters, 244
  problem formulation, 244
  software implementation, 245

elements, *see* mesh
equilibrium
  gas and liquid, 367
  liquid and solid, 368
  solid and gas, 369
error
  acknowledged, 140
  computer programming, 143
  debugging, 157
  defined, 140
  discretization, 142
    performing automated mesh convergence, 148
    performing mesh convergence manually, 148
    reduction through mesh convergence, 147
  global, 141
  iterative convergence, 142
  local, 141
  physical approximation, 141
  round-off, 142
  truncation, *see* error (discretization)
  types of, 140
  unacknowledged, 140
  usage, 143
  validation, *see* validation
errors, *see* numerical solution (finite element)
estimation of properties, *see* individual property data
evaporation, 349
explicit, *see* numerical methods (finite element)

fat
  thermal properties, 428
ferromagnetic heating, 392
Fick's law, 353
finite difference, *see* numerical solution
finite element, *see* numerical solution
formulation, *see* problem formulation
Fourier's law, 344
freezing, 346
  eutectic, 347
  tongue on a metal pole, 166

geometry, *see also* problem formulation
  1D, software implementation, 60
  2D, software implementation, 60
  3D, software implementation, 61
  bottom-up, 60
  COMSOL tools, 61
  in CAD software, 63
  in COMSOL, 60
  top-down, 60
  types, software implementation, 56
glaucoma, 164, 166
goals
  drug delivery, 4
  thermal therapy, 4

governing equations, 21, *see also* governing equations (continuity), *see also* governing equations (fluid flow), *see also* governing equations (heat transfer), *see also* governing equations (mass transfer)
  all, 373
  axisymmetry (conduction), 345
  axisymmetry (mass transfer), 356
  Cartesian, 373
  convection, software implementation, 65
  coupling, 360
  cylindrical, 373
  introduction, 23
  non-dimensionalization, 357
  software implementation, 65
  source terms, 379–413
  source terms, software implementation, 65
  spherical, 373
  terms to keep
    2D/3D/axisymmetric, 27
    convection, 25
    diffusion, 25
    generation (or source), 25
    transient, 24
  transient, software implementation, 65
  which ones, 23
governing equations (continuity)
  derived, 336
governing equations (fluid flow)
  derived, 338
  general form, 373
  non-dimensionalization, 360
  non-Newtonian fluid (implementation), 341
  software implementation, 65
governing equations (heat transfer)
  bioheat transfer, 349
  derived, 342
  evaporation, 349
  freezing, 346
  general form, 373
  generation term, 343
  lumped parameter, 346
  melting, 346
  non-dimensionalization, 357
  phase change, 346
  software implementation, 65
  storage term, 343
governing equations (mass transfer)
  derived, 353
  general form, 373
  multiple species, 356
  software implementation, 65

heart
  thermal properties, 428
heat equation, *see* governing equations (heat transfer)
heat source, *see* source terms
heat transfer
  boundary conditions, *see* boundary conditions (heat transfer)

governing equations, *see* governing equations (heat transfer)
heat transfer coefficients
    combined, 448
    convective, 444
    radiative, 448
Henry's law, 367
human body
    composition, 441
    reference man, 442
    reference woman, 442

ice formation, *see* freezing
implicit, *see* numerical methods (finite element)
infrared heating, 393
    clinical applicator, 393
    mechanism, 393
    software implementation, 393
    source term, 393
Initial conditions
    introduction, 31
    software implementation, 77
interphase equilibrium, 367
inverse problem, *see* optimization
    example
        detection of breast tumor, 173

kidney
    thermal properties, 428

laser heating, 395
    boundary conditions, 305
    clinical applicator, 397
    complexities, 400
    fluence, 397
    frequency effect, 399
    glaucoma, 164, 166
    governing equations, 303
    input parameters, 306
    irradiance, 397
    light diffusion, 400
    LITT, 397
    mechanism, 395
    problem formulation, 303
    software implementation, 307, 400
    source term (2D, axisymmetric), 398
    source term (simplified), 397
    very short pulse, 400
latent heat, *see* property data
liver
    diffusivity in, 440
    thermal properties, 428
lung
    thermal properties, 428

mass source, *see* source terms
mass transfer
    boundary conditions, *see* boundary conditions (mass transfer)
    governing equations, *see* governing equations (mass transfer)
    non-dimensionalization, 359

mass transfer coefficients
    convective, 444
material properties, *see* properties
melting, 346
mesh
    1D, software implementation, 91
    2D structured, 92
    2D unstructured, 92
    2D, software implementation, 92
    3D mixed, 97
    3D structured, 92
    3D unstructured, 97
    3D, software implementation, 92
    convergence, 98
    elements, 88
    free, 88
    introduction, 88
    mapped, 88
    moving, 317
    nodes, 88
    software implementation, 91
    structured, 88
    unstructured, 88
mesh convergence, *see* error(discretization)
metabolic heat, 420
    as source term, 380
metabolic rate
    basal, 420
    human activities, 420
    organs, 421
    tumor, 421
microwave heating
    clinical applicator, 386
    complexities, 389
    dielectric constant, 384
    dielectric loss, 384
    electromagnetic spectrum and, 383
    equivalent conductivity, 385
    frequency dependence, 385
    governing equations, 384
    heat source term, 384
    heat source term (simplified), 388
    ISM band, 383
    mechanism of, 385
    software implementation, 389
    temperature dependence, 386
    thermotherapy example, 388
model, *see* problem formulation
modeling
    academic preparations necessary, xx
    benefits of, xiii, xix
    design and, xxi
    improved understanding through, xx
    to optimize, xix
    userfriendliness of software and, xxi
    what is it, xix
molecular diffusivity, *see* diffusivity
Monte Carlo simulation
    software implementation, 234
moving mesh
    software implementation, 317

myocardium
    thermal properties, 428

Navier–Stokes equation, 338
Newtonian fluid, see property data
nitrogen elimination
    boundary conditions, 252
    governing equations, 251
    input parameters, 253
    problem formulation, 251
    software implementation, 253
nodes, see mesh
non-dimensionalization
    how to, 357
    fluid flow, 360
    heat transfer, 357
    mass transfer, 359
    what to substitute, 360
non-Newtonian fluid, see property data
    software implementation, 341
numerical solution
    finite difference, 458
        heat equation, 460
        stability, 464
    finite element, 465
        2D, 488
        boundary conditions, inclusion of, 475
        convergence, 487
        discretization of spatial term, 466
        discretization of transient term, 474
        errors, 485
        explicit and implicit, 474
        Galerkin, 466
        stability, 488
    flexibility, 456
    linear system solver
        choices, 483
        direct methods, 479
        iterative methods, 481
    non-linear systems, 483

objective function
    software implementation, 218
optimization, 155
    forward problem, 155
    how to, 155
    inverse problem and, 155
    objective function, 155
        RF heating, 169, 175
    software implementation, 218
organization of text, xxi
organs
    properties, see property data
overview of text, xxi
oxygen consumption
    organs, 421

pancreas
    thermal properties, 428
partition coefficient, 368
phase change, 346

physiological data, see property data, see property data
plot
    arbitrary functions, 129
    averages, 125
    contour, 117
    point, 113
    surface, 117
    surface (3D), 121
postprocessing, 53, 108–138
    animations, 129
    averages, 125
    contour plot (2D), 117
    data analysis guidelines, 133
    data at a point, 111
    dedicated postprocessing software, 131
    Ensight, 131
    exporting data in text format, 131
    general guidelines, 131
    in biomedical context, 108
    in Case Studies, 110
    plotting arbitrary functions, 129
    report writing, 133
    software implementation, 108
    surface plot (2D), 117
    surface plot (3D), 121
    Tecplot, 131
    transient data at a point, line or surface, 113
    variables in COMSOL, 110
preparations needed in using this book, xx
preprocessing
    guidelines, 54
    how to solve, 88
    software implementation, 55
    steps, 52
    what to solve, 52
problem formulation, 3–49
    application examples
        cryosurgery, 46
        drug delivery from a stent, 41
        drug delivery in brain, 39
        drug delivery using nicotine patch, 43
        drug delivery using therapeutic contact lens, 46
        laser refractive surgery, 47
        Laser-Interstitial Thermal Therapy, 41
        oxygen transport in alveoli, 46
        radiofrequency ablation, 44
        tooth drilling, 39
    biomedical transport, 3
    computational domain, 11
        1D implemented in 2D, 21
        1D, 2D or 3D, 16
        axisymmetry, 18
        connectivity between subdomains, 13
        how to decide, 11
        large, 14
        regions, 12
        semi-infinite, 14
        software implementation, 21
        solid-fluid, 12

subdomains, 12
      symmetry consideration, 18
   example, 5
   geometry, 11
   goals, 4
   goals, defining of, 9
   importances of, 7
   nitrogen elimination in alveoli, 41
   simplification, 11
   steps, 7
   therapeutic heating, 44
processing, 52
properties
   list, 32
   questions to ask, 33
   sensitivity analysis, 34
   simplification needed, 34
   software implementation, 79
   variable, software implementation, 81
   when not available, 34
property data, *see also* individual properties, 414–450
   air-water vapor mixtures, 437
   anatomical
      human body composition, 441
      reference man, 442
      reference woman, 442
   anatomical data, 417
   blood perfusion, 419
   density
      composition dependence, 433
      organs, 428
   diffusivity
      in biological fluids, 448
      in gases, 437
      in gels and tissues, 439
      in liquids, 437
      in water, 437, 445
      of chemicals, 439
      of glucose, 439
   emissivity, 434
   equilibrium vapor pressure, 435
   how accurate, 416
   metabolic heat, 420
   partition coefficient, 439
   permeability (diffusive), 439
   physiological
      blood, 427
   psychrometric, 436
   radiative, 434
   reaction rate constants, 443
   rheological, 424
   rheological properties
      Bingham plastic, 424
      blood, 425, 426
      blood plasma, 425
      Casson fluid, 424
      dilatant, 424
      Newtonian fluid, 424
      non-Newtonian fluid, 424
      pseudoplastic, 424

   viscosity, 424
   viscosity (apparent), 425
   sensitivity analysis, 417
   specific heat
      apparent, 431
      composition dependence, 429
      during freezing, 431
      organs, 428
   thermal conductivity, 427
      composition dependence, 429
      irreversible changes, 429
      organs, 428
      temperature dependence, 429
   thermal diffusivity, 433
   thermal properties
      air, 444
      cryoprotectants, 434
      density, 433
      latent heat, 434
      radiative, 434
      specific heat, 429
      water, 443
   transmissibility (diffusive), 439
   water activity, 435
   when not available, 416
   where, 414
   where available, 414
psychrometric chart, 449

Radiofrequency heating, 168, 169
radiofrequency heating
   clinical applicator, 390
   complexities, 392
   heat source term, 389
   heat source term (simplified), 390
   mechanism, 390
   software implementation, 392
   temperature dependence, 392
radioimmunotherapy
   boundary conditions, 273
   governing equations, 270
   input parameters, 273
   problem formulation, 270
   software implementation, 273
rheological properties, *see* property data

sensitivity analysis, 152
   how to, 153
   property data, 417
shrinkage of tissue, 422
simplification, *see* problem formulation
simulation
   not working, 158
skin
   diffusivity in, 439
   emissivity, 434
   thermal injury kinetics, 406
   thermal injury, quantification of, 407
   thermal properties, 428
software implementation
   analysis types, 58
   continuity, 337

geometry, *see* geometry
mesh, *see* mesh
moving mesh, 317
postprocessing, *see* postprocessing
preprocessing, *see* preprocessing
solution, *see* solution
tolerances, *see* tolerances
solution
  software implementation, 87
  solvers, 105
solvers
  direct, 105
  iterative, 105
source terms, 379–413
  blood flow contribution, 380
  heat
    electromagnetic heating, 382
    exponential decay, 380
    ferromagnetic heating, 392
    infrared heating, *see* infrared heating
    laser heating, *see* laser heating
    microwave heating, *see* microwave heating
    radiofrequency heating, *see* radiofrequency heating
    ultrasonic heating, *see* ultrasonic heating
  mass, 405
    first-order reaction, 405
    Michael–Menten reaction, 406
    oxygen consumption as, 406
    software implementation, 71, 408
    thermal injury as, 406
    zero-order reaction, 405
  metabolism, 380
  software implementation, 65
specific heat, *see* property data
  air, 444
  apparent, 348
  water, 443
spleen
  thermal properties, 428
stability, *see* numerical solution, *see* numerical solution (finite element)
surface area of human body, 417
surface forces, 339
swelling of tissue, 422
symbols, xxv

thermal ablation
  boundary conditions, 187
  governing equation, 186
  input parameters, 187
  problem formulation, 186
  software implementation, 188
thermal conductivity, *see* property data
  air, 444
  water, 443
thermal diffusivity, *see* property data
thermal properties, *see* property data
thermal therapy, 4
  examples, 183
  goals, 4
time steps
  fixed, 98

  introduction, 98
  software implementation, 101
  variable, 98
time-temperature equivalence, 408
tissue destruction
  temperature, 5
tissues
  destruction temperatures, 423
  diffusivity in, 439
  heat transfer, 349
  shrinkage, 422
  swelling, 422
tolerances
  absolute, 106
  relative, 106
  software implementation, 106
tooth
  thermal properties, 428
transfer coefficients, *see* heat transfer coefficients
transient
  software implementation, 65
tumor
  growth rate, 421
  metabolic rate, 421
  thermal properties, 428

ultrasonic heating, 401
  clinical applicator, 401
  complexities, 403
  mechanism, 401
  MRIgFUS, 401
  software implementation, 404
  source term, 403
  temperature dependence, 403
uncertainty
  defined, 140
  estimation of, 152
    how to, 153
  sources of, 141
undergraduate curriculum
  integration of simulation in, xiii
  use of this text in, xxiii
units, xxv

validation, 144–146
  against alternative solutions, 145
  against experimental data, 145
    cryosurgery, 170
  level of agreement, 144
  over large parameter range, 146
  qualitative checks, 144
viscosity, *see* property data (rheological properties)
  air, 444
  blood, 425
  water, 443

water
  diffusivity in, 445
  thermophysical properties, 443
web resources
  case studies, 183
  CFD codes, 52
  software implementation, 85

Printed in the United States
By Bookmasters